CRITICAL ACCLAIM FOR THIS BOOK

"The hazards of genetic engineering are among the most critical threats facing us. Knowledge is power. This book gives us power."

FRANCES MOORE LAPPÉ, author of *Diet for a Small Planet*

"Biotechnology is truly revolutionary. But it is far from a precise science and the alliance of government, industry and universities to profit from genetic engineering has resulted in extravagant claims of benefits that are highly speculative. This book provides a welcome perspective that is missing from all the biotechnology hype. For anyone concerned about the real implications and potential hazards of gene manipulation, it is an excellent starting point."

DR DAVID SUZUKI, geneticist and broadcaster

"This book explains why and how genetic engineering is adversely affecting our food, health, environment and human rights. It also shows what people are doing to counter the technology and the industry. It must be read by the public everywhere and by policymakers, especially in developing countries."

MARTIN KHOR, Director of Third World Network

"In this wide-ranging collection, scientists and activists discuss the pressing issues growing out of the wanton commercialization of the life sciences. The authors illustrate the dangers inherent in the unfettered manipulation of plant, animal and human biology for health and societal wellbeing. They also document the growing worldwide resistance to attempts to engineer life in all its forms. An excellent guide to the brave new world of genetic engineering."

RUTH HUBBARD, Professor Emerita of Biology at Harvard University, Board Member of the Council for Responsible Genetics

"The biotechnology industry has taken us beyond natural evolution into the unknown terrain of a never-to-be-natural-again world. We know nothing of the long-term biological, ecological, economic, public health and animal welfare consequences of this new technology. This book, with its excellent and wide-ranging coverage of this complex subject, will help us all awaken to the costs and potentially harmful, even catastrophic, consequences."

DR MICHAEL W. FOX, Senior Scholar, Bioethics, The Humane Society of the United States, Washington, DC

"We are involved in a global war called genetic engineering. Most of us are totally unprepared to fight this war. Reading *Redesigning Life?* is mandatory if you want you and your offspring to enjoy life as we have known it."

HOWARD F. LYMAN, President of EarthSave International, author of *Mad Cowboy: Plain Truth From the Cattle Rancher Who Won't Eat Meat*

"A cutting-edge critique of today's headlong rush toward genetic engineering and the ways in which a few corporate gene giants are dominating global food supplies and pirating the planet's biodiversity, the human genome and the keys to life itself."

HAZEL HENDERSON, author of *Beyond Globalization* and *Building A Win–Win World*

REDESIGNING LIFE?

The Worldwide Challenge to Genetic Engineering

EDITED BY
BRIAN TOKAR

McGILL-QUEEN'S UNIVERSITY PRESS
Montreal & Kingston · London · Ithaca

WITWATERSRAND UNIVERSITY PRESS
Johannesburg

SCRIBE PUBLICATIONS
Australia and New Zealand

ZED BOOKS
London & New York

Redesigning Life? was first published by Zed Books Ltd, 7 Cynthia Street, London,
N1 9JF, UK and Room 400, 175 Fifth Avenue, New York, NY 10010, USA in 2001.

Distributed in the United States exclusively by Palgrave,
a division of St Martin's Press, LLC, 175 Fifth Avenue, New York, NY 10010, USA.

Published in South Africa by Witwatersrand University Press,
P.O. Wits, Johannesburg 2050, South Africa.

Published in Canada by McGill-Queen's University Press,
3430 McTavish Street, Montreal, QC H3A 1X9, Canada.

Published in Australia and New Zealand by Scribe Publications Pty Ltd,
P.O. Box 287, Carlton North, Victoria 3054, Australia.

Copyright © Brian Tokar, 2001

Cover design by Andrew Corbett.
Typeset in Monotype Bembo by Illuminati, Grosmont.
Printed and bound in Malaysia.

A catalogue record for this book is available from the British Library

US CIP is available from the Library of Congress

Canadian Cataloguing in Publication Data
Main entry under title:
Redesigning life? : the worldwide challenge to genetic engineering
  Includes bibliographical references and index.
    ISBN 0–7735–2143–7 (bound). ISBN 0–7735–2144–5 (pbk.)
1. Genetic engineering—Social aspects. 2. Genetic engineering—Moral and
ethical aspects. 3. Agricultural biotechnology—Social aspects. 4. Agricultural
biotechnology—Moral and ethical aspects. I. Tokar, Brian
QH442.R44 2001   303.48'3
                                                       C00-901377-6

ISBNs
In South Africa 1 86814 3716 pb
In Australia and New Zealand 0 908011 60 1 pb
In Canada 0–7735–2143–7 hb 0–7735–2144–5 pb
In the rest of the world 1 85649 834 4 hb   1 85649 835 2 pb

Contents

INTRODUCTION

Challenging Biotechnology

BRIAN TOKAR

Perhaps once in a decade, a compelling new social or environmental concern will come to the forefront of public debate in the West, raising profound consequences for all life on earth, while thoroughly challenging our views of what kind of future is possible. The ensuing controversies provoke challenging questions about the very nature of our society and its institutions; they expose widespread myths and shatter foundational assumptions. In the 1950s, it was the problem of nuclear fallout and the looming threat of nuclear war. In the 1960s, citizens of the industrialized nations confronted their governments' participation in a genocidal war in Southeast Asia, while at the same time discovering the life-threatening effects of urban and industrial pollution, which made the air unhealthy to breathe and much of our water unsafe to drink.

In the late 1970s, a renewed antinuclear movement swept much of Europe and North America, focused on the proliferation of a new generation of weapons, and pledged to end the use of nuclear power to generate electricity. This movement halted the construction of nuclear power stations in all but a few Northern countries, and raised a profound challenge to militarism and the increasing centralization of political and economic power in society. In the late 1980s, a series of unusually hot summers – along with several major oil spills, crises over urban waste disposal, and a revival of environmental images in popular culture – helped bring the problem of global climate change to the centre of public concern, debate and action. The particulars vary considerably across geographic and cultural boundaries, but the broad impact of these issues on public consciousness has been a widespread, international phenomenon.

Today, at the start of a new millennium, questions about biotechnology and genetic engineering have come to occupy a central place in shaping public debates about the future. With profound implications for our health, the environment, the future of agriculture, and the relationship between human societies and the rest of nature, today's genetic technologies have aroused worldwide attention, provoked thousands of people to engage in political action, and galvanized a new international movement that has made the commercial use of genetically modified organisms (GMOs) one of the most contentious public issues throughout Europe, Asia and, increasingly, the Americas as well.

Less than a decade ago, the implications of biotechnology were largely an esoteric concern. The discussion was mostly limited to scientists, ex-scientists, and those who were moved to question the entire scientific enterprise. Biotechnology appeared to have emerged only recently from the annals of science fiction. Compared to more pressing concerns, from widespread hunger and the displacement of millions of impoverished peoples in Africa and elsewhere, the increasing militarization of Western societies, the extinction of countless living species and ecosystems, and worldwide assaults on basic human rights, concerns about biotechnology could safely be put aside for another day. Today, however, biotechnology and genetic engineering have become powerful symbols of what is most fundamentally wrong with our society: the unprecedented concentrations of political and economic power; economic globalization and the widening gap between the world's rich and poor; the worldwide loss of food security; and the rise of a new technological and financial elite who act as if the earth and all its diverse inhabitants are little more than chess pieces that can be endlessly played and manipulated to satisfy the insatiable wants of an extravagantly wealthy few.

To its many proponents and vocal supporters around the world, however, biotechnology represents something profoundly different. For some, it has come to represent the fullest realization of technological power for human benefit. Biotechnology, we are told, will usher in a more productive and less environmentally damaging agriculture, more effective medical treatments, cures for intractable genetic diseases and, ultimately, an epochal transcendence of human limitations such as infertility and ageing. Biotechnology, we are told, is the only possible solution to persistent problems of hunger, disease, population growth and environmental pollution. Not since the dawn of the nuclear age has a technology come to represent such diametrically opposite views of the future.

Indeed, commentators well across the spectrum of opinion often appear to agree that the twenty-first century will be the Age of Biotechnology. It was the American magazine *Business Week* that first coined the phrase "The Biotech Century." In an issue published shortly after the announcement by Scottish scientists of the first successful cloning of an adult sheep, *Business Week* celebrated the potential medical breakthroughs, and especially the investment opportunities, that biotechnology seemed to promise. Successes in mapping the human genome, for example, were, according to Stanford University geneticist Richard Meyers, "expanding people's imaginations, allowing them to think on a grand scale, asking and answering questions they would never have dreamed of before."[1]

World-renowned biotechnology critic Jeremy Rifkin also titled his 1998 book *The Biotech Century*. To Rifkin, who has initiated countless legal and political interventions against the unregulated development of genetic engineering in the United States, biotechnology embodies the most profound transformation in human consciousness since the European Renaissance. "Genetic engineering," proclaims Rifkin, "represents our fondest hopes and aspirations as well as our darkest fears and misgivings ... touch[ing] the core of our self-definition."[2] While recounting many of the hazards posed by genetically modified organisms, critiquing genetic reductionism, and exposing the core of eugenic ideology that lies behind much of today's research in human genetics, Rifkin's book also chronicles, with a persistent underlying fatalism, many of the biotech industry's extravagant promises of future miracles. Even the popular American news magazine *Time* would soon echo in with a special issue on "The Future of Medicine" – the introductory essay was titled, once again, "The Biotech Century."[3]

What is the nature of this technology that raises such a profound sense of awe among proponents and critics alike? What are the real consequences of current developments in genetic engineering, and what are their implications for our health, the environment, and society as a whole? How can we meaningfully evaluate the claims of biotechnologists and adequately challenge their underlying assumptions? Why have these technologies become such a lightning rod for public debate and action? These are only a few of the questions that this book seeks to address. With nearly 100 million acres of genetically engineered crops grown worldwide in 2000, and new biotech-based medical interventions being announced at an increasingly rapid pace, there is little doubt that these questions will have profound implications for our lives for many years to come.

"Biotechnology" has become the overarching term for a wide variety of new technologies, of which genetic engineering and the cloning of animals are only two of the best-known examples. What all these technologies have in common is the simulation and manipulation, whether in a laboratory or industrial setting, of fundamental life processes at the cellular and molecular levels. Modern biotechnologies include the selective culture of living cells and tissues to enhance the expression or selection of particular physical and biochemical traits, artificially stimulated fusion and fission of cells, chemical alteration of protein structures, the identification and mapping of unique genetic sequences, test tube (*in vitro*, literally "in glass") fertilization of eggs, and the artificial implantation of human and animal embryos, including embryos produced by cloning. All these technologies rely on basic discoveries about the structure and function of living cells that have emerged only in the last half-century, and have transformed the nature of biological research. In this volume we will focus on genetic engineering, cloning and the new reproductive technologies. Of all the new biotechnologies, these are by far the most controversial, and the ones with the widest implications for society as a whole.

Genetic engineering, or gene splicing, is the means by which segments of genetic material (DNA) from the cells of virtually any plant, animal or micro-organism – usually the genes that are believed to encode a distinct biological function – can be isolated in the laboratory and artificially inserted into the chromosomes of another, often unrelated, organism. First, plant and animal genes were inserted into bacteria, and the mass cloning of these genetically altered bacteria made it possible to produce large quantities of substances such as Bovine Growth Hormone and human insulin. More recently, bacteria and viruses that naturally infect living cells have been loaded up with particular genes of interest, and the new genetic material is inserted by these infectious agents into an organism's DNA. Today, in many cases, foreign genes are injected directly, using so-called "gene guns," which use microscopic pellets of gold or tungsten to insert foreign genes at essentially random locations in the host's chromosomes. Since the success rate of gene transfers by any of these methods is usually exceedingly low, "marker genes" for antibiotic resistance, herbicide tolerance, or other easily testable traits are included, making it possible for scientists or manufacturers to select only those cells that successfully incorporated the foreign, or "recombinant," DNA.

Whichever of these methods is used, the splicing of genes is radically different from more familiar processes, such as splicing magnetic tape, or cutting and pasting sentences in a word-processor. The regulation of gene expression, especially in multi-celled organisms, is a complex, interactive process, involving continual subtle alterations in the activities and inter-relationships of many different genes. Most biological traits involve the inter-play of several discrete genes, and individual genes are often assembled from DNA fragments that are dispersed along the chromosome. In recent years, scientists have observed that some genetic elements appear to jump around within a chromosome. There is sophisticated editing of genetic sequences, a wide range of hormonal effects, and numerous other phenomena that radically contradict simple, linear models of gene expression.[4] This is one reason why experiments in genetic engineering often yield wildly unpredictable results, such as the petunias whose colour genes were doubled in the hope of producing brighter flowers, but instead yielded growing numbers of white flowers, or the pig engineered to produce a human growth hormone, which turned out so weak, arthritic and overweight that it could barely stand up.[5] Genetic engineers are rarely able to predict how a foreign strand of DNA from any organism will interact with the subtle genetic regulatory processes in a given cell.

These fundamental biological realities contradict one of the biotechnology industry's most persistent claims: that genetic engineering is not essentially different from traditional, time-tested interventions such as the breeding of plants and animals, or using yeasts to make bread or beer. To use the same term, "biotechnology," to suggest a continuum from cultivating wheat to cloning sheep is a gross misrepresentation of both history and biology.

Ecologist Philip Regal of the University of Minnesota has emphasized the essential differences between traditional breeding and genetic engineering.[6] First, breeding only involves exchanges of genes between animals or plants that are able to mate naturally; they must either be of the same species or, in special cases, closely related ones. When a horse and a donkey are bred together to produce a mule, the offspring are unable to reproduce. This is a virtually inviolable natural barrier to the kinds of exotic combinations of genetic traits that are now possible in the laboratory.[7]

Second, genetic crosses in nature involve the natural recombination of analogous DNA fragments that lie in the same location on the same chromosome of each parent. Plants might exchange traits for different coloured flowers, and two animals might produce an offspring that shares some of the more noticeable traits of both parents. But breeding, as Dr Regal emphasizes, "has *not* in fact involved simply 'moving genes around' and introducing new and functionally proven genes to chromosomes or genomes" in random locations.[8] Further, successive breeding to enhance a particular genetic trait almost always involves a loss in some other important traits, ultimately reducing the offspring's fitness to survive in the wild. This is often not the case with organisms that have been genetically altered using recombinant DNA.

Finally, genetic engineering has the potential to access portions of the genome that are not usually subject to the processes of natural selection. A plant is no more likely to acquire the ability to secrete a bacterial toxin, than a breed of dogs is likely to grow an extra eye, or a goat is likely to sprout an elephant's trunk. Yet these are just the kinds of monstrosities that today's genetic engineers would aspire to create. For Regal, these three qualities underscore what is potentially the most dangerous aspect of genetic engineering: "the potential to create types of organisms that can interact with particular ecosystems and biological communities in novel competitive or functional ways."[9]

More recent evidence suggests that the effects of gene splicing on the processes of genetic regulation in living cells may be even more disruptive than previously realized. With "success" rates in the range of one in ten thousand to one in a million, genetic engineers attempt to leave as little to chance as possible. In addition to the functional genes of interest, genetic engineers introduce an entire "construct" of promoters, marker genes and other functional DNA fragments to improve the success rate of their experiments. To genetically engineer plants, for example, scientists use powerfully disruptive promoter sequences from the DNA of viruses (usually a cauliflower mosaic virus) to override the host cell's regulatory genes and literally force the plant to express the artificially inserted genes.[10]

These added viral genes increase the instability of the host plant's genome and improve the odds of hijacking the plant's metabolism to express the imported genetic trait. This practice may have serious consequences, however. Dormant viruses may be activated, genes essential to the normal functioning of the plant may be shut off or "silenced," and these viral vectors may open up the possibility of further gene transfers to other, unrelated organisms, inducing new, unpredictable genetic recombinations that would not otherwise be possible.[11] A group of European scientists has speculated that the rapid spread of genetically engineered organisms in the environment may be one factor in the emergence of so many new, highly virulent disease pathogens in recent years, many of which are simultaneously resistant to several different classes of antibiotics.[12]

Indeed, even the relatively well defined genetic modifications that have been developed and commercialized to date have introduced a plethora of unanticipated problems. Genetically engineered crops have been shown to harm beneficial insects such as ladybugs/ladybirds, lacewings and monarch butterflies, to cross-pollinate at higher rates than their non-engineered counterparts, and to be more susceptible to the effects of environmental stresses. Consuming these foods has been associated with unusual allergies, irritations of the digestive tract, the uncontrolled spread of antibiotic resistance, and possible distortions in the growth and development of vital organs. Biotechnology companies are seeking to develop plant varieties whose seeds are sterile, and to alter commercially important species of trees and fish to grow dramatically faster than their non-engineered counterparts. Not only have sheep and cows been cloned, but they have simultaneously been genetically engineered to produce useful proteins in their milk, with unknown effects on their metabolism and their health.

The social and ethical consequences of these technologies may prove to be even more disruptive than their ecological effects. Farmers face an unprecedented concentration of ownership in the seed and agrochemical industries, a problem that has very closely paralleled the development of genetically modified crop varieties. Some US farmers have been punished with large fines for carrying on the ages-old practice of saving seeds for replanting. Biotechnology companies are seeking out and patenting genetic information from plants, animals, and even humans from some of the remotest corners of the earth. These activities threaten traditional agricultural practices, and defy indigenous cultural norms limiting the uses of living materials. The agendas of medical research are being transformed by a narrow genetic reductionism that undermines research on the environmental causes of disease, while research on inheritable forms of gene therapy is ushering in a revival of eugenics, the once-discredited "science" of "perfecting" the human genetic stock. All these diverse concerns, and numerous others, will be explored in detail in the pages of this book.

Perhaps the most significant, overarching impact of biotechnology is this industry's overwhelming drive to commodify all that is alive: to bring all of life into the sphere of commercial products. From microorganisms that lie deep within the boiling hot geysers of Yellowstone National Park – found to be the subject of a secret agreement between the US National Park Service and a San Diego-based biotechnology company called Diversa[13] – to millions of human DNA sequences being mapped by both public and private agencies, all of life on earth is being reduced to a set of objects and codes to be bought, sold and patented.

This process of commodification takes a number of different forms. First, and foremost, biotechnology seeks to alter the fundamental patterns of nature so as better to satisfy the demands of the commercial marketplace. Wherever the patterns of nature are not well suited to continued exploitation, biotechnology offers the promise of redesigning life forms to satisfy the demands of the economic system. Where soil fertility and plant health are undermined by monocropping and chemical fertilizers, biotechnologists make crops tolerant to herbicides so growers can use more noxious chemicals to destroy weeds, and also try to make cereal grains fix nitrogen like legumes. Where industrial-scale irrigation lowers the water table and makes the soil saltier, they offer to make food crops more resistant to drought and to salt, instead of addressing the underlying causes of these problems.

Where marketable fish species like salmon have difficulties surviving year round in far northern hatcheries, genetic engineers try to splice in frost resistance from cold-water species such as flounder, and also make them grow dramatically faster. If naturally bred livestock cannot satisfy the demand for ever-increasing profit margins, commercial breeders might instead offer clones of their most productive animals. Instead of addressing the effects of excessive pulp and paper production on the biological integrity of native forests, timber companies will seek to raise plantations of genetically engineered trees that grow faster, and have an altered chemical makeup that may be more amenable to processing. In each instance, biotechnology helps perpetuate the myth that the inherent ecological limitations of a thoroughly nature-denying economic and social system can simply be engineered out of existence.

The biotechnology industry is also in the forefront of patenting living beings. They have brought the agenda of life patenting into the European Parliament, as well as to global institutions such as the World Trade Organization. The US government has threatened trade sanctions against countries such as India that have resisted the patenting of life. Meanwhile, corporate bioprospectors are surveying the entire biosphere, from the arctic to the tropics, in search of DNA sequences to study, manipulate and patent. The identification and patenting of human genes is also proceeding at a staggering pace, despite successful campaigns on behalf of three indigenous nations to overturn the patenting of their genes by the US National Institutes of Health.[14] At least 1,000 human genes have been patented to date, and the US Patent Office has been flooded with literally millions of requests for patents on small fragments of DNA that have become a tool for researchers seeking to accelerate and simultaneously privatize the mapping of the human genome.

At the same time, the biotechnology industry is broadening the range of living "materials" that are available to be bought, sold and marketed as commodities. Genes and gene sequences are only one step. Hundreds of for-profit fertility clinics in the US and elsewhere are purchasing human eggs and sperm from willing "volunteers" and offering them for sale. Recent breakthroughs in cloning have suggested the very real possibility that an above-ground market in human cells, tissues and even laboratory-created organs may soon complement the shadowy but lucrative international trade in human organs for transplantation.[15] Will our consumer society have the ethical fortitude to resist a future where human embryos, selected for particular genetic traits, will become available on the so-called "free market" as well?

In institutional terms, the biotechnology industry represents an unprecedented concentration of corporate power in two areas that are central to human survival: our food and our health. The late 1990s saw a heretofore unimaginable wave of corporate mergers and acquisitions in virtually every economic sector, and now the three pivotal areas of seeds, pharmaceuticals and agricultural chemicals are increasingly dominated by a small handful of transnational giants, all centrally committed to the advancement of biotechnology. The first company to join the global top five in all three of these areas was Novartis, formed in 1996 through a merger of the Swiss companies Ciba-Geigy and Sandoz.[16]

Today, with recurring waves of new mergers in the pharmaceutical and agricultural biotech industries, the concentration of corporate power in these areas has grown to truly staggering proportions. By 1999, five companies – AstraZeneca, DuPont (owner of Pioneer Hi-Bred, the world's largest seed company), Monsanto, Novartis and Aventis – controlled 60 per cent of the global pesticide market, 23 per cent of the commercial seed market and nearly all of the world's genetically modified seeds.[17] Aventis – formed from the merger of the chemical giants Hoechst and Rhone Poulenc – was also the world's largest pharmaceutical company, and Novartis and AstraZeneca were numbers four and five, respectively, in global pharmaceutical sales.[18]

Many factors helped create the economic climate for these mergers, some technological and some purely financial. One key factor, though, was the extent to which diverse areas of plant, animal and human biotechnology rely on common laboratory methods and the use of huge, often proprietary databases of genetic sequences. In the late 1990s, many corporate managers believed that companies able to control key technologies in several of these areas would be in a position to extend their technological benefits across disciplines, leading to faster product development and faster commercialization of new discoveries.[19] The move towards greater corporate consolidation in agriculture and medicine is thus significantly driven by developments in biotechnology, while at the same time profits from the sale of herbicides and other chemicals are channeled towards the development of new genetically modified life forms.[20]

A particularly staggering result of this trend has been the increasing domination of the worldwide seed industry by companies that specialize in chemical production and biotechnology. In 1999, ten companies controlled a third of the world's seed trade, and three companies best known for their chemical and biotechnology products – DuPont, Monsanto and Novartis – accounted for nearly 20 per cent of all seed sales. During the late 1990s, Monsanto, the world's most aggressive promoter of genetic engineering, bought several of the most important commercial seed companies, including DeKalb Genetics, Asgrow and Holden's in the United States, Brazil's Sementes Agroceres, and Unilever's Plant Breeding International, which was once a public institution based at Cambridge University.[21] The company spent over a year trying unsuccessfully to acquire the Mississippi-based Delta and Pine Land Company, developer of the notorious Terminator sterile seed technology. This rash of corporate takeovers in the seed industry has helped many farmers see that biotechnology may represent the greatest threat yet to the independence and stability of agricultural producers all over the world; a recent opinion piece in the *Wall Street Journal* suggested that farmers were about to become "more like Detroit's auto parts makers," mere subcontractors to a tiny handful of global corporations.[22]

The most significant obstacle to this strategy of corporate consolidation over the life sciences has been the worldwide opposition to genetically modified foods. The genetic engineering controversy in Europe escalated to the point where, by the summer of 1999, several companies were beginning to divest their agricultural divisions to protect lucrative profits in the pharmaceutical sector. Major players on New York's Wall Street and in Europe's financial capitals helped create the phenomenon of "life science" mergers, and were now suggesting that it may have gone too far. With genetically engineered crops rapidly becoming an economic "liability to farmers," Germany's Deutsche Bank anticipated "a major change in the market's view of GMOs": "We predict that GMOs, once perceived as the driver of the bull case for [the chemical] sector, will now be perceived as a pariah."[23] Monsanto was pressed to renounce the use of "Terminator" sterile seed technology, even though they did not own the notorious Delta and Pine Land Terminator patent. The *Wall Street Journal* soon announced that Monsanto, once the seemingly invincible world leader in biotechnology, would be worth significantly more to investors if it were simply to be broken up.[24] Monsanto's merger with the pharmaceutical giant Pharmacia and Upjohn, announced at the end of 1999, may be only the beginning, as the merged company had to pledge to sell off 20 per cent of its agricultural division (which retains the Monsanto name) to avoid further depressing the value of its stock.[25]

Still, biotechnology represents an estimated $13 billion in corporate revenues worldwide, with over 75 per cent of corporate investments based in the United States. There are some 1,300 companies involved, and just over 150,000 people employed in the biotechnology industry. The biotech industry's overall market capitalization skyrocketed from $97 billion in 1998 to a staggering $350 billion in early 2000.[26] But unlike the computer and telecommunications industries, where new startup companies can generate almost unbelievable short-term returns for investors, small biotech companies are often unable ever to bring a product to market. Developments in biotechnology are far more speculative, and new drugs and foods far more difficult to bring to market, than innovations in computer software or Internet applications. Many new biotechnology companies either disappear or become subcontractors to the transnational "life science" giants. The trend is clearly towards "fewer small organizations and more larger ones," Nobel prizewinning biologist Phillip Sharp told the magazine *Technology Review* at the end of 1999, with smaller entrepreneurial firms devoting increasing attention to cultivating and maintaining ties with their corporate patrons.[27] The *New York Times*, reporting the views of numerous venture capitalists and Wall Street investment analysts, described this phenomenon, with no intended irony, as an example of "corporate Darwinism."[28]

This unprecedented concentration of corporate power in the agricultural and pharmaceutical industries is only one of many factors driving today's intense debates over biotechnology. Public reactions to genetic engineering, cloning and other recent developments have been aroused by a staggeringly wide range of health, environmental, ethical and political concerns. First and foremost, millions of people worldwide see products of genetic engineering as a serious threat to their health and the health of their families. After decades of corporate scandals over unsafe pharmaceuticals and pesticides, tainted beef, and other such outrages, the potential health consequences of genetically engineered foods are one more risk that people simply do not wish to have forced on them. In a 1999 report, the British Medical Association validated the growing public concern, urging more comprehensive health studies and a moratorium on the commercial planting of engineered crops, until there is a scientific consensus on the potential long-term effects.[29]

Environmental concerns have also been in the forefront of the public debate over GMOs. The discovery in 1999 of the deadly effects of pollen from genetically engineered corn on immature monarch butterflies – with an almost 50 per cent mortality rate for caterpillars that were exposed to the altered pollen – dramatized the environmental consequences of genetic engineering in an easily comprehensible fashion. Effects on other beneficial insects, the threat of "super weeds," genetic contamination from engineered trees and fish, and the surprising death of soil microbes exposed to an experimental genetically engineered bacterium in an Oregon laboratory[30] have all contributed to raising the level of environmental concern. Ultimately, no one can predict the full effects of releasing countless millions of new, reproducing, genetically manipulated organisms on the earth's diverse natural ecosystems.

Ethical and religious considerations have played a central role in debates over biotechnology as well. Some opponents are moved by ethical concerns for the integrity of nature, as well as the threat to human identity posed by cloning, human embryo selection and the intensification of research into inheritable (germ-line) gene therapies. Others speak from a religious commitment to protecting God's creation from interventions that are harmful at best, and diabolical at worst. Genetic engineering also violates religious strictures against consuming certain foods or combinations of foods, especially where such foods cannot be clearly identified before they are eaten. The proper relationship between our human communities and the rest of the natural world is a subject of philosophical reflection and debate in many diverse cultures. Whether one is motivated primarily by secular or religious concerns, the profound ethical implications of genetic engineering and other new biotechnologies have proved impossible to ignore.

Farmers around the world have played a distinctly important role in debates about the effects of engineered organisms on nature and society. In India, hundreds of thousands of farmers have demonstrated against the corporate ownership of seeds. In southern France, farmers and cheesemakers concerned about the effects of US trade sanctions dumped truckloads of rotten fruit and manure at the doors of McDonald's restaurants. American farmers were in the forefront of early campaigns against the introduction of genetically engineered recombinant Bovine Growth Hormone (rBGH) for dairy cows, and are in the untenable position of having first been sold on the purported benefits of genetically engineered seed and only later informed that these crops are of lesser value on the world market than traditional varieties. "GMOs have become the albatross around the neck of farmers on issues of trade, labeling, testing, certification, segregation, market availability and agribusiness concentration," said Gary Goldberg, the head of the American Corn Growers Association, in response to agribusiness giant Archer Daniels Midland's 1999 announcement that the company would no longer accept genetically engineered crop varieties that had not been approved in Europe.[31]

All these concerns, and numerous others, have helped inspire powerful grassroots movements against genetic engineering and other biotechnologies around the world. Activists in Europe have pressured their governments to limit imports of engineered crops from the US, won key concessions from major supermarket chains and food processors, and taken direct action against experimental plots of genetically engineered crops. In India, some farmers have uprooted or burned test plots of Monsanto's pesticidal cotton varieties, while others have reasserted the importance of traditional seed saving and convinced the Indian Supreme Court to consider whether the planting of GMOs represents a violation of fundamental constitutional rights. Canadian activists joined with sceptical government scientists to pressure their government successfully to prohibit the use of engineered Bovine Growth Hormone. Indigenous activists throughout the world have objected to the appropriation of their crops, their medicinal wild plants and their own chromosomes for genetic information that can be patented and sold by transnational companies. In the United States, where a virtual media blackout helped delay the GMO debate by several years, food companies have begun promising to avoid genetically engineered ingredients, and public demonstrations against genetic engineering have become more and more visible.

This book is divided into four parts: the first three concentrate on particular facets of genetic engineering and other biotechnologies, highlighting the wide-ranging consequences of these technologies for people and the environment; the fourth brings together voices from the growing worldwide opposition.

First we examine the effects of genetic engineering on our health, our food, and the environment. We challenge the biotech industry's persistent claim that their technology is necessary to feed a growing world population, and then address the specific impacts of genetic engineering on food safety and the environment. Part II looks at the controversies surrounding medical genetics and the genetic manipulation of humans. We consider the implications of cloning, gene therapy, new reproductive technologies, and other developments; show how recent discoveries in human genetics have enabled the emergence of a disturbing new, market-oriented eugenics; and see how the agendas of medical research and health care have been distorted by the exaggerated claims of genetic engineering proponents.

Part III examines the institutional roots of genetic engineering, specifically the corporations and government agencies that have created today's biotechnologies and forced its products into our food supply. This section also examines the impacts of biotechnology and genetic research on people around the world, and the ways in which the biotech industry's agenda of patenting life specifically threatens the world's remaining land-based traditional peoples. Finally, Part IV highlights the growing resistance to genetic engineering by people all around the world. Activists from North America, India and Europe tell their stories, examine their successes (and shortcomings), and discuss the movement's prospects for the future.

A brief word on terminology is in order before we proceed. In this book, the terms "genetic engineering," "genetic modification" and "genetic alteration" are used interchangeably. Activists in the United States are increasingly adopting the European usage, "GMOs," for genetically modified organisms. There is something troubling about this term however. In the 1970s and early 1980s, proponents and opponents alike referred to either "genetic engineering" or to "genetic manipulation." This began to change around 1988, when the European Commission adopted the term "genetic modification" in its first proposed directive on the release of engineered organisms.[32] It was clear from the beginning that this usage was intended to soften the public impact of discussions of this technology. Of course, "engineering" implies a degree of certainty and predictability that this technology has never even approached, and today the term "GMO" has come to represent all the potential horrors of genetic manipulation. Clearly this is why the Cartagena Protocol on Biosafety, adopted in Montreal in January of 2000, sought to institutionalize an even more absurd and meaningless euphemism, LMO, for "living modified organism."

A book of this magnitude clearly depends on the cooperation and active engagement of a large number of people. All the contributors are to be thanked effusively for taking time out of their inordinately busy schedules of activism, research, teaching, lecturing, lobbying and other important activities to pull together the very best of what their work has to offer. Special thanks are due to Zoë Meleo-Erwin, who played a central role in the initial development of this book, and to Beth Burrows, who helped me track down several of the contributors, and kept insisting on the compelling need for such a book whenever the obstacles seemed insurmountable. Sidney Solomon and Heidi Freund helped shape the eventual format of the book, and Robert Molteno of Zed Books offered much encouragement and some helpful suggestions.

Finally, I wish to thank two larger groups of people without whose support and encouragement this book would never have happened. First, I am indebted to all my incredibly dedicated colleagues, students and friends at the Institute for Social Ecology and Goddard College in Plainfield, Vermont, for allowing me to take the necessary time away from my other responsibilities in order to see this project through. Second, I need to thank all the grassroots activists in Vermont, New England, across the United States and around the world whose tireless labours and unsurpassed inspirations have made the worldwide campaign against genetic engineering the incredibly dynamic and inspiring movement it truly is.

"It is not inconceivable," wrote the New York bureau chief for Britain's international business magazine *The Economist*, in an editorial on the European GMO debates, "that in a decade's time people will look back on the current rows about food as a turning point for both globalization and what used to be called the Western alliance."[33] By all accounts, this is a very apt observation. People throughout the world have become disenchanted with a globalized economy that consistently undermines democracy, stifles people's aspirations, and threatens to absorb every remaining inch of the earth's surface, every surviving traditional culture, and ultimately everything that is alive into its insatiable sphere of speculative markets, exploitable assets and tradable commodities. Debates over food, biotechnology and the commodification of life have clearly become key flash points for a growing worldwide resistance to corporate globalism and its horrific designs for our future.

Like many aspects of the global economy, the biotechnology industry tries to paint itself as institutionally invincible, and as the very embodiment of human progress and enlightenment. Biotech companies are closely allied with national governments and international financial institutions, and, for much of the last quarter century, the biotechnology industry has been a darling of the world's financial markets. However, the speed with which Monsanto's top officials were demoted from prophets to pariahs, in response to widespread public rejection of their biotech food products, offers hope that this industry's appearance of invincibility will have been a very short-lived phenomenon.

Just over two decades ago, another new technology was being promoted as the very key to prosperity and human progress. It, too, represented the commercialization of dramatic new scientific discoveries, and its advocates foresaw limitless improvements in human well-being. This technology had the full political and economic clout of the world's military establishments underwriting its development, and was also supported by many of the world's most powerful corporations. That technology was nuclear power.

The United States government was confidently predicting that by the year 2000 several hundred nuclear power stations would be providing most of the country's heat and light, and sustaining the development of a limitless array of new electronic wonders. Today, we may indeed have more new electronic devices than we know what do with, but nuclear power in the US peaked at just over 20 per cent of the nation's supply of utility-generated electricity, and its contribution has been essentially flat since the early 1990s.[34] Nuclear technology continues to play an important but limited role in a few key areas of medical diagnostics, but there has not been a single new commercial nuclear reactor ordered in the United States since the Three Mile Island accident in 1979.

Less than two weeks after the aforementioned commentary on globalization and GMOs was published in the New York Times, The Economist published an extended editorial on the increasing difficulties facing genetically engineered crops in the global marketplace. They quoted a seed industry consultant, formerly affiliated with Monsanto's subsidiary Calgene: "These [biotechnology] companies have great faith in their technology," he said. "They see themselves as the semiconductor industry of the 21st century." "But given the size of public opposition," The Economist's editorialist cautioned, "proponents of GM foods could be risking the fate of a rather different technology that once looked high-tech and futuristic – nuclear power."[35]

Today's activists against genetic engineering have inherited a great deal from the antinuclear campaigns of twenty years ago, including a decentralized approach to organizing, a commitment to nonviolent direct action, and an overarching belief that even the most technically daunting issues are not beyond the reach of an engaged and empowered citizenry. There is much cause for hope that the movement against genetic engineering and the commodification of life will continue to achieve comparable successes.

NOTES

1. Quoted in John Carey et al., "Special Report: The Biotech Century," Business Week, March 10, 1997, p. 80.

2. Jeremy Rifkin, The Biotech Century: Harnessing the Gene and Remaking the World, New York: Putnam, 1998, p. xii.

3. Walter Isaacson, "The Biotech Century," Time, January 11, 1999, pp. 42–3.

4. John Rennie, "DNA's New Twists," Scientific American, March 1993, pp. 88–96.

5. Andrew Kimbrell, The Human Body Shop: The Engineering and Marketing of Life, San Francisco: HarperCollins, 1993, pp. 175–6. The petunia experiments are described in Ricarda Steinbrecher's chapter in Part 1.

6. P. J. Regal, "Scientific principles for ecologically based risk assessment of transgenic organisms," Molecular Ecology 3, 1994, pp. 5–13.

7. Even so-called "wide crosses" of unrelated plants require a recent evolutionary history of close genetic interaction. See Michael K. Hansen, Genetic Engineering is Not

an Extension of Conventional Plant Breeding, New York: Consumer Policy Institute, February 2000.

8. Regal, "Scientific Principles," p. 7. Emphasis in original.

9. Ibid.

10. Hansen, "Genetic Engineering."

11. Mae-Wan Ho et al., "Gene Technology and Gene Ecology of Infectious Diseases," *Microbial Ecology in Health and Disease* 10, 1998, pp. 33–59; Mae-Wan Ho, Angela Ryan, and Joe Cummins, "Cauliflower Mosaic Virus Promoter – A Recipe for Disaster," *Microbial Ecology in Health and Disease* 11, 1999, at http://www.scup.no/mehd/ho.

12. Ho et al., "Gene Technology and Gene Ecology of Infectious Diseases."

13. Jim Robbins, "Yellowstone's Microbial Riches Lure Eager Bioprospectors," *New York Times*, October 14, 1997, p. B10; Christopher Smith, "Park Deal: Some Call it 'Biopiracy'," *Salt Lake Tribune*, November 9, 1997; see also http://www.edmonds-institute.org.

14. The NIH obtained patents on genetic material collected from the Guaymi of Panama, the people of the Solomon Islands, and the Hagahai of Papua New Guinea. See the collected papers in *Cultural Survival Quarterly*, vol. 20, no. 2, Summer 1996.

15. Kimbrell, *The Human Body Shop*.

16. RAFI (Rural Advancement Foundation International) Communiqué, "The Life Industry," September 1996, at http://www.rafi.org.

17. RAFI, "Seed Industry Giants: Who Owns Whom?" September 1999.

18. RAFI Communiqué, "The Gene Giants: Masters of the Universe?" March/ April 1999, p. 9. In 2000, Novartis, Aventis and AstraZeneca began to spin off their agricultural divisions, concerned that controversies over GM food were threatening pharmaceutical profits. Aventis Crop Science took the name Agreva, reminiscent of parent company Hoechst's original agbiotech division AgroEvo, which is referred to in Chapters 3 and 6.

19. Ann M. Thayer, "Living and Loving Life Sciences," *Chemical and Engineering News*, November 23, 1998, pp. 17–24.

20. For example, sales of glyphosate herbicides such as Roundup accounted for half of Monsanto's operating income in 1996, even before the company divested its industrial chemicals division. See Kenny Bruno, "Say it Ain't So, Monsanto," *Multinational Monitor*, vol. 18, no. 1–2, January/February 1997; Mark Arax and Jeanne Brokaw, "No Way Around Roundup," *Mother Jones*, January-February 1997.

21. See Brian Tokar, "Monsanto: A Checkered History," *The Ecologist*, vol. 28, no. 5, September/October 1998, p. 259.

22. Holman W. Jenkins, Jr., "Fun Facts to Know and Tell About Biotechnology," *Wall Street Journal*, November 17, 1999.

23. "Appendix 2: GMOs are Dead," in Deutsche Bank Alex Brown investor's report on DuPont Chemical: *Ag Biotech: Thanks, But No Thanks?*, July 12, 1999, p. 18. This Appendix was apparently released by Deutsche Bank as an independent report to investors on May 21, 1999.

24. Scott Kilman and Thomas M. Burton, "Monsanto Feels Pressure from the Street," *Wall Street Journal*, October 21, 1999. In the fall of 2000 the now-diminished Monsanto was re-established as a separate company, but with Pharmacia retaining 15 per cent ownership.

25. Robert Langreth and Nikhil Deogun, "Investors Cool to Pharmacia Merger Plan," *Wall Street Journal*, December 21, 1999.

26. Riku Lahteenmaki, "Investment Indicators Show US is Still Ahead," *Nature Biotechnology* 16, February 1998, p. 149; Rifkin, *The Biotech Century*, p. 15; "Top 50 Biofirms and 10 Leading Drug Companies," *Genetic Engineering News*, vol. 20, no. 6, March 15, 2000.

27. Stephen Hall, "Biotech on the Move," *Technology Review*, November–December 1999, p. 68.

28. Andrew Pollack, "Weed-Out Time in Biotechnology: Once-Hot Industry Feels the Impact of Corporate Darwinism," *New York Times*, December 16, 1998, p. C1. For a somewhat rosier view, see Justin Gillis, "Wall Street Makes it Official: Biotech Has Arrived," *Washington Post*, August 22, 1999, p. H1.

29. "The Impact of Genetic Modification on Agriculture, Food and Health: An Interim Statement," British Medical Association, Board of Science and Education, May 1999.

30. These results are discussed in the chapters by Martha Crouch, Beth Burrows and Ricarda Steinbrecher in Part I of this volume.

31. American Corn Growers Association press statement, "Corn Growers Call on Farmers to Consider Alternatives to Planting GMOs if Questions Are Not Answered." August 25, 1999.

32. Les Levidow and Joyce Tait, "The Greening of Biotechnology: From GMOs to Environment-Friendly Products," Occasional Paper 21 of the Open University Technology Policy Group, Milton Keynes, August 1990.

33. John Micklethwait, "Europe's Profound Fear of Food," *New York Times*, June 7, 1999.

34. US Department of Energy, "Nuclear Power Plant Operations," at http://www.eia.doe.gov/pub/energy.overview/monthly.energy/mer8–1. In 1989 this represented 6.6 per cent of total US energy use. *Nucleus*, vol. 13, no. 1, Spring 1991, p. 2 (Cambridge, MA: Union of Concerned Scientists).

35. Editorial: "Food for Thought," *The Economist*, June 19, 1999, p. 21.

PART I

Our Health, Our Food
and the Environment

Few subjects generate as much public passion and outrage as threats to the safety of the food we eat. This is especially true in Europe, where people have weathered repeated food safety scandals, from "mad cows" in Britain, to dioxin-laden animal feed in Belgium, and persistent attempts by the United States to force countries to accept unwanted imports of hormone-treated beef. In the United States, people are more accustomed to the idea of food as an industrial product, often heavily laden with synthetic ingredients. But even Americans pay very close attention when a new revelation casts doubt on the safety of their food. The widespread rejection of irradiation as a means to lengthen the shelf life of fresh fruits and vegetables is one important example.[1] Another was the successful effort to ban the chemical alar, once widely used to make apples ripen more uniformly, after it was discovered in the late 1980s that one of its breakdown products is highly carcinogenic.[2]

So it is hardly a surprise that the food safety aspects of genetic engineering have been in the forefront of public debate. In many countries, revelations about the health implications of genetically engineered Bovine Growth Hormone (rBGH) in the mid-1990s led either to a complete ban on the use of the drug or, as in the United States, a considerable curtailment in its use by dairy farmers. In Britain, the implications of genetically engineered food came to the centre stage in February of 1999 when scientists from across Europe spoke out in defence of Dr Arpad Pusztai, who had been fired from a senior position at Scotland's Rowett Research Institute for publicly announcing his findings that genetically engineered potatoes can seriously damage the internal organs of laboratory rats.

Since then, most of the largest supermarkets in Britain, and many across Europe, have removed genetically modified ingredients from their store brand products, and the largest food distributors including Nestlé and Unilever have followed suit. In the US, companies ranging from Gerber Baby Foods (ironically a subsidiary of the biotechnology giant Novartis) to the Whole Foods chain of natural foods supermarkets have also sworn off genetically modified inputs. The growing consumer rejection of genetically engineered foods has helped convince many farmers, even in agribusiness-dominated regions of the American Midwest, that it is in their best interest to return to non-genetically engineered varieties.

Having failed to convince consumers in the industrialized world quietly to accept the genetic engineering of our staple crops, the biotech industry has tried to recast itself as the saviour of the world's hungry. "How will we ever feed a rapidly growing population," scientists and corporate spokespersons echo, "if we cannot use the newest production-enhancing technologies?" In 1998, Monsanto tried this argument in a series of national advertisements in the British press, and it continues to be raised in public forums all across the US. In Britain, people from various international relief agencies, and from countries facing severe food shortages, were able quickly to deflate Monsanto's campaign, and it had to be withdrawn before it became a national laughing stock. Dr Tewolde Egziabher of Ethiopia offered a characteristically outraged response:

> There are still hungry people in Ethiopia, but they are hungry because they have no money, no longer because there is no food to buy.... We strongly resent the abuse of our poverty to sway the interests of the European public.[3]

Where people are displaced from the land in huge numbers, compelled to grow luxury crops for global markets, or driven to urban areas where they must rely on the market economy to satisfy all their basic needs, the false promises of the biotechnology industry are a cruel hoax at best. At the worst, genetic engineering threatens to perpetuate a never-ending cycle of famine, hopelessness and impoverishment.

The experiences of farmers who are actually growing genetically engineered varieties of corn (maize) and soybeans in the United States further discredit the industry's glowing claims. Studies in both experimental and commercial settings all across the American Midwest affirm what biotech critics have suspected all along: that genetically modified varieties do *not* help farmers to reduce chemical use, and they only rarely demonstrate significant improvements in yield.[4] The first point comes as no surprise, given that agricultural biotech companies still rely upon the sales of pesticides and other agricultural inputs for much of their operating income. The second point is also a predictable result: as the metabolic energy of engineered plants is diverted to express exotic traits such as herbicide tolerance and pesticide release, it is little surprise that both the yields and the quality of the crop can suffer considerably. Cotton farmers across the southern US have discovered this the hard way: their widespread difficulties with biotech crops – misshapen cotton bolls, significant crop losses, and so on – led to a series of lawsuits, arbitrations and out-of-court settlements.[5] As corporate consolidations in the seed industry continue at a staggering pace, large commercial growers in the United States are increasingly at the mercy of corporations determined to further the development of genetic engineering at all costs.

The implications of genetically engineered crops reach far beyond the often rather technical arguments about the track record of specific products. Genetic engineering serves to perpetuate an entire system of globalized agriculture that many careful observers view as the main underlying cause of hunger and poverty. Contrary to the claims of development specialists and agricultural scientists in the employ of institutions such as the World Bank and the International Rice Research Institute, the technical achievements of "Green Revolution" industrial agriculture have done little to stem poverty and hunger in the so-called developing world.

The work of development critics such as Vandana Shiva, Amartya Sen, Frances Moore Lappé, Walden Bello, Peter Rosset and many others clearly demonstrates (1) that hunger is largely a *political* problem resulting from the maldistribution of food and the displacement of people from productive land, and (2) that the imposition of "modern" agricultural practices at best brings only short-term increases in the yields of individual crops. "Improved" Green Revolution crop varieties have helped further to enrich the few at the expense of the many, and introduced Northern, industrial approaches to crop cultivation and pest control. This comes at the expense of groundwater supplies, soil integrity and traditional patterns of land tenure that, for all their shortcomings, have kept people on the land and supported traditional polycultures of locally well-adapted staple crops.[6] Where the market economy succeeds in thoroughly eclipsing the traditional subsistence economy, there is little recourse for those left behind.

"Agriculture is war by other means," trumpeted *The Economist* as the debates over genetically modified food began to heat up all across Europe during the summer of 1998.[7] Genetic engineering may have little to offer the hungry, or those concerned about the survival of family- and community-scaled farms, but it has become a crucial weapon in the continuing war of corporate agribusiness on the land, the soil, and the food security of people worldwide. Time and again, proponents of organic agriculture have shown that it is possible to grow more food by working with natural patterns rather than against them, by building up the soil and sustaining healthy agro-ecosystems.

Agribusiness seeks to solve each successive technical problem with another generation of technology, to develop newer, more sophisticated chemical inputs and ever more complex genetically engineered adaptations. But the work of pioneering organic researchers, from the Rodale Research Institute in Pennsylvania to Dr Elaine Ingham at Oregon State University, shows that a careful mix of traditional organic methods and a sophisticated scientific understanding of how soil fertility is naturally sustained is far more likely to solve the practical problems faced by today's farmers. Dr Ingham, for example, has

developed a method for replenishing depleted soil microbes on a rather large
scale, enabling even some of the largest commercial growers in the Western
United States to stop using chemical fertilizers and pesticides while dramatically increasing yields.[8]

The chapters in this section attempt a comprehensive overview of the diverse
issues that have emerged from the worldwide debate over genetically modified
foods. Martha Crouch begins by placing the discussion of GM food in
the context of ongoing debates about the causes of hunger and poverty and
the failings of the Green Revolution. She exposes the myths that underlie the
current generation of genetically engineered crops, describes her research on
the consequences of Terminator seed technology, and explains why the anticipated "next generation" of engineered crops is no more likely to "cure"
world hunger. Vandana Shiva follows up on this discussion, explaining why
the much-touted vitamin A-enhanced rice varieties that are currently under
development will not cure blindness among the world's poor, but will only
serve to exacerbate the problems that cause blindness and other symptoms of
malnourishment. Next, Sonja Schmitz reflects on some of her experiences as
a molecular biologist formerly on the research staff at DuPont, and explains
the agendas that underlie corporate research on herbicide tolerance and other
genetic traits. Jennifer Ferrara and Michael Dorsey then review our current
knowledge about the health effects of genetically engineered foods, from
rBGH to today's genetically engineered crops.

The discussion of the environmental consequences of genetic engineering
begins with Beth Burrows's account of the emergence of international debates
over the safety of engineered organisms released into the environment. These
discussions eventually led to the adoption in early 2000 of the Cartagena
Protocol on Biosafety under the auspices of the UN Convention on Biological Diversity. Then Ricarda Steinbrecher offers a comprehensive analysis of
the state of our scientific knowledge on the ecological effects of engineered
organisms. While studies of the consequences of genetic engineering have
only begun to catch up with twenty-five years of corporate research narrowly
focused on product development, recent findings have confirmed nearly all of
the likely effects that critics of genetic engineering have predicted since the
early 1990s. Jack Kloppenburg and Beth Burrows next review the wide array
of reasons – political, technical, philosophical and practical – why genetic
engineering is incompatible with the goal of a sustainable agriculture.

This section concludes with a pioneering discussion by Orin Langelle of
the increasing use of genetic engineering and other biotechnologies by the
forest products industry. While the consequences of genetic engineering of
trees are not yet as widely understood as the problems of genetically engineered food, Langelle's discussion helps illuminate the longer-range goals –
and the consequences – of the biotechnological view of all of nature as
merely a set of objects to be controlled, manipulated and harvested for profit.

The discussion of genetically engineered trees highlights the profound consequences for all life on earth of the thoroughly life-denying world-view of biotechnology.

NOTES

1. For an up-to-date chronicle of the campaigns against food irradiation in the US, see the *Food & Water Journal*, published in Walden, Vermont, and its predecessor, *Safe Food News*.

2. Peter Montague, "How They Lie," *RACHEL's Environment & Health Weekly* 503, July 18, 1996.

3. Quoted in CornerHouse Briefing no. 10, *Food? Health? Hope? Genetic Engineering and World Hunger*, Sturminster Newton: CornerHouse, October 1998, p. 3.

4. These studies have been catalogued most thoroughly by agricultural consultant Charles Benbrook of Idaho, via his website http://www.biotech-info.net.

5. For example, "Cotton Growers Blame New Seed for Crop Losses," *Augusta Chronicle* (Georgia), January 1, 1999; "Monsanto to Pay Cotton Farmers," *Financial Times* (US Edition), February 25, 1998; Kurt Kleiner, "Monsanto's Cotton Gets the Mississippi Blues," *New Scientist*, November 1, 1997.

6. See, for example, Vandana Shiva, *Monocultures of the Mind*, London: Zed Books, 1993; and Frances Moore Lappé, Joseph Collins, and Peter Rosset, *World Hunger: Twelve Myths*, New York: Grove Press, 1998.

7. *The Economist*, June 6, 1998.

8. "Dr. Ingham Puts Soil Life to Work," at http://www.soilfoodweb.com/mainfarmer.html.

From Golden Rice to Terminator Technology: Agricultural Biotechnology Will Not Feed the World or Save the Environment

MARTHA L. CROUCH

Imagine a university forum on the problems of biotechnology in agriculture. After an hour of carefully constructed arguments against further modernization of agriculture via biotechnology, and in favour of strengthening traditional ways of growing food, I wrap up my speech and wait for questions. A hand shoots up and the inevitable response is hurled towards the podium:

> You have not mentioned the population explosion. It might be nice to go back to the good old days before fertilizers, pesticides and high-yielding monocultures, but to do so would condemn millions of people to starvation. We have no choice but to forge ahead with the most advanced technologies we can devise. How else are we going to feed the world?

By now, I have heard this comment over and over again, and always the invocation of the population explosion and our inexorable race against time overshadows all other logic, every alternative.

The so-called "life science" companies that are developing biotechnology for agriculture have taken to citing the population problem as a main rationale for their products. For example, Monsanto trotted out the hunger scenario in their 1998 advertising campaign in Europe.[1] Several African luminaries signed on to a Monsanto letter claiming that protests against genetic engineering in the North are an obstacle to the development of foods that could save the South from certain starvation. The implication? Opposing biotechnology is a selfish, ethnocentric stance that only rich Europeans and North Americans can afford. This is a clever argument. Most people who actively work against biotechnology are also concerned about overpopulation, want to end hunger, and try to be sensitive to the needs of the Third World. It is easy for activists to become confused, and to lose momentum for opposing biotechnology when the issues are framed in this way. Furthermore, while advertisements can be brushed aside as cynical manipulations by self-interested agribusiness, many of the editorials and essays praising biotechnology are written by basic research scientists, liberal leaders and progressive activists.[2] I believe that they are sincere. Clearly, until the population-crisis-as-imperative-for-biotechnology rationale is decentred, it will be difficult for those of us who oppose biotechnology to make headway.

THE POPULATION BOMB IS A DANGEROUS DISTRACTION FROM IMMEDIATE PROBLEMS

Almost every defence of biotechnology in agriculture begins with a litany of statistics. These are crafted to scare us into welcoming any means necessary to allay the impending chaos of hungry hordes seething at our doorsteps. For example, former US president Jimmy Carter states that

> the global population is expanding by 100 million people each year.... It took some 10,000 years to expand food production to the current level of about 5 billion tons per year. In the next 35 years we will have to double that amount. The world's farmers will not meet this challenge unless they have access to ... continuing breakthroughs in agricultural science and technology.[3]

Many environmentalists use similar scare tactics, and conclude that "[n]othing short of massive intervention into the forces causing environmental deterioration will be adequate.... We are in fact at war with ourselves and our future, and only a similarly strong counter-response can save the day."[4]

Are there so many people in the world, and are the numbers increasing so rapidly, that all other concerns pale by comparison? Should we declare war on ourselves to try to defuse the population bomb before the explosion destroys the planet? First, let's look at the numbers. There are a lot of people in the world, and humans as a whole have increased rapidly in the last few hundred years, from about one billion in 1800 to over six billion today.[5] Estimates vary, but a middle projection is that the human population will stabilize at just over ten billion by the year 2200.[6] These numbers have a ring of certainty and objectivity. However, reducing the faces of humanity to a single number hides what is most important for determining whether panic is indeed in order.

Try to imagine a billion of anything and you will discover that it is impossible to visualize them, except as an abstract mass. Thus, conclusions based on such a large number have little meaning where life is lived, on the local level. The numbers are silent about how many are urban as opposed to rural, in deserts or in the tropics. They say nothing of which people are having a negative impact on the earth, ruining soil with salt and toxins, polluting rivers, and depleting the ozone layer; and which are having a positive impact by enriching the soil, planting and protecting trees, and maintaining good relationships with other species. Still, we are told, the equations prove that there is a population crisis, and that the ills of the world can be logically traced to the rapidly expanding number of people. Experts divide the total amount of some "resource" by the total number of people in the world, to determine whether there will be enough of that resource to go around. How much arable land is there to support each person? How much grain is produced per capita? How much petroleum does each individual get? In this way, the total number of people who can be supported by the earth's resources, and the impact of a person on the earth, is estimated.[7]

Such "carrying capacity" equations have understandable appeal. They acknowledge that we are subject to the laws of nature, just like all of the other creatures. However, to a greater degree than most other organisms, human populations are amazingly diverse in how they interact in the world. It is exceedingly difficult to pin down what impact a specific individual will have, and what they will need to live a healthy life. As an exercise, try to determine how much wood an "average" person needs. Do they cook with it, use it for building, shipping pallets, paper? Is paper a "basic need"? Are we calculating carrying capacity based on the unlimited desires of a capitalist consumer culture, the basic needs of a human animal, or something in between?

More sophisticated population models do try to account for some of the differences between populations, by correcting for affluence and technological capabilities.[8] However, none of these comes close to describing the real world, because they leave out social and political realities, such as who is able to buy food and who isn't, which land cannot be cultivated due to disruptions of war, impacts of global as opposed to local economy, and so on. The equations are biased in the assumption that all humans are in direct competition with each other and with the rest of nature for scarce resources. For example, ecologists have calculated that humans commandeer almost 50 per cent of the energy captured by all of the photosynthesizing plants on earth, leaving only half for the other millions of species.[9] In these calculations, they assumed that if an acre was planted in a crop, that land and the plants on it were out of bounds for the rest of nature. This may be largely true for industrial monocultures. But in a traditional peasant's polycultural plot, the diversity of birds, soil microorganisms and insects may be almost as high as in the surrounding forests.[10] The farmer may even enhance conditions for some species by creating richer soil, making more water available, and may encourage diversity by providing plants for beneficial insects or special fruits for the birds.[11] The complexities of interactive webs of species are unknown, little appreciated, and not easy to quantify.

We may never be able to compute how many people can live well on the earth. The diversity of each region of the earth and population of humans must be considered, and limits change with political, social and technological contingencies. Although the current population of 6 billion sounds high, and the projected ten billion seems overwhelming, how these people live may be far more important than the numbers.

HUNGER TODAY IS CAUSED BY POVERTY, NOT BY OVERPOPULATION

Let us consider the problem of hunger. Nina Fedoroff, a prominent molecular biologist, argues:

> To keep increasing food production in the face of population growth and the ac-
> celerating degradation of earth's productive land will take some miracles of knowl-
> edge and genetic engineering.... In new knowledge lies our only hope for wresting
> more productivity from increasingly impoverished land.[12]

Typically, the number of acres in production is divided by the amount of
grain each person will need times the number of people. The alarming result
is offered as proof that unless yield per acre increases dramatically, there will
either be widespread hunger, or all remaining wild lands will go under the
plough. Thus it would be criminal for people concerned about human welfare or the environment to stand in the way of biotechnologies designed to
make agriculture more efficient.[13]

We live in a world where hunger amongst the poor increases as agriculture
becomes commercialized, even when yields improve.[14] The Green Revolution
of the 1960s and 1970s introduced higher-yielding varieties of staple food
crops, new tilling methods and increased use of chemical inputs. Along with
these technical innovations, modernizers promoted commercial, export-based
agriculture using loans, technical advisors, aid programmes, tax incentives,
advertising and military support. Farming with fossil fuels, fertilizers, pesticides, herbicides and scientifically bred seeds has been subsidized and encouraged at the expense of peasant subsistence and local-market-based agricultures.
These successors of colonialism continue to convert the best land to export
agriculture and ranching. They produce cattle and other livestock, grain for
processing and animal feed, luxury foods such as coffee and chocolate, fruits
for Northern markets, and so on.[15]

Peasants formerly used this land to grow food, fibre, forage, building
materials and medicines. They have been moved to marginal land, and have
a harder time growing what they need. When they can't make it, they end up
as squatters on the edges of large cities, or they migrate to work for low
wages on the bigger farms. It is difficult to make enough money to buy the
food that they only recently grew for themselves. The food they can buy is
lower in nutrients; it is refined, high in fat, low in vitamins. In study after
study of poor children before and after commercialization of agriculture in a
region, researchers have found an increase in malnutrition at the very time
that overall yields per acre have increased.[16]

In the United States, no one can say that there is too little food to feed
the population. There are large agricultural surpluses. However, hunger is a
serious problem, with charities reporting a burgeoning need for food in both
urban and rural settings. How can this be, when the stock market is at an all-
time high, the economy is booming, and the harvests are record-setting? It is
clearly the concentration of wealth in the hands of fewer people that is
resulting in poverty for more people today.[17]

Who would believe that the best way to decrease poverty in the United
States is simply to increase the total amount of money available? Dividing the
total amount of money by the number of people to arrive at a measure of
how well the majority is doing is meaningless. By analogy, when food is a

commodity, entirely within the market realm, it is unrealistic to assume that an increase in the amount of the commodity will translate into an even distribution of food or profits from agriculture. In fact, both will flow to and concentrate around the people and places with the most money. The disparity between rich and poor will intensify. And it has, in every region of the world.

In Brazil, soybeans have become king, making corporate farmers rich and replacing the black beans important to peasants. The final products – soya oil, animal feed, and industrial chemicals – are sold to people who already have plenty to eat, while the displaced farmers go hungry. In Africa, many countries have up to half of their arable land planted in luxury export crops, from cacao to peanuts. In Malaysia, oil palm plantations replace rainforests where local people once made their livings. In Colombia, poor workers toil to grow flowers for the North. In the Caribbean, bananas and sugar are grown with toxic pesticides and slave-like employment practices. Global prices for these goods are influenced by International Monetary Fund decisions and the whims of traders and speculators, and can fluctuate wildly. The instability of the market favours the largest players, who can withstand such temporary set-backs.[18] It also accelerates the flow of resources, products and money to the middle class and elites from the poor people around the world.

What if most of the arable land were being sustainably farmed in direct food crops that were available to everyone? How many people could be fed? Contrary to the bleak statistics on the productivity of peasant agriculture that are cited by proponents of modernization, less biased studies find that traditional ways of growing food can be very effective at providing for the needs of a dense population. Studies of non-industrial agriculture are misrepresentations, because they tend to measure the yields of just the seeds, only one of the crops grown in a polyculture, thus underestimating how much else is contributed. The investigators ignore the complexity of traditional agricultural systems. The farms they study are often already operating suboptimally, on marginal land, or under disrupted cultural conditions. The negative costs of modernization of agriculture are not included in the equations: environmental degradation, depletion of non-renewable resources such as petroleum and fresh water, pollution, and social disruption.[19]

Recent studies indicate that well-functioning polycultures based on traditional systems of cultivation are more energy-efficient, maintain healthier soils, keep pests and pathogens below epidemic levels without harmful chemicals, are more resilient during adverse conditions, and result in a more balanced diet than the Green Revolution alternative. All this is accomplished with less dependence on exploitative relationships with institutions and corporations, and without the need for large amounts of money, while employing many more people and maintaining a vibrant rural culture. Enough food could be grown on existing agricultural land to feed the projected population without invoking a new Green Revolution, as long as food, and not money, is the primary goal.[20] If any technology is going to help reduce hunger, it must relieve poverty and must not threaten subsistence farming.

BIOTECHNOLOGY DOES NOT AND WILL NOT
RELIEVE POVERTY

Agricultural biotechnology leaders are more open about their motives when speaking to the industrialized farmers who share their world-view. In a full-page advertisement in *Farm Journal* (January 1998), Monsanto tells farmers that "It takes millions of dollars and years of research to develop the biotech crops that deliver superior value to growers. And future investments in biotech research depend on companies' ability to share in the added value created by these crops." Agribusiness leaders make it sound as if they just want their fair share, and that everyone else will win if they get it. Unfortunately, they don't really want to share. These transnational corporations want it all, and their genetically engineered products speak for themselves. The most commonly planted engineered crops are either tolerant toward specific herbicides, or have built-in insecticides. In both cases, environmental benefits are short-lived, leaving the fundamental problems of industrial agriculture unsolved. Making matters worse, the new technologies have potentially dangerous consequences of their own. A close look at the results from the early years of growers' experiences with these crops will illustrate my points.

Herbicide-tolerant soybeans

More land is planted in Monsanto's Roundup Ready soybeans than in any other genetically engineered crop. Between 1995 and 1999, the proportion of Roundup Ready soy varieties in the United States rose from zero to almost 50 per cent, more than 18 million acres. The plants have been engineered with an altered enzyme that allows them to grow when sprayed with Monsanto's Roundup herbicide. Roundup is a broad-spectrum herbicide that kills most leafy plants except for varieties that have been made "ready" for it by biotechnology.

How is herbicide-tolerance construed to be a boon to humanity? Scientists give several reasons. First, they assume that farmers are either ploughing weeds under, or are using herbicides to kill everything before planting seeds, in a practice called "burndown." Both ploughing and burndown can enhance soil erosion. With Roundup Ready plants, the herbicide is supposed to be applied after the crop is holding down the soil, and without ploughing, thus reducing chances of soil loss. Second, they claim that Roundup is one of the least toxic herbicides, and that farmers using Roundup Ready soybeans are switching from more dangerous herbicides, thus reducing pollution. Lastly, proponents of herbicide-tolerant plants contend that if weeds are not controlled, crop yields will decrease, causing hunger to increase in the world. Therefore, herbicide-tolerant plants will help curb starvation.[21]

The story from the fields is more complex and less optimistic. The first assertion, that readiness for Roundup will do away with other herbicides or with ploughing, is false. Nature is adaptable, and always finds ways to circumvent a system that is based on a single strategy. In this case, not only soybeans but also many other crops are being engineered to rely on one substance, Roundup. Already, the few weeds that grow well in the presence of Roundup, such as convolvulus, are becoming more common, in a phenomenon called "weed shift." In addition, some weeds are developing genetic resistance to Roundup.[22]

The appearance of Roundup-tolerant weeds means that some farmers are now spraying with up to three different herbicides to control weeds. Others are either ploughing or using a burndown herbicide even with Roundup Ready soybeans, because weed seeds germinate at different times in the season, and one application of Roundup is not enough. The effectiveness of Roundup will continue to erode in the future.

Farmers are trying the Roundup Ready varieties because they hope to save money and simplify weed management. Some will use less total herbicide on their land in the short term. They will continue to use Roundup Ready plants as long as the savings in chemicals outweigh the increased costs of seeds and Monsanto's technology fees, and yields are comparable to conventional varieties. However, in the long run, the treadmill of chemical weed control is accelerated by increased reliance on Roundup, so that the pattern of using more and different chemicals will continue. The prospects for a sustainable reduction in herbicides are not good.

The idea that Roundup is a benign or "friendly" herbicide is also overstated. It is true that Roundup is less acutely toxic to humans than are many of the other commonly used herbicides. However, it is still dangerous. Roundup is sold as an active ingredient, glyphosate, mixed with chemicals that help glyphosate function better in the field. These so-called inert ingredients do not kill weeds, but they can poison humans and other organisms. Farm workers have reported serious ailments from using Roundup. Also, Roundup has been shown to harm beneficial fungi and other soil organisms, and has been implicated in increased outbreaks of some plant diseases.[23]

Of course, Roundup is designed to kill plants, so it is not at all friendly as far as wildflowers, hedgerows, and ditch plants are concerned. These oases of non-cultivated plants on the edges of agricultural land are important for the biodiversity of organisms. Farmers I have talked to are of the opinion that glyphosate breaks down in the soil within a few days of spraying, whereas it actually can persist for more than a year. Also, sprays often drift from cultivated land into adjacent areas. What will happen if all of the major crops are ready for Roundup and are sprayed with it year after year? Without a diversity of plants around the fields, the spiders, bees, warblers, and wasps that live on them will disappear, leaving agriculture more vulnerable to pests and pathogens.

All of these disadvantages might be tolerable if having fewer weeds in soybeans helped starving people to survive. Currently, Roundup Ready soybeans are producing lower yields, on average, than the non-engineered elite varieties.[24] However, it is safe to say that if a few extra bushels of soybeans per acre in the US were grown, they would not feed hungry poor people, anywhere in the world. The Roundup Ready soybean varieties are used for the most part to be fed to animals, processed for oil or other ingredients in packaged foods, or converted into industrial chemicals.[25] The starving masses are not going to buy these products.

More surprisingly, in times of famine, it is the weeds, not the crops, that keep people alive.[26] These famine foods are high in nutrients and are able to withstand stresses that their cultivated cousins cannot handle. Weeds are also important sources of medicines and fodder, provide habitat and food for beneficial insects, and so on. In most parts of the word, weeds are managed so as not to compete too heavily with the crops, but people do not want to eradicate weeds or to reduce their diversity. Use of herbicides like Roundup makes people more vulnerable to disasters, not less so.

Insect-resistant crops

Maybe a better argument can be made for engineering crops with their own internal insecticides. After all, any technology that reduces pesticide use must be a clear environmental boon. As with Roundup Ready soybeans, the reality is different than the advertised scenario. Today, the only crops in the fields that have been genetically engineered to resist insects contain variants of the same gene, called "Bt."[27] "Bt" is an abbreviation for *Bacillus thuringiensis*, a bacterium that makes insecticidal proteins. These bacteria have been used for decades in spray or dust formulations to control various insects. Molecular biologists isolated the bacterial genes that contain instructions for making the insecticides, modified the genes, and then put them into corn, cotton, potatoes, and dozens of other crops still awaiting approval. These engineered plants make the Bt insecticidal proteins in their own tissues, and thus do not have to be sprayed to be protected against the same insects that Bt bacteria kill.

The environmental benefits seem obvious. Insects are usually controlled by spraying a crop with a mix of pesticides which may kill beneficial insects as well as pests; contaminate air, soil, and water; drift into wild areas, killing other animals; and work their way throughout the food chain, with various health consequences. With Bt-engineered plants, the insecticide is within the plant tissues, eliminating the expense and imprecision of spraying. Also, Bt is fairly specific for particular classes of insects, minimizing damage to beneficial or non-target organisms, including people. No wonder many farmers in the US and elsewhere have adopted Bt varieties.

However, it is the very success of Bt crops that is shaping up to be their downfall.[28] Insects may be even better at developing genetic resistance to insecticides than weeds are to herbicides. Plant-internalized insecticides are no exception. Without safeguards, and with the large-scale use of Bt crops, significant insect resistance is anticipated within five years or less. In laboratory experiments, several kinds of insects are already resistant to Bt toxins. If insects become resistant in the field, then the conventionally sprayed bacterial mixtures used by organic farmers, orchardists, foresters, and others will be rendered useless.

Besides the problem of insect resistance, other side-effects of widespread use of engineered Bt are appearing.[29] Some beneficial organisms, such as ladybirds and lacewings, are being poisoned as they eat insects that have fed on the Bt-containing leaves. Monarch butterflies are poisoned by the pollen from Bt crops. Also, the Bt toxins contained in decaying plant tissues are not breaking down in the soil as quickly as expected, with unknown consequences. Food safety issues are unresolved, as well. The Bt-toxins in engineered plants are in a different form than the original bacterial proteins, and they are also in new combinations with the naturally occurring chemicals in foods, such as potatoes. These Bt-containing foods have not been thoroughly tested for health effects.

CAUTIONARY TALES OF UNANTICIPATED FAILURES

So far, I have only argued that genetic engineering will not solve the basic problems of modern agriculture. However, there is also a danger of creating new problems. With genetic engineering, genes can be moved between any species, whether or not they would normally be able to mate – for example, from fish to flowers, from humans to pigs, and from bacteria to frogs. The use of biotechnology removes all species boundaries.

By transferring genes between very distantly related organisms, scientists create varieties with radically new capabilities. Unfortunately, we know very little about the complex interactions between species in nature, and are thus unable to predict how their creations will behave. The results can be surprising. For example, a genetically engineered bacterium in Oregon, designed to solve an air pollution problem, turned out to be able to kill almost any kind of plant it came in contact with.[30] For years, the farmers who grow grass seed for the turf industry have burned their fields at the end of the season to destroy the disease-containing stubble. This is an increasing problem for air quality. Scientists decided to try a new approach. They transferred genes from one bacterium, *Xanthomonas*, into another common soil bacterium, *Kebsiella planticola*, giving the recipient bacterium a new ability to break down dead plant tissue into alcohol. Grass farmers would collect plant residues into vats instead of burning them. They would ferment the material with the engineered bacteria and sell the resulting alcohol as a fuel additive, reducing air pollution from vehicles as well as from crop burning. In addition, the residue from the fermentation would be used as fertilizer when spread on the soil.

Before the researchers could field-test their engineered bacteria, the EPA performed various experiments to ensure the safety of releasing them into the environment. To everyone's surprise, when wheat seedlings were planted in soil containing the engineered bacteria, the plants died in about a week. *Klebsiella planticola* normally lives around the roots of many species of plants. With its new genes, it now used secretions produced by the plants and organic matter in the soil to make alcohol, and thus poisoned the seedlings. The unsettling part of the story is that the EPA could easily have missed this danger if they had used their standard tests. A soil ecologist from Oregon State University just happened to suggest a different experimental design, which caught the problem before the bacteria were let out of the laboratory.

This illustrates one of my concerns about genetic engineering in agriculture. Once the bacteria were on the farms, it would have been impossible to recall them. This concept was captured well in a recent bumper sticker: "Genetic Engineering: Giving Pollution a Life of its Own." Organisms are not machines, nor are they merely chemicals. They grow, reproduce, interact and change. Scientists accelerate the rate of change by moving genes between different species, shaking up long-standing relationships in unforeseen ways. In nature, novelty is usually disruptive.

These problems are more than simply glitches in the system to be worked out in future applications. It is an entire way of thinking about agriculture that is faulty. Unless future scientists formulate the basic problems facing farming in a different way, the same issues will resurface.

Plants that make plastics or other industrial feedstocks directly in their cells, cows with drugs in their milk, fruit containing vaccines, and pigs with human antigens making their organs more suitable for transplants all sound great at first blush. However, in each case the possible good accrues to those who can pay and the likely harm is borne by the environment and by people who cannot afford to pay. Biotechnology is designed to make a profit for some of the largest corporations in the world, and if it doesn't do that, by their own admission, it won't be pursued. These same companies have been responsible for intentionally demolishing local food self-sufficiency, for preferentially polluting poor countries, and a whole host of other unethical behaviours.[31] How likely are they to change their stripes with biotechnology, and suddenly set about sharing with the poor? Their public relations campaigns involve token gestures of goodwill, but are they sincere in their concerns?[32]

TERMINATOR TECHNOLOGIES ARE A WINDOW ON THEIR WORLD

A chilling example of the actual intentions of the biotechnology industry is contained in a group of patents designed to cripple or kill crops under specific conditions. In March of 1998, the first of these plans, a patent designed to give self-destruct instructions to proprietary seeds, was brought to public attention by the Rural Advancement Foundation International (RAFI).[33] They dubbed it "Terminator Technology," a term that reflects the technology's implications for farmers and plants. The industry prefers to call it a "Technology Protection System." The strategy was laid out in a patent awarded to the Delta and Pine Land Company in collaboration with the US Department of Agriculture: Patent Number 5,723,765: Control of Plant Gene Expression.[34] One of the main applications of the patent is a scheme to engineer crops to kill their own seeds in the second generation.

Further research by RAFI turned up more than two dozen patents that are either awarded or pending to accomplish the same goal of incapacitating the regenerative abilities of proprietary seeds. Most of the companies involved in agricultural biotechnology have their own versions of this technology. A 1999 RAFI Communiqué explains:

> The new generation of Terminator patents goes beyond the genetic neutering of crops. The patents reveal that companies are developing suicide seeds whose genetic traits can be turned on and off by an external chemical "inducer" – mixed with the company's patented agrochemical. In the not-so-distant future, we may see farmers planting seeds that will develop into productive (but sterile) crops only if sprayed with a carefully prescribed regimen that includes the company's proprietary pesticide, fertilizer, or herbicide.
>
> ... Chemically dependent suicide seeds are a dazzling technological achievement and a brilliant marketing strategy, but it's grim news for farmers, the environment and global food security.[35]

Grim news, indeed. I have examined the first Terminator patent in detail and have found serious potential problems.[36] The plethora of related patents means that the final applications are likely to differ in detail from the one I critiqued. However, the general way of thinking about agriculture is dangerous and is shared by all of these schemes.

The system has three key components: (1) A gene for a toxin that will kill the seed late in development, but that will not kill any other part of the plant. (2) A method for allowing a plant breeder to grow several generations of plants, already genetically engineered to contain the seed-specific toxin gene, without any seeds dying. This is required to produce enough seeds to sell for farmers to plant. (3) A method for activating the engineered seed-specific toxin gene after the farmer plants the seeds, so that the farmer's second generation of plants will be killed.

The various patents describe different ways of doing this. In the Delta and Pine Land/USDA version, the toxin gene has to be rearranged by an enzyme, whose synthesis may be triggered by applying the antibiotic tetracycline to the seeds before they are planted. Until tetracycline is added, the plants reproduce normally. It is amazing to me that these scientists should propose such a risky process, which on a large scale would be expensive, dangerous to people with allergies to antibiotics, unlikely to be completely effective, and increase the risk of bacteria becoming resistant to an important medication. Because of these problems, other methods will probably be used. Some of the patents outline methods whereby plants require continuous or repeated treatment with proprietary chemicals in order to grow normally. Others use insect hormones, temperature shocks, or even the company's pesticides to trigger toxins at will. All these traits are transferred permanently to the plant, so that they are passed on via the plant's normal reproduction. If combined with some commercially valuable trait – for example, a Roundup Ready gene –

the Terminator prevents farmers from replanting saved seeds, thus "protecting" the company's patents.

With increasing concentration in the seed industry, there is no telling how many varieties of seed may someday carry a Terminator gene. In the future it may be difficult to buy so-called elite, high-yielding varieties of any crop without also accepting Technology Protection Systems. The consolidation and vertical integration in the food industry is resulting in near-monopolies in seeds for some crops.[37]

CAN THE TERMINATOR TECHNOLOGY KEEP GENETICALLY ENGINEERED ORGANISMS UNDER CONTROL?

Terminator technologies are being promoted as environmentally friendly, because they are supposed to slow the spread of genetically engineered crops, or the genes from such crops, out of agricultural fields into nature. This feature is meant to appeal to those who worry that herbicide-tolerant or insect-resistant genes will flow through pollen or seeds to relatives of crops to create super-weeds; or that toxins will become incorporated into other species with uncertain results. Also, there is evidence that organic farmers are suffering from "genetic pollution" when their fields of non-genetically engineered varieties are pollinated by adjacent genetically engineered plants.[38] The resultant seeds cannot then be saved or harvested as organic, thus losing the value of the crop. If Terminator works as planned, it should prevent these transgressions.

Terminator cannot be trusted to contain the spread of engineered genes, however. If a gene or plant would be dangerous on the loose in nature, then it should not be released from the laboratory into agriculture. No matter how fancy the engineering, plants are not machines. Their behaviour will be more complex and changeable than predicted. Agricultural landscapes are integral parts of nature, where organisms interact, exchange genes, move about and evolve. Terminator-engineered plants will be no exception. If the toxin gene did not get activated properly, fertile seeds would be produced from the defective Terminator plants, and they would be able to multiply. Pollen would also be able to fertilize non-terminator plants without killing them. The traits being protected could spread by other means that bypass sex and seeds, including DNA transfer by viruses, bacteria, insects, or fungi.[39] Even a rare event could be important in spreading the unwanted genes, including Terminator, if those combinations of genes provide unique advantages.

Even if Terminator works reliably, there may be unforeseen problems. Pollen from functional Terminator plants would carry toxin genes, able to make proteins if they participated in seed formation. Any seed pollinated by activated Terminator plants would thus be killed. Although it is often claimed that the technology will only be used in self-pollinating crops, the patent covers all plants, and self-pollination is only a matter of degree. Cotton, for example, is described as self-pollinating, and has been mentioned as a likely first-candidate for Terminator insertion. But there can be as high as 40 per cent cross-pollination, depending on the variety, the plant's stage of growth, the weather, bee populations, and so on; in some cases fields would have to be more than ten miles away to be isolated from contact with pollen from other varieties.[40] Self-pollination is more accurately viewed as a tendency, rather than a condition of a species.

Seeds that have been killed by cross-pollination with Terminator plants would not reproduce, so the effect might be limited to the first generation. But the farmer would not know that the seeds were sterile until after they were planted, and even then the decrease in germination percentage would be very difficult to pin on cross-pollination with Terminator plants.

People often wonder if Terminator could spread into successive generations, causing a wave of sterilizations. This is indeed possible, because of a phenomenon called gene silencing. In some genetically engineered plants, previously active (introduced) genes can suddenly stop working. If this phenomenon occurred with seeds containing the Terminator gene, plants with the silenced toxin gene could grow and reproduce, perhaps for several generations. Thus, Terminator and other engineered genes could be carried into the future, to be expressed unexpectedly at some later time. In other words, future generations of seeds could suddenly die, without warning. No one can predict how likely such a scenario would be, because gene silencing is poorly understood.[41]

Although new schemes may overcome some of these technical objections, there is an on/off, machine-like logic, assuming the ability to control life precisely, inherent in all of the genetic engineering approaches I have seen. They are unlikely to work as planned, once they are incorporated into living organisms in the real world, where local conditions vary and are often unpredictable. Seeds hold the potential of future generations, both for themselves and for us. Why engage in the certain folly of trying to control the cycle of seed regeneration for uncertain, short-term, and private gain?

FROM TERMINATOR TECHNOLOGY TO GOLDEN RICE

Even some proponents of genetic engineering are disturbed by the consequences of Roundup Ready, Bt and Terminator technologies, and are distancing themselves from these "first generation" products. However, they argue that we might miss out on the clear benefits of the next generations of bioengineered crops if we slow down the process of research and development at this point.[42]

A shining example of the good things coming in the future is "golden rice," so-called because of the yellow colour of the grain. The rice is yellow because it has been engineered to have high levels of carotenoids (which make carrots orange), a precursor of vitamin A. It also has more iron.[43] The scientists who developed golden rice are giving the technology to an international rice breeding centre to be distributed to the Third World, and they hope that seeds will provide an antidote to malnutrition. Golden rice is perhaps the best-intentioned product of agricultural biotechnology, and is therefore the best example of why genetic engineering is doomed to fail the poor. Basically, a perfect food is not the answer to hunger. Packing all of the necessary nutrients into one food plant instead maintains the conditions that create poverty, for both people and the land.

To illustrate this, imagine yourself as a peasant in Ireland a few years before the great potato famine in the 1840s.[44] As an adult, you would be eating around ten pounds of potatoes each day, supplemented with a bit of milk. Ten pounds! I can barely force myself to eat two baked potatoes at dinner, a mere pound. We humans are omnivores and both require and desire a diverse diet. Potatoes are an unusually good source of a variety of nutrients, but they lack sufficient vitamin A and iron to be the sole food. Now what would you say to an international aid worker who suggested that he could provide you with nutritionally superior potatoes to cure your deficiencies? I would be outraged. As an emergency measure, of course, I would accept immediate help. But why would someone who genuinely cared about my well-being propose that I continue to eat only potatoes? Why didn't they bring emergency supplements, now available for pennies today from a variety of sources? My problem would not be fixed without more democracy, land reform, a shorter work day and week, and so on, with the result that I could grow my own vegetables and livestock (and keep them to eat), and have more money to buy food.

In the rice belt of Asia and Africa, rice is mixed with an amazing diversity of different species of plants, animals and fungi to create curries, soups, stews and stir-fries. What desperation must lead a person to eat so much rice, and so little else, that they suffer from vitamin-A deficiency, particularly when a cup of greens, a mango, a sweet potato or a melon is all that one needs? If the hundreds of millions of dollars and dozens of years that have gone into the development of genetically engineered rice had been spent promoting community and urban gardens to re-establish a diverse diet and more control over food resources, how many poor children would be healthier today?

A diet restricted to a few species leads to an agriculture devoid of biodiversity. Even if rice is engineered to have all of the nutrients needed for human health, it will never have all of the characteristics necessary to support a diverse community of organisms. Agriculture is part of nature, and the number of species used as food by humans is an important aspect of biodiversity, providing food, shelter and other functions for many organisms. A diverse diet ensures a varied landscape, which promotes stability in the ecosystem. No magic pill can solve the ills of the world, and to search for one wastes valuable time and energy.

HOW I LEARNED TO STOP WORRYING AND LOVE THE POPULATION BOMB

It is my opinion that the best thing we could do to promote harmony in the world is to learn to love ourselves as a species, to understand that humans are capable of being good and of having a beneficial interaction with nature. We are part of nature. When we wage war against ourselves, we also battle the world in which we are embedded, and there are many casualties. To dwell on the large number of people on earth, and to make decisions in panic mode, is not only misguided, but also dangerous. Draconian measures are taken against false targets while the real perpetrators get away free. With an emphasis on population as the villain, for example, women in the Third World are blamed and punished for participating in the cycle of life by having children. It is time to point a finger at perpetrators of economic systems that spread greed and fear, and to embrace the fecundity of nature, including our own species, as something sacred.

By defusing the population bomb, we have plenty of time to think about biotechnology, and to determine whether it is good or bad for society. No one is going to starve because we take time to debate the issues. On the contrary, a biotech slowdown would be more likely to relieve some of the pressure on small farmers, thus feeding more hungry people. The environment is not going to deteriorate more rapidly because we stand in the way of biotechnology. Biodiversity cannot be saved with biotechnology. Only a commitment to diversity at all levels can nurture other species; monocultures are part of the problem, not the solution. Agricultural biotechnology promotes intensification of monocultures, and is thus more likely to erode the environment than to heal it. Owning and patenting organisms, reducing them to their genes, and thinking of them as machines is not the way to engender respect for life.

I thus encourage everyone freely and energetically to oppose the use of genetically engineered crops, and to do so with confidence that you are improving our chances of creating a truly green and abundant future.[45]

NOTES

1. K. Bruno, "Monsanto's Failing PR Strategy," *The Ecologist*, vol. 28, no. 5, 1998, pp. 287–93.

2. For examples of scientists and liberal politicians writing in support of agricultural biotechnology, see P.H. Abelson and P.J. Hine, "The Plant Revolution," *Science* 285, 1999, pp. 367–8; R.N. Beachy, "Facing Fear of Biotechnology," *Science* 285, 1999, p. 335; Jimmy Carter, "Forestalling Famine with Biotechnology," *Washington Times*, July 11, 1997; N.V. Fedoroff, "Food for a Hungry World: We Must Find Ways to Increase Agricultural Productivity," *Chronicle of Higher Education*, June 20, 1997, p. B4; C. Mann, "Reseeding the Green Revolution," *Science* 277, 1997, pp. 1038–43.

3. Carter, "Forestalling Famine with Biotechnology."

4. T.E. Lovejoy, "The Problem of Third World Development," *Journal of Environmental Health*, Spring 1990, pp. 31–2.

5. United Nations, Population Division, Department of Economic and Social Affairs: Revision of the World Population Estimates and Projections. http://www.popin.org/pop1998/.

6. United Nations, Population Division, Department of Economic and Social Affairs: Long-Range Projections Based on the 1998 Revision. http://www.popin.org/longrange/keyfindings.htm.

7. Examples of the use of statistical averages to calculate human needs and impacts can be found in publications of the WorldWatch Institute and Carrying Capacity Network. For example, L.R. Brown, "Can We Raise Grain Yields Fast Enough?" *WorldWatch*, July/August 1997, pp. 8–17; R.F. Preiser, "Living Within Our Environmental Means: Natural Resources and an Optimum Human Population," *Carrying Capacity Network Clearinghouse Bulletin*, vol. 4, no. 6, 1994, pp. 1–2.

8. P. Ehrlich and A. Ehrlich, *The Population Explosion*, New York: Simon & Schuster, 1990; M. Wackernagel and W. Rees, *Our Ecological Footprint: Reducing Human Impact on the Earth*, Gabriola Island, BC: New Society Publishers, 1996.

9. P.M. Vitousek, P.R. Ehrlich, A.H.Ehrlich and P.A. Matson, "Human appropriation of the products of photosynthesis," *BioScience*, vol. 36, no. 6, 1986, pp. 368–73.

10. J. Benitez and I. Perfecto, "Efecto de diferentes tipos de manejo de cafe sobre las comunidades de hormigas," *Agroecologia Neotropical*, vol. 1, no. 1, 1989, pp. 11–15. Also, the Smithsonian Migratory Bird Center has educational materials concerning coffee and cocoa production and biodiversity, http://www.si.edu/smbc/coffee.htm.

11. F. Apffel-Marglin, ed., with PRATEC, *The Spirit of Regeneration: Andean Culture Confronting Western Notions of Development*, London: Zed Books, 1998.

12. Fedoroff, "Food for a Hungry World," 2.

13. D.T. Avery, *Saving the Planet with Pesticides and Plastic: The Environmental Triumph of High-Yield Farming*, Indianapolis: Hudson Institute, 1995. While its title makes the book sound like a parody, it is a serious, widely quoted contribution to the pro-development and modernization literature.

14. F.M. Lappé, J. Collins and P. Rosset, *World Hunger: Twelve Myths* (2nd edn), New York: Grove Press, 1998. Related updates and books from the Institute for Food and Development Policy can be found at http://www.foodfirst.org.

15. P. Goering, H. Norberg-Hodge and J. Page, *From the Ground Up: Rethinking Industrial Agriculture*, London: Zed Books, 1993; and other books from the International Society for Ecology and Culture in Berkeley, California; V. Shiva, *The Violence of the Green Revolution*, Dehra Dun, India: Research Foundation on Science, Technology and Ecology, 1989.

16. K.G. Dewey, "Nutrition and Agricultural Change," in C.R. Caroll, J.H. Vandermeer and P.M. Rosset, eds, *Agroecology*, New York: McGraw-Hill, 1990, pp. 459–80.

17. A. Mittal and P. Rosset, eds, *America Needs Human Rights*, Oakland, CA: Food First Books, 1999.

18. W. Bello, *Dark Victory: The United States and Global Poverty* (2nd edn), Oakland, CA: Food First Books, 1999; R.G. Williams, *Export Agriculture and the Crisis in Central America*, Chapel Hill: University of North Carolina Press, 1986.

19. P. Rosset, "Small is Bountiful," *Ecologist*, vol. 29, no. 8, 1999, pp. 452–6.

20. United States Department of Agriculture, *A Time to Act: A Report of the USDA National Commission on Small Farms*, USDA Miscellaneous Publication no. 1545, 1998. http://www.reeusda.gov/agsys/smallfarm/report.htm.

21. M. Potlak, "Designer Seeds," in the series *Beyond Discovery: The Path from Research to Human Benefit,* a project of the US National Academy of Sciences, 1998. http://www2.nas.edu/bsi.

22. C.M. Benbrook, "World Food System Challenges and Opportunities: GMOs, Biodiversity, and Lessons from America's Heartland," paper presented January 27, 1999, as part of the University of Illinois World Food and Sustainable Agriculture Program. http://www.pmac.net/IWFS.pdf; also see Benbrook's extensive compilation of articles and data at http://www.biotech-info.net.

23. C. Cox, "Glyphosate (Roundup): Responding to a Chemical Goliath," *Journal of Pesticide Reform*, vol. 18, no. 3, 1998, p. 2; C. Cox, "Glyphosate (Roundup): Herbicide Fact Sheet," *Journal of Pesticide Reform*, vol. 18, no. 3, 1998, pp. 2–16. Also at http://www.pesticide.org/factsheets.html.

24. *Journal of Pesticide Reform*, vol. 18, no. 3, 1998, p. 22.

25. D. Imhoff and P. Washall, "Soybean of Happiness: A 3,000-Year History of Our Most Modern Oilseed," *Whole Earth*, Summer 1999, pp. 75–9.

26. C.E. Sachs, "Gendered Fields: Rural Women, Agriculture, and Environment," Boulder, CO: Westview Press, 1996.

27. M. Mellon and J. Rissler, eds, "Now or Never: Serious New Plans to Save a Natural Pest Control," Cambridge, MA: Union of Concerned Scientists, 1998, at http://www.ucsusa.org.

28. Ibid. I do not mean to imply that Bt crops are working as designed. There have been several crop failures with Bt cotton and soybeans, with compensation awarded to affected farmers, as described in "Consumers Union's comments on Docket No. 99N-4282, Biotechnology in the Year 2000 and Beyond; Public Meetings," at http://www.consumer.org/food/fdacpi100.htm.

29. This is an outline of some of the problems with Bt crops. For details and up-dates, see http://www.biotech-info.net.

30. M. Holmes and E.R. Ingham, "Ecological effects of genetically engineered *Klebsiella planticola* released into agricultural soil with varying clay content," *Applied Soil Ecology*, vol. 3, 1999, pp. 394–9; M.T. Holmes, E.R. Ingham, J.D. Doyle and C.W. Hendricks, "Characterization of genetically engineered *Klebsiella planticola* and effects on plant growth," *Applied and Environmental Microbiology*, forthcoming. Also see http://www.soilfoodweb.com, and the Edmonds Institute's *Manual for Assessing Ecological and Human Health Effects of Genetically Engineered Organisms* at http://www.edmonds-institute.org.

31. D. Fagin, M. Lavelle and the Center for Public Integrity, *Toxic Deception: How the Chemical Industry Manipulates Science, Bends the Law, and Threatens Your Health*, Monroe, ME: Common Courage Press, 1999; R. Mokhiber and R. Weissman, *Corporate Predators: The Hunt for Mega-Profits and the Attack on Democracy*, Monroe, ME: Common Courage Press, 1999; J.C. Stauber and S. Rampton, *Toxic Sludge is Good for You! Lies, Damn Lies and the Public Relations Industry*, Monroe, ME: Common Courage Press, 1995.

32. J. Greer and K. Bruno, *Greenwash: The Reality Behind Corporate Environmentalism*, Penang: Third World Network and the Apex Press, 1997; B. Tokar, *Earth for Sale: Reclaiming Ecology in the Age of Corporate Greenwash*, Boston, MA: South End Press, 1997.

33. The Rural Advancement Foundation International has communiqués, press releases, books and other publications about Terminator Technology and other problems of industrial agriculture at http://www.rafi.org.

34. United States Patent Number 5,723,765: Control of Plant Gene Expression, issued on March 3, 1998, to Delta and Pine Land Co. and The United States Department of Agriculture. Inventors: M.J. Oliver, J.E. Quisenberry, N.L.G. Trolinder, and D.L. Keim.

35. Rural Advancement Foundation International Communiqué, March 30, 1999, "The Gene Giants: Masters of the Universe?" at http://www.rafi.org.

36. M.L. Crouch, "How the terminator terminates: An explanation for the non-scientist of a remarkable patent for killing second generation seeds of crop plants," Edmonds Institute Occasional Paper, 1998, at http://www.edmonds-institute.org.

37. J.R. Kloppenburg, Jr., *First the Seed: The Political Economy of Plant Biotechnology, 1492–2000*, Cambridge: Cambridge University Press, 1988; also see RAFI Communiqué July 30, 1998, "Seed Industry Consolidation: Who Owns Whom?" at http://www.rafi.org.

38. M.-H.R. Martens, "Is Your Organic Farm Safe? Protecting Your Crops from Genetic Contamination," *Acres*, vol. 30, no. 1, 1999, pp. 1, 8–10; C.L. Moyes and P.J. Dale, "Organic Farming and Gene Transfer from Genetically Modified Crops," MAFF Research Project OF0157, John Innes Centre, UK, 1999, at http://www.gmissues.org/orgreport/gmissues[1].htm. A review of the propensity of crops to cross with weeds is:

N.C. Ellstrand, H.C. Prentice and J.F. Hancock, "Gene flow and introgression from domesticated plants into their wild relatives," *Annual Review of Ecology and Systematics* 30, 1999, pp. 539–63.

39. M.-W. Ho, *Genetic Engineering: Dream or Nightmare?*, Bath: Gateway Books, 1998; J. Rissler and M. Mellon, *The Ecological Risks of Engineered Crops*, Cambridge, MA: MIT Press, 1996.

40. S.E. McGregor, "Insect Pollination of Cultivated Crop Plants," reprinted from the original USDA version, 1976, and updated on-line by *BeeCulture Magazine*. Cotton pollination is discussed in chapter 9. At http://bee.airoot.com/beeculture/book/chap9/cotton.html.

41. E.T. Ben-Ari, "The Silence of the Genes," *BioScience*, vol. 49, no. 6, 1999, pp. 432–7; R.A. Steinbrecher, and P.R. Mooney, "Terminator Technology: The Threat to World Food Security," *The Ecologist*, vol. 28, no. 5, 1998, pp. 276–9.

42. D. DellaPenna, "Nutritional Genomics: Manipulating Plant Micronutrients to Improve Human Health," *Science* 285, 1999, pp. 375–9; T. Gura, "New Genes Boost Rice Nutrients," *Science* 285, 1999, pp. 994–5.

43. There is another problem with engineering foods to have specific nutrient levels: scientists do not know very much about nutrition. It is a young science, and very complex. Although it is almost universally believed that low iron is a major health problem worldwide, extensive recent research leads to the conclusion that too much iron in the diet is a more serious problem. See the Iron Disorders Institute Newsletter, at http://www.irondisorders.org; and E.D. Weinberg, "Iron loading and disease surveillance," *Emerging Infectious Diseases*, vol. 5, no. 3, 1999, pp. 346–52.

44. R. Salaman, *The History and Social Influence of the Potato*, Cambridge: Cambridge University Press, 1949. This is an excellent, detailed history of the potato in Europe, with eyewitness descriptions of life in Ireland before and after the introduction of the potato, from the fifteenth to the nineteenth centuries.

45. For additional critiques of the presumed role of biotechnology in ending hunger, see *Food? Health? Hope? Genetic Engineering and World Hunger*, CornerHouse Briefing no. 10, October 1998, via cornerhouse@gn.apc.org; B. Kneen, *Farmageddon: Food and the Culture of Biotechnology*, Gabriola Island, BC: New Society Publishers, 1999; M. Lappé and B. Bailey, *Against the Grain: Biotechnology and the Corporate Takeover of Our Food*. Monroe, ME: Common Courage Press, 1998; and P.R. Mooney, "The Parts of Life: Agricultural Biodiversity, Indigenous Knowledge, and the Role of the Third System," *Development Dialogue* 1–2, 1996 (Uppsala: Dag Hammarskjöld Foundation).

2

Genetically Engineered "Vitamin A Rice": A Blind Approach to Blindness Prevention

VANDANA SHIVA

Genetically engineered "vitamin A rice" has been proclaimed a miracle cure for blindness – "a breakthrough in efforts to improve the health of billions of poor people, most of them in Asia." More than $100 million has been spent over ten years to produce a transgenic rice at the Institute of Plant Sciences at the Swiss Federal Institute of Technology in Zurich. The Zurich research team headed by Ingo Potrykens and Xudong Ye introduced three genes taken from a daffodil and a bacterium into a rice strain to produce a yellow rice with high levels of beta-carotene, which is converted to vitamin A within the body. The rice is being promoted as a cure for blindness since vitamin A deficiency causes vision impairment and can lead to loss of sight. According to the UN, more than 2 million children are at risk due to vitamin A deficiency. The work in Zurich was funded by grants from the Rockefeller Foundation, the agency that had launched the age of chemical agriculture in Asia through the Green Revolution, which led to the erosion of biodiversity and diverse sources of nutrition for the poor. The Swiss government and the European Community also supported the research. It will take millions more in dollars and another decade of development work at the International Rice Research Institute (IRRI) to produce vitamin A rice varieties that can be grown in farmers' fields.

Is the "golden" rice a miracle that is the only means of preventing blindness in Asia, or will it introduce new ecological problems like the Green Revolution did, and create new health hazards like other genetically engineered foods?

The genetic engineering of vitamin A rice deepens the genetic reductionism of the Green Revolution. Instead of millions of farmers breeding and growing thousands of crop varieties to adapt to diverse ecosystems and diverse food systems, the Green Revolution reduced agriculture to a few varieties of a few crops (mainly rice, wheat and maize) bred in one centralized research centre

(IRRI for rice and CIMMYT for wheat and maize). The Green Revolution led to massive genetic erosion in farmers' fields and knowledge, and the breakdown of farming communities, besides leading to large-scale environmental pollution due to use of toxic agrichemicals and wasteful use of water.

Genetically engineered rice, as part of the second Green Revolution, is repeating the mistakes of the Green Revolution while adding new hazards in terms of ecological and health risks. The "selling" of vitamin A rice as a miracle cure for blindness is based on blindness to alternatives for eliminating vitamin A deficiency, and blindness to the unknown risks of producing vitamin A through genetic engineering.

ECLIPSING ALTERNATIVES

The first deficiency of genetically engineering rice to produce vitamin A is the eclipsing of alternative sources of the vitamin. Per Pinstrup Anderson, head of the International Rice Research Institute, has said that vitamin A rice is necessary for the poor in Asia, because "we cannot reach very many of the malnourished in the world with pills." However, there are many alternatives to pills for vitamin A. Vitamin A is provided by liver, egg yolk, chicken, meat, milk and butter. Betacarotene, the vitamin A precursor, is provided by dark green leafy vegetables, spinach, carrot, pumpkin and mango. Women farmers in Bengal use more than a hundred plants that are green leafy vegetables.

Sources of vitamin A in the form of green leafy vegetables are being destroyed by the Green Revolution and genetic engineering, which promote the use of herbicides in agriculture. The spread of herbicide-resistant crops will further aggravate this erosion of biodiversity with major consequences in terms of increase in nutritional deficiency. For example, bathua, a very popular leafy vegetable in North India, has been pushed to extinction in Green Revolution areas where intensive herbicide use is a part of the chemical package.

The lower-cost, accessible and safer alternative to genetically engineered rice is to increase biodiversity in agriculture. Further, since those who suffer from vitamin A deficiency suffer from malnutrition generally, increasing the food security and nutritional security of the poor – by increasing the diversity of crops and therefore diets of poor people, who suffer the highest rates of deficiency – is the reliable means of overcoming nutritional deficiencies.

ENVIRONMENTAL COSTS

Vitamin A from native greens and fruits is produced without irrigation and wastage of scarce water resources. Introducing vitamin A in rice implies a shift from water-conserving alternative sources of vitamin A to a water-intensive system of production, since so called high-yielding rice varieties are highly water demanding. Vitamin A rice will therefore lead to mining for ground water or intensive irrigation from large dams, with all the associated environmental problems of waterlogging and salinization.

Further, as in the case of other genetically engineered crops, rice with vitamin A will have an impact on the food chain. The ecological impact on soil organisms and other organisms dependent on rice in the food chain should be part of the biosafety analysis of genetically engineered rice before it is released for production. Research has already shown that indigenous rice varieties support far more species than do Green Revolution varieties. How will genetically engineered rice impact upon biodiversity and the potential for disease and pest vulnerability?

SAFETY RISKS

Since rice is a staple eaten in large quantities in Asian societies, vitamin A rice could lead to excessive intake of vitamin A, especially among those who do not suffer from vitamin A deficiency. Excess vitamin A can lead to hypervitaminosis A, or vitamin A toxicity. Such toxicity is known to occur due to overingestion of vitamin A-rich food – for example, polar bear liver, or as the result of food faddism encouraged by oversolicitous parents, or as the side-effect of inappropriate therapy. Vitamin A toxicity can lead to abdominal pain, nausea, vomiting, dizziness, popillidena and bulging fontanelle. Chronic toxicity can occur after ingestion of large quantities of vitamin A for protracted periods. Chronic toxicity is characterized by bone and joint pain, hyperotosis, hair loss, dryness and fissures of lips, a nausea intraeranial hypertension, low-grade fever, pruritis, weight loss and hepatosplenomegaly.

While plant sources of vitamin A through its precursor betacarotene do not normally lead to vitamin A toxicity, safety analysis needs to be carried out to see whether the artificially introduced betacarotene in rice has the same impact on human health and other species as betacarotene in plants. Such safety analysis is necessary because there is now ample evidence to show that the assumption of substantial equivalence does not hold; naturally occurring organisms are not equivalent to genetically engineered ones.[1]

Natural sources of vitamin A are consumed seasonally and in small quantities as greens, relishes and fruits, and hence do not carry the risks of vitamin A toxicity. Rice-eating regions have been found to be associated with higher malnutrition than wheat-eating regions, especially after the Green Revolution, which destroyed the fish and plant biodiversity necessary for a balanced diet. These regions also have a higher prevalence of water-borne diseases like diarrhoea, amoebiasis, hepatitis A and E, dysentery, and vector-borne diseases like malaria, which increasingly comes in a more virulent form.[2] These health problems are known to involve damage to the liver. The additional risks of vitamin A toxicity under these conditions of vulnerable health endured by the poor in Asia needs to be assessed with care before a large-scale push is given to genetically engineered rice.

Further, the globalization of agriculture is leading to an increase in malnutrition in the Third World, as the most fertile ecosystems are diverted to the production of luxury export crops, and as domestic markets are destroyed due to the dumping of subsidized agricultural commodities. In India, per-capita consumption of cereals has declined by 12 per cent in rural areas over the past two decades. The shift from policies based on the "right to food" to free-trade policies will push millions into hunger and poverty.

Genetically engineered rice is part of a package of globalized agriculture that is creating malnutrition. It cannot solve the problems of nutritional deficiency, and it can introduce new risks of food safety. Since the vitamin A in rice is not naturally occurring and is genetically engineered, novel health risks posed by vitamin A rice will need to be investigated before the rice is promoted by IRRI and aid agencies or commercialized.

The risk assessment of vitamin A rice, as required under Annex II of the Convention on Biological Diversity's Biosafety Protocol, should therefore involve the following steps: (1) identification of any novel genotypic and phenotypic characteristics associated with the vitamin A rice that may have adverse effects on biological diversity in the likely potential receiving environment, taking into account also the risks to human health; (2) evaluation of the likelihood of these adverse effects being realized, taking into account the level and kind of exposure of the likely potential receiving environment; and (3) evaluation of the consequences should these adverse effects be realized. The risk assessment also needs to take into account the vectors used, the insects, the ecological differences between transgenic vitamin A rice and conventional rice varieties. The diverse contexts in which the rice is potentially to be introduced also need to be taken into account. This includes information on the location, geographical, climatic and ecological characteristics, including relevant information on biological diversity and centres of origin of the likely potential receiving environment.

It is these potential risks that have put a question mark over genetic engineering in agriculture. The genetically engineered vitamin A rice is now being used as a Trojan horse to push genetically engineered crops and foods. Mr Pinstrup Anderson, the IRRI director, has suggested that the "vitamin A rice could provide a public relations boost for plant biotechnology, which has been criticized by some environmentalists and consumer activists for promoting 'Frankenfoods'." It has yet to be established that genetically engineered rice is not a Frankenfood. Yet one thing is clear: promoting vitamin A rice as a cure for blindness while ignoring safer, cheaper, available alternatives provided by our rich agrobiodiversity is nothing short of a blind approach to blindness control.

NOTES

1. Erik Millstone, Eric Brunner and Sue Mayer, "Beyond 'Substantial Equivalence'," *Nature* 401, October 7, 1999, pp. 525–6.
2. In earlier years, a less hazardous form of malaria was caused by *Plasmodium vivax*; today the disease is increasingly becoming *Falciparum* malaria.

3

Cloning Profits: The Revolution in Agricultural Biotechnology

SONJA A. SCHMITZ

Gold beads blast from the barrel of a gun at 1,000 m.p.h. Their target: soft plant tissue nestled in a sterile Petri dish. The golden bullets blast their way through thick cell walls, membranes and cytoplasm of the plant cells. Finally, they penetrate the nuclear membrane and deliver the information with which they have been coated; cloned genes that insert themselves randomly along the chromosomes. Only a fraction of the cells will survive the bombardment. Only one in a million will express the new genetic information correctly. That cell will be grown to maturity and eventually, after years of nurturing in the hands of plant breeders, will produce fields of genetically engineered plants. Vandana Shiva, director of the Research Foundation for Science, Technology and Natural Resource Policy in India, describes the genetic engineering of plants as the latest manifestation of colonization: this time, invasion of the seed.[1] The process of colonization uses weapons in order to exploit another culture. The gene gun, known as Bioblaster, is the weapon of biotechnology. To many scientists it represents a technological advance that will revolutionize agriculture.

In recent years, up to 50 million acres of genetically engineered crops have been planted on US farmland. Genetically engineered foods are for sale at the supermarket. To the agricultural industry these events herald a revolution in biotechnology.[2] The public is concerned about use of genetically engineered organisms in agriculture because of the incompatibility of biotechnology with sustainable agriculture, the environmental impacts of releasing engineered organisms, and the safety of genetically engineered foods, among other issues. In drafting organic standards for the US Department of Agriculture, the question arose as to whether engineered foods should be considered organic. Nearly 300,000 people wrote letters and postcards to the USDA, largely in response to the suggestion that genetically engineered products might qualify as organic.

These issues are important. They are also difficult to resolve because they require going beyond the facts of science to debate ethics and values. Such a debate impinges on issues of democracy, the access to technology, and its effect on social structure and community. It requires us to have agency and agree upon a vision for the future. Currently, decisions surrounding the use of technology are driven primarily by economics and the corporations that dominate the market. Agricultural biotechnology is promoted as a sustainable means of addressing the issues of food security and safety, but, in fact, biotechnology is generating products whose sole purpose is to benefit and sustain industrial agriculture.

A CAREER IN BIOTECHNOLOGY COMES TO AN END

Science has always been a part of my life. As a child, I set up a laboratory bench in my bedroom and bred parakeets to see what colour feathers would result from the cross of recessive and dominant traits. I later cultivated an interest in other subjects – writing, literature, feminism and religion – but I turned to science as a way to make a living. My career began as a medical laboratory technician and I have continued working as a scientist ever since. When I graduated in the late 1980s with a Master's degree in molecular biology, biotechnology was an industry in its infancy. Although this was not a particularly socially conscious period of my life, the rhetoric that pervaded the university I attended clearly spoke to the hope that biotechnology would benefit society by revolutionizing agriculture and medicine.

Feeling proud of my educational accomplishment and hopeful that as a scientist I could make a valuable contribution to society, I accepted a lucrative offer from DuPont and began working as a molecular biologist in their agricultural biotechnology department. During my employment there, I worked for many good people and dedicated scientists. I learned that agriculture was no longer synonymous with the notion of the small family farm. Small to medium-size farms were quickly being replaced by industrial farms, and companies like DuPont played an important role in this transition.

Job satisfaction became an issue when I realized the projects on which I worked did not address food security or safety. For instance, there was the "novel starch" project: genetically engineering corn to produce altered starches. Although scientifically challenging, the end product was destined for the food processors who fill grocery shelves with such items as instant puddings, gravies and frozen dinners. Then there were the "fartless" soybeans, a project we jokingly called cloning the "FART 1" gene. These soybeans contained smaller amounts of the sugars which normally result in human flatulence. My colleagues were working on equally "socially beneficial" projects. How was this work part of the agricultural revolution that biotechnology was supposed to spur?

My ever-growing disillusionment culminated in the realization that biotechnology is big business. This issue gets lost in the debate between an industry that tells us we need their technology to ensure abundant, safe and nutritious food, and activists who focus on the health and environmental risks. At the core of the industry's argument for developing biotechnologies is the forever impending human population boom and the growing demand for more food. Some activists are quick to point out that food security is more an issue of political economy than productivity. There is sufficient food produced globally; it just doesn't get to everyone because of political and economic inequities.

Agricultural biotechnology is producing commodities whose sole purpose is to profit the industry that makes them. These new products do not address the social problems of hunger, nutrition, environmental safety and health to which they make claim. Aside from our recently acquired capacity to clone genes and transfer them between unrelated species, agricultural biotechnology is not generating anything unique or new. I no longer believe that the use of biotechnology heralds the dawn of an agricultural revolution.

CLONING THE VALUE OF FOOD

"Value-added" is a value-laden term. It is one of several buzz words adorning the overheads of corporate presentations. Whether they are called value-added, value-enhanced, identity-preserved or specialty grains, the terms refer to fruits and vegetables that (1) are designed for delayed ripening; (2) are engineered for increased yields of the starch, protein, fibre or oil extracted from them; or (3) contain genetically altered starch, oils or protein.

Value-added grains are not new to agriculture; they have existed since the 1930s when plant breeding produced high-yielding varieties of hybrid corn. Many of today's specialty grains, like corn seed that contains 8 per cent oil instead of the standard 4 per cent, are produced through plant breeding. The latest value-added grains are the result of genetic engineering: canola (oil-seed rape) that has elevated levels of unsaturated fats, or corn that may contain essential amino acids. Several biotech companies are developing "novel" starches. A closer look at starch technology allows us to examine the role that biotechnology plays in generating products that are questionably "novel."

Starches are extracted from a variety of different crops including corn, rice, wheat and potatoes. Each starch behaves differently during the cooking process, resulting in a variety of thicknesses that add texture to our food. Anyone who cooks knows that starch can be used "as is" to thicken soups, gravies and puddings. The food industry, however, is interested in adding functional properties to foods by chemically altering starch: food labelled with "modified starch" as one of its ingredients is made with a chemically modified starch. For example, many of the fat-free products on the market today are the result of replacing fats and oils with a chemically modified corn starch that creates the "mouth feel" of fats.

Current research efforts are aimed at developing "novel" starches through genetic engineering. For biotechnologists, this means experimenting with the pathways for starch metabolism in an effort to produce a "naturally" modified starch that behaves like a chemically modified starch. There is a fascination in exploring the relationships among enzymes to see how manipulating their genes can alter the structure of starch. They seek to replace chemically modified starches with genetically engineered ones, provided of course that instant pudding still looks and feels like instant pudding. The "typical" consumer will never know the difference between pie filling made with genetically altered starch and one that is chemically modified. What is so novel about that?

RESISTING HERBICIDE-TOLERANT CROPS

Herbicide-tolerant crops (HTCs) are another commodity produced by the so-called revolution in agricultural biotechnology. Unlike genetically engineered foods whose primary value to the industry is in cost savings, HTCs add profitability by creating a new market niche. HTCs and their corresponding herbicides are sold together as treatment systems and are marketed as sustainable, environmentally friendly additions to chemical agriculture. Monsanto's Roundup Ready soybeans and Aventis/AgroEvo's Liberty Link corn were among the first to reach the market. While the industry claims environmental benefits from the use of HTCs, the problems created by their use, especially the emergence of herbicide-resistant weeds and the transfer of genes to wild relatives of crops, invite just the kinds of solutions the chemical industry is eager to solve.

The industry cannot guarantee that HTCs will reduce the application of herbicide to farmland. Their careful use of qualifiers such as "has the potential to…" and "may lower the use of…" in their propaganda suggests doubt. Critics point out that rapid degradation of herbicide could result in the need to apply herbicide more often, as weeds emerge after it degrades. Also, since these herbicides are sprayed directly onto crops, as opposed to direct soil application, they are more likely to drift onto neighbouring fields. This could induce other farmers to switch to the same herbicide to protect their crops, thereby increasing overall herbicide use.[3]

HTCs do not exclude the use of other herbicides. A herbicide treatment regimen typically requires several applications of several different herbicides to combat all weeds. Although broad-spectrum herbicides eliminate most weeds, they are not effective on all of them. For instance, Liberty, a broad-spectrum herbicide sold by Aventis/AgroEvo, is not effective on the roots and rhizomes of quack and Johnson grasses.[4] Therefore farmers who plant Liberty Link corn and apply Liberty herbicide have to purchase additional herbicides to control these weeds.

Organisms have a natural compulsion to overcome barriers that stand in the way of survival. From bacteria to plants, organisms develop resistances after repeated exposure to deleterious substances whether they be antibiotics, viruses or herbicides. The emergence of herbicide-resistant weeds has plagued agriculture for decades. In the last ten years the number of weed populations resistant to an array of herbicides has increased.[5] Whether herbicide-resistant weeds emerge spontaneously or are the result of genetic engineering, the transference of resistance genes to neighbouring populations of the same species is a matter of environmental concern. More research needs to be done to assess adequately the impact of unwanted herbicide-resistant plants on surrounding habitats. The industry's response to the emergence of herbicide-resistant weeds is to implement an integrated weed management programme. This entails rotating HTCs and herbicides. For example, if Roundup-resistant soybeans are grown one year, they should be rotated with Liberty Link corn the following year. They remain silent on the issue of gene flow to natural populations.

Who should bear responsibility for the ecological risks posed by genetically engineered plants in the environment? Federal officials at the USDA believe that because chemical companies have an incentive to protect their investments, they will naturally bear the responsibility. Companies risk losing a portion of the herbicide market if resistance genes are transferred to wild relatives because it renders the herbicide obsolete. But, consider this: if new weeds are created by the transfer of herbicide-resistant genes, it forces the farmer to purchase other herbicides. The chemical companies can rely on the sale of traditional herbicides or offer the farmer a different seed/herbicide system for purchase. Why else would Monsanto and other companies invest billions of dollars in research that would appear to lose money eventually? The agricultural chemical companies have an army of chemists constantly developing new herbicides and just as many molecular biologists probing the metabolism of crop plants for genes that will confer resistance. The problems posed by the introduction of seed/herbicide systems only serve to invite another technological fix. Herbicide technology, whether "traditional" or genetically engineered, has a built-in obsolescence. Seed/herbicide systems are the logical solution for companies who operate in a system that depends upon technological fixes to keep generating new markets.

FALLING FOR THE BIOTECHNO-FIX

Beginning with the scientific revolution of the seventeenth century, we have increasingly sought answers through science and technology. There is now a pervasive belief that solutions to our technological problems lie in generating new technologies. We have become so enamoured of science that it is difficult to recognize its limitations and distinguish between its real economic power and the illusion of its power to provide sustainable solutions. This illusion protects us from confronting the more difficult issues that underlie the social problems science pretends to address. It is perhaps useful to remind ourselves that science is an art, a palette of techniques, in our relentless quest to understand our place in nature. It is much easier (and more fun!) to go into the lab and clone a gene than it is to tackle the social and economic restructuring required to create a society that is in sustainable harmony with the planet.

Our love affair with science and technology is not the only reason we maintain faith in their power to provide answers. Technology is directly related to generating capital. It has become a mechanism for generating new markets and commodities. While technology creates many modern innovations, it does not always represent the best solution, particularly in agriculture. Western economies have historically relied upon a built-in obsolescence to their products while maintaining the ability to create replacements. That is why companies have research and development departments. Biotechnology fits this agenda by generating products that will, for example, replace chemically modified foods with genetically engineered ones and by adding herbicide-resistant crops to their product lines.

Genetically engineered products are sold as the progeny of an agricultural revolution that will improve the quality, nutrition and abundance of our food supply. At best, the notion of an "agricultural revolution" reflects a social ignorance and arrogance that is typical of the scientific community. At worst, it is an intentional lie generated by an industry trying to promote itself. The insidiousness of these products lies in their role as technological fixes. When markets become saturated, as with chemically modified food starches, or problems arise, such as weed resistance to herbicides, new commodities must be created in order to realize profit. Scientists employed in research and development are under continual pressure to create new products or design technologies to decrease production and processing costs. Given the market pressure to produce new commodities, it is naive to believe that the agricultural industry will create solutions that are environmentally sustainable. When they use the term "sustainable" they mean able to sustain their share of the market economy.

Our current economic system is incompatible with an ecological society, one that can sustain itself within the environmental limits of our planet. Sustainable agriculture, when defined as regenerative, low-input, diversified, and decentralized, is the antithesis of industrial agriculture. Industrial agriculture depends upon obsolescence, high inputs, monocultures, and the centralization of power. While not yet a monopoly, industrial agriculture is increasingly controlled by just a few companies. Such centralization of power threatens democracy. We live in an era where the meaning of democracy has been diminished. The majority of us have lost the power to make decisions about issues that impact on our communities and environment. In order to live in ecological harmony within the limits of our environment, we must recreate an egalitarian, political forum for making decisions.

At a lecture given by Vandana Shiva, I asked which she perceived as the more imminent danger: the potential environmental and health risks of releasing genetically engineered organisms, or the insidious role the products of biotechnology play in promoting the growth of capitalism. She replied that while the two should not be separated, potential environmental risks may not be fully realized for another ten or fifteen years, but the social and economic effects of globalization are affecting peoples' lives today, particularly in developing countries where agribusiness strategies, including intellectual property rights over genetically engineered crops, threaten to devastate indigenous farmers. It is important to expose the role of agricultural biotechnology as another technological fix rather than a method that will revolutionize agriculture.

For those of us who work to create an ecological society, it is important to understand the role science and technology play. Only then can we begin to address the larger questions about what role we want technology and science to play in an ecological society. Would we choose to use biotechnologies if they were divorced from their role in sustaining the chemical agriculture industry? The application of technology should be determined by the needs of an ecological society, as opposed to the needs of a market economy, an economy on a collision course with environmental disaster.

NOTES

1. Vandana Shiva, *Biodiversity: Social and Ecological Perspectives*, London and New Jersey: Zed Books, 1991.

2. Howard Rudnitzky, "Another Agricultural Revolution," *Forbes Magazine*, May 20, 1996, pp. 159–61; Harlander, Susan, "Social, Moral, and Ethical Issues in Food Biotechnology," *Food Technology*, May 1991, pp. 152–9.

3. GRAIN, "Round Up Ready or Not," *Seedling*, vol. 14, no. 1, 1997, pp. 18–23.

4. L. Grooms, "Liberty Rings Throughout the US," *Seed World*, September 1996, pp. 10–14.

5. M. Jasnieniuk, A. Brule-Babel, and I. Morrison, "The Evolution and Genetics of Herbicide Resistance in Weeds," *Weed Science*, 44, 1996, pp. 176–93.

4

Genetically Engineered Foods: A Minefield of Safety Hazards

JENNIFER FERRARA AND MICHAEL K. DORSEY

Long before the first genetically engineered food product ever hit super-market shelves, public relations officials of agribusiness corporations were out in force to sell the idea of food bioengineering to American citizens, farmers, food processors, media and lawmakers. The corporate sales pitch was simple: genetic engineering was the latest, and the most monumental, in a long and fruitful line of agricultural breeding innovations that began with Gregor Mendel and his peas. Now that scientists could directly alter the genetic make-up of life forms to change their functions and characteristics, anything was possible. Transgenic agricultural products would make the food supply more nutritious and farms more prosperous and sustainable. It was the key to unlock an agricultural cornucopia. It would feed a starving world. And, most importantly, the genetically engineered products that were coming to market were completely safe.

Agribusiness and biotechnology corporations still use the same sales pitch today, even though their promises have always been false. Some genetically engineered foods have proven hazardous to human, animal and environmental health, and all pose unnecessary health and environmental risks. A mere quarter of a century after genetic engineering was first made possible, these products have become part of the world's food supply.

Health and safety risks posed by genetically engineered foods include: the spread of antibiotic resistance; new or increased levels of toxins in foods; hidden allergens; higher incidence of food contamination due to more factory-style farms; nutrient depletion resulting from reduced genetic diversity in the food supply; potential food shortages caused by the loss of biodiversity; and possible ecological disruption when genetically altered organisms are released into the environment. Those most at risk from engineered foods are the most vulnerable – children, the elderly, the chronically ill and the poor. And these are just the risks that we know to exist. Scientists warn that genetic engineering often has unexpected, and even unforeseeable, results. Every bioengineered plant, animal or product introduced into the food supply is like a wild card that could cause any number of possible health and environmental consequences.

Genetically engineered foods are not necessary, and their development could have been stopped. Indeed, that is what the US public wanted. Environmentalists, consumer groups, and social justice and sustainable agriculture advocates adamantly protested government approval and corporate marketing of engineered foods. Opinion polls showed that average citizens were wary of genetically engineered foods and did not want them.[1] But in the rush to develop genetic engineering, the public had virtually no say in determining the scientific and social agenda. So today genetically engineered foods are a part of the food supply, and the public is plagued with yet another food safety hazard. The context in which these foods were developed, approved, and marketed poses as much of a food safety hazard as the actual engineered products themselves. We are faced with these new food safety threats because agribusiness corporations and the US government were, and still are, banking on biotechnology.

THE RUSH FOR PROFIT

The first bioengineered agricultural products to be developed, approved and marketed did not improve food nutrition or enhance human, animal, and environmental health. They were designed specifically to increase industrial agricultural production and facilitate industrial food processing. They include products like the Monsanto corporation's recombinant Bovine Growth Hormone (rBGH), which artificially increases a cow's milk production, Roundup Ready soybeans designed to survive applications of Monsanto's own toxic herbicide Roundup, and Calgene's Flavr Savr tomato, engineered to ripen on the vine, to withstand packing and shipping, and to have and a long shelf life. In a 1997 report, 93 per cent of the genetically engineered food crops undergoing field tests were intended to make food processing more profitable, while 7 per cent were engineered for nutritional or flavour enhancement.[2] By developing, owning and heavily marketing genetically engineered seeds that would produce novel products for use in the industrial food system, agribusiness corporations would ensure the expansion of that system and tighten their monopoly over it.

Beginning in the 1970s, agribusiness and chemical corporations made large investments in biotechnology, reflecting corporate projections that biotechnology would be the money-maker in industrial agriculture in the coming decades.[3] Corporations like Monsanto and Eli Lilly began buying out small biotechnology companies and building their own biotechnology research centres.[4] DuPont built a $120 million "life sciences" research centre,[5] and Monsanto spent millions on a Molecular Biology Center in St Louis.[6] They also began granting substantial funds to US universities for biotechnology research.[7] At about the same time, corporations like Ciba–Geigy and Sandoz (which merged in 1996 to create the Novartis corporation), Shell and Monsanto began buying agricultural seed companies, an industry worth tens of billions of dollars.[8] By 1984, more than a thousand new and established corporations were involved in agricultural genetic research.[9]

Market predictions explained oil and chemical corporations' rush into agricultural biotechnology. In the 1980s it was believed that the worldwide market for "agrigenetic" products would be ten times that of genetically manufactured medical and pharmaceutical products.[10] One report stated that the market for bioengineered agricultural products would reach $50 to $100 billion by the year 2000.[11] Monsanto predicted that during the 1990s they would market genetically engineered wheat and soybean products that would generate $300 to $500 million in revenue.[12] In 1984, investments in genetic engineering totalled over $2.5 billion.[13] Six years later, biotechnology companies sold $17.7 billion in new stock in a five-month period.[14]

BGH: AN UNCONTROLLED EXPERIMENT

Recombinant BGH (rBGH) is a drug injected into dairy cows that mimics their naturally occurring growth hormone and is reported to increase milk production by 10 to 20 percent. As the first product of genetic engineering to significantly impact the US food supply, its consequences are perhaps the best understood. This artificial cow hormone is produced by splicing the growth hormone gene from cows into E. coli bacteria to generate commercial quantities of the drug. Using research generated by the Monsanto corporation and Monsanto-funded university scientists, the US Food and Drug Administration (FDA) approved the general use of the drug in November 1993, ostensibly to increase the nation's already overabundant milk supply.

Before critical data concerning the human health effects of rBGH milk was available, and eight years before the FDA ruled that rBGH was safe for cows, the agency declared that milk from rBGH-treated cows was safe for human consumption[15] and permitted milk from test herds involved in rBGH research studies to be sold to the general public in states including California, Wisconsin[16] and Vermont.[17] The FDA reasoned that there was no difference between genetically engineered rBGH and BGH as it naturally occurs in cows, nor between rBGH-milk and milk produced naturally. But rBGH, in fact, has a different molecular structure to the natural hormone, and there are significant differences between milk from rBGH-treated cows and milk produced without rBGH.[18] Scientific evidence showing that rBGH is not safe for humans or cows arose during the FDA's review of the drug and is still emerging today. The FDA continues to ignore the safety questions and allow Monsanto to profit from the sale of this dangerous drug.

It is well documented that rBGH has adverse effects on the health of cows, and this fact is cause for human concern. As rBGH increases the incidence of udder infections (mastitis) and numerous other diseases in cows, farmers utilize increasing amounts of legal and "extra-label" antibiotics.[19] This in turn raises antibiotic residues in milk. Some people are allergic to antibiotics and can be affected by antibiotic residues. Though the dairy industry claims that the milk supply is adequately tested for drug residues, the FDA analyses only 500 milk samples a year nationwide for residues of just twelve antibiotics approved for use on dairy cows. A 1992 General Accounting Office (GAO) report indicated that eighty-two drugs which may leave residues in milk were known to be or suspected of being used on dairy cows. Of these, only thirty are approved for use on dairy cows.[20] The medical establishment relies on antibiotics to treat diseases in people, and antibiotic-resistant bacteria – which can proliferate rapidly where antibiotics are used in excess – can render antibiotics ineffective against illness.

Cows treated with rBGH must also eat a high-protein diet to be able to produce more milk. This diet might appear healthier, but rendered protein from animals, including cows, is a major ingredient in high-protein feeds. Independent scientists have expressed concern that feeding "rendered" cow-derived feed to cows might increase the risk of spreading the deadly disease known as bovine spongiform encephalopathy – also known as BSE or "mad cow disease."[21] In 1996 Britain faced a public health crisis when at least ten people died from the human variant of BSE, Creutzfeldt–Jakob Disease (CJD). It is believed that they contracted the disease after eating beef from BSE-infected cows. Since CJD is always fatal, the European Union was prompted to ban the importation of British beef and the British government ordered millions of beef cattle to be destroyed. Scientists, including Dr Michael Hansen of the Consumers' Union, warn that if US cattle are infected with BSE and the practice of feeding rendered animal proteins to dairy cows continues, rBGH could hasten the spread of the disease and increase the chances that BSE-infected meat would enter the human food supply.[22]

Yet recombinant BGH fits perfectly into the industrial food system. Large, highly mechanized farms with hundreds, even thousands, of cows are more likely to use the drug "successfully" than smaller farmers who do not have the extra money to buy the drug and high-protein feed, pay the veterinary bills, and accept that their cows will have a significantly shorter productive lifespan.

BGH AND CANCER

Of all the health risks associated with rBGH, Monsanto and the dairy industry have tried hardest to quell public concern about the drug's possible link to cancer. Though the FDA says there is no real difference between rBGH-produced milk and non-rBGH milk, levels of a growth hormone called Insulin-like Growth Factor (IGF-1) are increased in rBGH-produced milk. IGF-1 in cows is chemically identical to IGF-1 in humans, an active hormone that causes human cells to divide and has been linked to cancer.[23]

Recombinant BGH stimulates increased production of IGF-1, which, in turn, is the hormone that actually stimulates milk production. The FDA acknowledges that studies show a 25–70 per cent increase in IGF-1 in rBGH-produced milk,[24] and other rBGH proponents acknowledge that IGF-1 levels are at least doubled.[25]

The FDA and the dairy industry maintain that IGF-1 cannot survive the human digestive system and that increased levels of IGF-1 in milk will have no effect on human health. The FDA based this conclusion on unpublished IGF-1 toxicity studies conducted by Monsanto and Eli Lilly.[26] Independent scientists have criticized these studies, citing the lack of sufficient controls, the short study periods, and the fact that tests were conducted on adult rather than infant rats. Though test subjects showed statistically significant growth increases, the FDA concluded that IGF-1 had no effect on the rats.[27] Studies show that IGF-1 may indeed survive digestion, and evidence that high levels of IGF-1 in milk could result in serious human health effects continues to build. A study published in the *Journal of Endocrinology* in August 1995 shows that IGF-1 in the presence of casein, the principal protein in cows' milk, is not destroyed by digestion.

If IGF-1 survives digestion, as studies indicate, the question remains as to what effects artificially raised levels of IGF-1 will have on the human body. We already know that people who produce excessive amounts of IGF-1 in their own bodies suffer from a condition called acromegaly, characterized by enlarged hands, feet, face and head. Studies within the last decade suggest that IGF-1 may also encourage tumour production.[28] A 1994 study published in the British medical journal *Lancet* showed that IGF-1 treatment of human small intestine cells promoted cell division (cancer is uncontrolled cell division). A 1995 study in *Cancer Research* found that IGF-1 promotes cancer tumour growth in laboratory animals and humans by preventing cells from dying in their normal life cycle.[29] A 1998 study published in *Science* magazine found that men with elevated, but still normal, levels of IGF-1 in their blood were four times more likely to develop prostate cancer. Further, a May 1998 study in the *Lancet* found that among pre-menopausal women those with the highest levels of IGF-1 were seven times more likely to get breast cancer.[30]

Analysing these studies, Peter Montague, director of the Environmental Research Foundation and editor of *Rachel's Environment and Health Weekly*, pointed out that IGF-1 is associated with larger relative risks for common cancers than any other factor yet discovered. "Thus," Montague writes, "these latest cancer findings raise important public health questions about the safety of milk from cows treated with bovine growth hormone."[31]

GENETICALLY ENGINEERED CROPS

Knowing the story of rBGH makes corporate assurances about the safety of bioengineered foods and agricultural crops even harder to swallow. According to the Union of Concerned Scientists, genetically engineered canola, corn, cotton, papaya, potato, soybean, squash, tomatoes and other crops have been approved for sale in the United States, with many more awaiting approval.[32]

Six different bacteria bioengineered to control insect pests or increase crop yields have also been approved. Of the thirty-eight approved biotech crops at the end of 1999, almost all are among the country's major crops. Fifteen of these are engineered to withstand toxic herbicide application, and six are engineered to make industrial-scale marketing or food processing easier. Most of the remainder secrete a toxin from *Bacillus thuringiensis* (Bt) to control insect pests. Engineered crops compound the safety problems already caused by industrial agriculture as well as creating new ones.

Industrial agriculture promotes genetic uniformity in crops, as farmers plant fewer varieties of crops bred for higher yields and other characteristics conducive to food processing. History has demonstrated the danger of a uniform food system. The uniformity of the potato crop was a factor in the Irish potato famine in the 1800s. In 1916, a wheat disease in the US destroyed 200 million bushels of wheat.[33] Around 1950 corn breeders began developing varieties whose cytoplasm (the liquid medium surrounding each cell's nucleus) expedited the breeding process by rendering plants unable to self-pollinate. In 1970, farmers planted 46 million acres of corn that had been bred with this male-sterile breeding system. Unknown to the plant breeders and the farmers, T-cytoplasm also carried a mitochondrial gene that made the corn plant susceptible to a particular fungus. That fungus duly infected the crop, causing the US corn blight of 1970.[34] In 1972, the National Academy of Sciences issued a report warning that most of the country's major food crops were "impressively uniform genetically and impressively vulnerable," but US agriculture has not changed. In 1985, 43 per cent of US corn crops were from just six inbred strains of corn plants.[35] In 1996, the first commercial genetically engineered crops were planted in the United States.

The industrial food system thrives because of uniformity, which is precisely one of the goals behind genetically engineered crops. These crops will be used primarily on large, highly mechanized, monoculture farms. As Jack Doyle describes, in these "simplified environments" a small genetic change in a crop can produce great and swift consequences.[36] These consequences could involve human and ecological safety hazards and destruction of a country's food security.

One potentially harmful health consequence of expanding US agriculture's uniformity is a degradation of the nutritional value of the food supply. Agribusiness's track record indicates that genetic engineering may decrease nutritional content. Prior to genetic engineering, some crops bred through traditional methods to fit the industrial food system lost nutritional value. Tomato varieties bred for mechanical harvesting lost 15 per cent of their vitamin C content. Potatoes bred for food processing lost 50 per cent of their riboflavin and 40 per cent of their potassium. As it happens, some genes that enhance characteristics good for mechanical harvesting and processing are inversely related to genes for good nutrition.[37]

Despite this record, agribusiness corporations attempt to sell the idea of biotechnology by arguing that it will make the food supply more nutritious, and there is currently a big push within the agribusiness industry to try to engineer foods for increased nutrition. Companies see this as an opportunity to gain acceptance for genetically engineered foods among health-conscious consumers. Even if these attempts are successful, there is no guarantee that engineered high-nutrition foods will be healthy or safe. Altering a plant's (or an animal's) nutrient composition affects the organism's metabolism, which will likely create other changes in the organism. Engineered organisms are not tested for these secondary effects before going to market. Whatever trait a plant (or other organism) has been genetically engineered to express, disruptions in essential metabolic pathways and processes of genetic regulation are not only highly likely, but extremely difficult to detect.

ALLERGIES

One corporation's attempt to engineer a food product for higher nutritional content confirmed the warnings of biotechnology opponents that a single intentional change in a plant's DNA can cause other, unintended health and safety problems. In an attempt to increase usable protein levels in soybeans, researchers at Pioneer Hi-Bred International (now wholly owned by DuPont) inserted a brazil nut gene into soybeans. Blood tests conducted for Pioneer at the University of Nebraska found that the brazil nut gene produced proteins that caused strong, potentially deadly allergic reactions in samples from people allergic to brazil nuts.[38] Pioneer had intended to use its engineered soybeans for high-protein animal feed, but dropped plans to market the crop.[39]

Millions of people in the US have food allergies. Food allergies are confirmed in 8 per cent of children and 2 per cent of adults, and one-quarter of Americans believe that they or their children suffer from food allergies. According to Dr Marion Nestle of New York University, many more people suffer from food sensitivities. Nestle's editorial in the *New England Journal of Medicine* reported that this trend matches the rising proportion of processed foods in the food supply.[40] Allergic symptoms range from discomfort and swelling to death. Allergies cannot be cured, and the only way for people with food allergies to stay healthy is to avoid the foods to which they are allergic.

Though the potential for genetic engineering to transfer allergenicity from one food to another had been suspected, the brazil nut study was the first to confirm the suspicion. The FDA only requires labelling and safety testing of engineered foods containing genes from about ten common sources of food allergens, such as nuts and shellfish. Yet it is not only those who suffer from known allergies who are in danger from bioengineered foods. As engineered foods containing genes from known but not common sources of food allergies do not have to be labelled, people with uncommon allergies may unknowingly eat a genetically engineered food with an allergy-causing gene that they thought was safe. Genetically engineered foods are also produced with genes from non-food sources, such as non-food plants or bacteria. Researchers were able to determine that the brazil nut/soybean caused allergic reactions because they were able to test people with known brazil nut allergies. It is much harder to determine whether foods spliced with genes from non-food sources will cause allergic reactions in people. There is, however, growing evidence that problems are on the horizon.

In early 1999, British researchers found that allergic reactions to soybeans had increased by 50 per cent from the previous year. The study, by researchers at the York Nutritional Laboratory, was reported as offering "real evidence that GM [genetically modified] food could have a tangible, harmful impact on the human body."[41] John Graham, managing director of the York Labora-tory, was quoted as saying that the study "raises serious new questions about the safety of GM foods because it is impossible to guarantee that the soya used in the tests was GM-free."[42] For the first time in seventeen years of testing soya was one of the laboratory's ten most allergenic foods.

Arguably these data are inconclusive, and Graham subsequently denied ever making such a claim;[43] it is a warning sign nonetheless. Indeed, the allergenic potential of new GM proteins, writes Marion Nestle, "is uncertain, unpredictable, and untestable." Though the public got lucky with the transgenic brazil nut soybeans, Nestle warns, "The next case could be less ideal, and the public less fortunate."[44] The appearance in mid-2000 of a potentially allergenic variety of Bt corn in the US food supply added new fuel to these well-founded fears.

DR PUSZTAI'S POTATOES

Allergenicity tests on transgenic soybeans were the first to affirm critics' concerns about the safety of genetically engineered foods. Many questions remain, in light of the wide variety of genetic manipulations that are cur-rently being explored. There may be other effects on human health, less expected than allergies, from the consumption of transgenic foods. One study conducted at the Rowett Research Institute (RRI) in Scotland sought to investigate the potential health effects of potatoes that had been engineered to express an insecticidal compound (lectin) from the snowdrop plant, by feeding them to rats and evaluating their effects on metabolism and organ development. The study, conducted under the direction of renowned lectin specialist Dr Arpad Pusztai, compared transgenic potatoes with their parental (non-altered) potatoes, testing for nutritional composition and effects on laboratory animals. Pusztai found that the amounts of protein, starch and sugars in the genetically altered potatoes varied from unaltered varieties by as much as 20 percent.[45] This directly contradicts the claim of "substantial equivalence" for transgenic foods and offers evidence of possible gene silencing, suppression and/or alter-ation of neighbouring genes as a result of gene insertion (see Chapter 6).

This alone was a significant and crucial finding, sufficiently damaging to halt further work on these potatoes, according to Pusztai. But as the study continued, it found that when these transgenic potatoes were fed to laboratory rats in controlled studies the result was a in significant reduction in the weights of the rats' vital organs. Many organs, including the intestine, pancreas, kidneys, liver, lungs and brain underwent highly significant changes in weight.[46] Further, the immune system response was significantly depressed, and there was evidence of serious intestinal inflammation and infection.[47] Since potatoes directly sprayed with the same lectin had no such effects on the rats, Pusztai concluded that the effects were associated with other elements of the DNA construct that is used to increase the odds of successful genetic engineering. These include such components as the viral promoter sequence from cauliflower mosaic virus that is commonly used to carry the lectin genes into the host.[48]

On August 10, 1998, Dr Pusztai announced his research results on a television talk show. Within days a flurry of news articles and other media in the UK denounced GM foods. Two days later, the RRI's director denied the existence of Pusztai's experiments. Then, marked as a whistle-blower, Pusztai was suspended from the RRI. The Institute's director questioned Pusztai's character and his credibility as a researcher. As the initial media frenzy ebbed, Pusztai distributed his findings to colleagues around the world. On February 12, 1999, he called a press conference where he presented a joint statement from twenty-one scientists from thirteen different countries stating that the study's conclusions were justified, and that his methods and data were legitimate. This was supported by an Audit Report published three months after Pusztai's suspension, which confirmed that despite previous false claims by the RRI's director, Dr Pusztai's team had in fact performed several feeding experiments on rats with transgenic potatoes, and that the experimental approach and methods were correct. Yet the attacks against Pusztai did not cease. In May 1999, four major reports lauded the safety of GM foods and further condemned Pusztai's efforts to vindicate his work. Perhaps significantly, all of the reports were released within two days of each other.[49]

While the attacks and efforts to denigrate Pusztai and his work continued, the British Medical Association (BMA) helped validate the importance of his findings with a public call for a moratorium on the planting of genetically modified crops. The Association cited numerous potential hazards of engineered foods that have not been studied sufficiently, supported full labelling of products containing GMOs, and called the use of antibiotic-resistant marker genes "a completely unacceptable risk, however slight, to human health."[50] The BMA's announcement came in the immediate aftermath of Pusztai's initial announcement and subsequent attacks on him by colleagues at the RRI and by members of the British Royal Society.

The British Medical Association helped draw attention to another key problem with genetically engineered foods that had already raised significant concern throughout Europe. Virtually all genetically engineered crops are produced with an antibiotic resistance marker, which serves to facilitate detection of the altered gene. A genetically engineered food supply containing marker genes heightens the chances that antibiotic resistance will be transferred to bacteria in the environment, including bacteria that live in the guts of

animals and humans. Bacteria are already prone to becoming antibiotic resistant through direct exposure to high levels of antibiotics used on farm animals
and elsewhere, and the presence of marker genes in the environment may
exacerbate antibiotic resistance. The health risk is that as more bacteria become resistant to antibiotics, antibiotics will become increasingly ineffective in
treating people, especially children, for bacteria-related diseases and infections,
contributing to a general slow deterioration of public health. No studies have
been done to determine the long-term effects of bioengineered foods containing marker genes on human health.

ARMED PLANTS

Over one billion pounds of insecticides, herbicides and fungicides are used on
US food crops every year. These chemicals are all classified as pesticides –
toxic chemicals that kill insects, weeds and fungi, respectively. But these toxins
don't just harm what they are targeted to kill; pesticides are also toxic to farm
workers, citizens who live near sprayed fields or drink from contaminated
water supplies, consumers who eat pesticide-laden foods, and wildlife. Some
pesticides that are legally used in the US are known or suspected to cause
reproductive damage, cancer and other serious health problems in humans.
Pesticides are often used in combination with each other, and studies show
that combinations of chemicals have a greater effect on human health and the
environment than chemicals used in isolation.

Industrial agriculture relies heavily on pesticides. Crops grown in large
monocultures are extremely susceptible to insect and weed infestation and
disease. Instead of changing the way crops are grown, US agribusiness has
steadily increased its use of toxic chemicals since the 1950s. As pesticide use
has increased, insects and weeds have adapted to these toxins. Over five hundred species of insects are now resistant to insecticides,[51] and over a hundred
weed species are herbicide-resistant.[52]

The public is increasingly concerned about the effect of pesticides on the
food supply and on their health. Agribusiness hit a sensitive public nerve
when it marketed genetic engineering as a way to reduce pesticide use and
make US agriculture more "sustainable." But instead of freeing farmers from
chemical dependency, genetic engineering is giving them cause to use more
toxins. Chemical and biotechnology corporations are engineering crops that
are herbicide-tolerant. In other words, crops have been engineered to withstand application of certain herbicides in doses they could not survive naturally. Companies that manufacture particular brands of herbicide have designed
crops to be sprayed with that herbicide; the genetically engineered crops and
their matching herbicides are sold to farmers as "packages." Agribusiness corporations aim to increase sales of already high-selling herbicides and reap new
profits in the market for genetically engineered crops.

For example, Monsanto has aggressively marketed its soybeans that are
genetically engineered to withstand high doses of Roundup, the trade name

for Monsanto's brand of herbicide that contains the active ingredient glyphosate.[53] Monsanto requires farmers who choose to plant Roundup Ready soybeans to sign affidavits pledging to use only the company's brand of glyphosate, though farmers can still use several additional herbicides on their crop. By engineering Roundup-resistant beans and inserting this stipulation into agreements with farmers, Monsanto has been protecting itself from likely profit losses as the patent on Roundup, their greatest profit making product, has now run out. Roundup has accounted for up to 17 per cent of Monsanto's annual sales,[54] and approximately 50 per cent of the company's operating income.[55]

Roundup Ready soybeans are increasing farmers' use of Roundup, while applications of other pesticides are not reduced in proportion. Though Monsanto advertised glyphosate as a tool that can "eradicate weeds and un-wanted grasses effectively with a high level of environmental safety," research indicates otherwise.[56] Studies show that glyphosate causes long-term toxic and reproductive effects, and glyphosate-containing products cause genetic damage. Though the EPA has deemed it a non-carcinogen, the Oregon-based Northwest Coalition for Alternatives to Pesticides has analysed studies for glyphosate and determined that the carcinogenicity of glyphosate-containing products remains unknown pending more long-term studies.[57] Glyphosate is also acutely toxic. A 1993 study at the University of California at Berkeley's School of Public Health found that glyphosate was the most common cause of pesticide-related illness among landscape maintenance workers in California, and the third leading cause among agricultural workers.[58]

Plants can take up glyphosate into parts used for food. Glyphosate has been found in strawberries, wild blueberries and raspberries, lettuce, carrots and barley after crops or wild areas have been sprayed. The herbicide has been found in food plants a year after fields have been sprayed. When glyphosate is used before harvest on grain crops to dry out the grain, it can leave residues in the grain. Glyphosate can contaminate surface water when the soil particles it binds to are washed into rivers, streams, lakes and ponds. The herbicide kills beneficial insects and worms, and has indirect hazardous effects on birds and small mammals. Glyphosate and glyphosate-containing products are also acutely toxic to fish.[59]

Clearly, the use of Roundup on genetically engineered soybeans and other crops poses immediate health threats to farmworkers, the public and the environment. And as is often the case with genetically engineered foods, there are also questions about secondary, unintended health effects of these genetically engineered soybeans. Marc Lappé and Britt Bailey at the Center for Ethics and Toxics (CETOS) in Gualala, California, raised questions about various changes in these soybeans as a result of genetic engineering in their book *Against the Grain: Biotechnology and the Corporate Takeover of Our Food*.[60]

The authors describe how in order to make a plant resistant to Roundup, Monsanto's genetic engineers boost the activity of a gene that produces critical amino acids. The higher levels of amino acids inside plant cells change the plant's metabolism. A byproduct of soybean metabolism is a class of compounds known as phytoestrogens, named for their similarity to human oestrogenic hormones. Oestrogens have many roles within the human body, including controlling sexual function and development, immune function, and carcinogenesis. An independent, non-industry study cited by Lappé and Bailey found that Roundup Ready soybeans may have increased levels of phytoestrogens. Unrelated data about the already high activity of phytoestrogens in soy products and their hormonal effects on infants is causing the medical community to re-evaluate the wisdom of a heavy soy diet for very young children. The question of whether or not Roundup Ready soybeans have even higher levels of phytoestrogens than non-engineered soybeans is still unanswered. Considering the possible consequences, it is a question that warrants serious consideration.

Monsanto is also marketing canola (rapeseed), corn and cotton crops that are resistant to glyphosate. Other companies are marketing corn and canola crops resistant to the herbicide glufosinate.[61] Rhone–Poulenc (now Aventis and its planned spin-off Agreva) developed a variety of cotton resistant to bromoxynil, an acute toxin known to cause birth defects and a possible human carcinogen.[62] Bromoxynil is already used on US food crops, including cereals, corn and sorghum. The increased use of herbicides on genetically engineered crops may cause weeds to become resistant to them at a faster rate. In turn, herbicide-resistant weeds can spread, making it more likely that other farmers who use these herbicides on non-engineered crops – some of which are food crops – will increase their use of toxic chemicals to kill virulent weeds. This points to a twofold health threat from herbicide-resistant crops. First, there are the possible health problems caused by the direct increased use of herbicides on genetically engineered crops. And second, other crops may receive heavier doses of herbicides, which can end up on food, in water supplies and in the environment. It is still unknown whether or not some of these crops will cause secondary, unintended health effects.

Crops have also been engineered to resist viruses. For example, the Asgrow seed company – a Monsanto subsidiary – markets a virus-resistant yellow crookneck squash. To make plants resistant to viruses, genetic engineers take genes from viruses that cause plant disease and insert them into the plants, in a sense inoculating them. Little is known about how the parts of viruses spliced into plants will behave over time, but studies have shown that these can combine with other viruses in the environment to create new, even more virulent viruses that may attack plants.[63] Nevertheless, industry apologists say the threat to the environment from virus-resistant genetically engineered plants is "vanishingly small," and that the benefits of such crops outweigh the risks.[64] Yet those who will benefit most from these new crops are undoubtedly the agribusiness corporations that sell them; the risks to everyone else remain. Though plant viruses don't pose direct human health threats, these experiments indirectly threaten food security.

CONCLUSION

There are many genetically engineered agricultural and food products in the corporate pipeline, and, just like those already on the market, all carry the risk of unknown or suspected health effects. Thousands of unmonitored field tests of genetically engineered organisms have occurred in almost every state in the US.[65] For instance, potatoes genetically engineered with genes from chickens, moths, peas, soybeans and viruses have all been field tested.[66] Studies show that transgenic plants can transfer engineered traits to neighbouring crops and to wild relatives. Others show that pollen can escape from fields of genetically engineered crops and fertilize plants in other locations.[67]

Fish genes have been inserted into tomato and tobacco plants. Human genes have been inserted into mice, cattle, pigs, salmon and sheep.[68] The American Cyanamid corporation has inserted a toxin gene from a scorpion into an insect virus to be used as a pesticide on lettuce, cabbage and tobacco without knowing the possible ecological and human health effects.[69] US genetic engineers are also working to develop edible vaccines for common yet deadly illnesses like cholera.[70] In theory, foods such as bananas can be genetically engineered to produce antigenic proteins to stimulate the human body's immune system. The hope is to market these edible vaccines in Third World countries where cholera rates are high, perhaps selling the engineered plants to communities to grow themselves. But if a plant is genetically altered to carry a vaccine, other effects on plant metabolism are likely to occur, with unknown results. In addition, no one knows what would happen if vaccine-carrying banana trees, for example, crossed with or outcompeted non-engineered trees, or if people accidentally ate vaccine-carrying bananas. Notably, the primary market for these genetically engineered vaccine-foods will not be in the United States.

Genetic manipulation often has unexpected and hazardous results, disrupting the tight genetic control and metabolic balance with which living organisms have evolved. As corporate laboratories spill over with new product ideas, agribusiness corporations have increased the risks of public exposure to health and safety hazards immeasurably. And the US government is right behind them, clearing the way with lax regulations for a genetically engineered food supply (see Chapter 2 in this volume).

Proponents of genetically engineered foods say that opponents are spreading an irrational fear of the unknown that will only hold back so-called progress. But people are fighting bioengineered foods for very real – and very serious – reasons. Evidence is growing that genetically engineered foods are not safe. At least one tragedy has occurred, when a genetically engineered dietary supplement approved by the FDA, L-tryptophan, killed twenty-seven people and injured 1,500 in 1990.[71] Even without the help of genetic engineering, agribusiness has already gained excessive control over the world's food system, compromised nutrition, poisoned the environment, and contributed to rising cancer rates. The full impact on human and environmental health of a genetically engineered food system is simply and undeniably unforeseeable. The public must refuse to be blind-sided by this so-called agribusiness miracle.

NOTES

1. Michael K. Hansen, *Biotechnology and Milk: Benefit or Threat?* Mount Vernon, NY: Consumers Union of the United States, 1990, pp. 17–18.

2. Kristin Dawkins, *Gene Wars: The Politics of Biotechnology*, New York: Seven Stories Press, 1997, p. 30.

3. Jack Doyle, *Altered Harvest*, New York: Viking Penguin, 1985, p. 27.

4. Ibid., p. 109; Steven M. Gendel, A. David Kline, D. Michael Warren and Faye Yates, eds, *Agricultural Bioethics: Implications of Agricultural Biotechnology*, Ames: Iowa State University Press, 1990, p. 182.

5. Doyle, *Altered Harvest*, p. 26.

6. Ibid., p. 109.

7. Gendel et al., *Agricultural Bioethics*, p. 186.

8. Ibid., p. 183; Doyle, *Altered Harvest*, p. 97.

9. Doyle, *Altered Harvest*, p. 27.

10. Ibid.

11. Ibid.

12. Ibid., p. 109.

13. Michael Fox, *Superpigs and Wondercorn: The Brave New World of Biotechnology and Where It All May Lead*, New York: Lyons & Burford, 1992, p. 6.

14. Ibid.

15. Hansen, *Biotechnology and Milk*, p. 22.

16. Ibid., p. 3.

17. Tim Peek, "Cow Growth Hormone Has Farmers Upset," *Addison County Independent*, May 11, 1989.

18. Hansen, *Biotechnology and Milk*, p. 3.

19. Under the FDA's Extra-Label Use Policy, the agency does not take enforcement action against farmers who use a drug that is not approved for cows but legal for some other animal. This may include antibiotics that are suspected carcinogens, or are responsible for other health problems in animal toxicity tests. See Hansen, *Biotechnology and Milk*, p. 10.

20. Michael K. Hansen, "Testimony before the Veterinary Medicine Advisory Committee on Potential Animal and Human Health Effects of rBGH Use," New York: Consumers Union, March 31, 1993.

21. Ibid.; Sheldon Rampton and John Stauber, *Mad Cow U.S.A.: Could the Nightmare Happen Here?*, Monroe, ME: Common Courage Press, 1997, p. 3.

22. Hansen, *Biotechnology and Milk*.

23. Peter Montague, "Milk Safety," *Rachel's Environment & Health Weekly* 454, August 10, 1995 (Annapolis, MD: Environmental Research Foundation); Samuel S. Epstein, "Unlabeled Milk from Cows Treated with Biosynthetic Growth Hormones: A Case of Regulatory Abdication," *International Journal of Health Services*, vol. 26, no. 1, 1996, pp. 173–85; Joel Bleifuss, "Mucking with Milk," *In These Times*, January 10, 1994; Peter Montague, "Breast Cancer, rBGH, and Milk," *Rachel's Environment & Health Weekly*, no. 598, May 14, 1998.

24. Epstein, "Unlabeled Milk."

25. Montague, "Milk Safety," p. 25.

26. Epstein, "Unlabeled Milk."

27. Ibid.; also Bleifuss, "Mucking with Milk," p. 12.

28. Montague, "Milk Safety."

29. Mariana Resnicoff et al., "The Insulin-like Growth Factor I Receptor Protects Tumor Cells, *Cancer Research* 55, June 1, 1995, pp. 2463–69.

30. Peter Montague, "Milk, rBGH and Cancer," *Rachel's Environment & Health Weekly* 593.

31. Ibid.

32. Union of Concerned Scientists, "Foods on the Market," *Foodweb: A Public Voice on Food, Farming and the Environment,* February 2000, pp. 12–14; also at http://www.ucsusa.org.

33. Doyle, *Altered Harvest,* p. 177.

34. Ibid., p. 13.

35. Ibid., p. 15.

36. Ibid., p. 241.

37. Jack Doyle, "Potential Food Safety Problems Related to New Uses of Biotechnology," in *Biotechnology and the Food Supply: Proceedings of a Symposium,* Washington, DC: National Academy Press, 1988, pp. 52–3.

38. Warren E. Leary, "Genetic Engineering of Crops Can Spread Allergies, Study Shows," *New York Times,* March 14, 1996; Julie A. Nordlee et al., "Identification of a brazil-nut allergen in transgenic soybeans," *New England Journal of Medicine,* vol. 334, no. 11, March 14, 1996, pp. 688–92.

39. Rebecca Goldburg, "Pioneer Drops Allergenic Soybeans," *Gene Exchange,* August 1994 (Washington, DC: Union of Concerned Scientists), p. 5.

40. Marion Nestle, "Allergies to Transgenic Foods – Questions of Policy," *New England Journal of Medicine,* vol. 331, no. 11, March 14, 1996, pp. 726–7.

41. Mark Townsend, "Why Soya is a Hidden Destroyer," *Daily Express,* March 12, 1999; also Rex T. Linao, "Who's Afraid of Genetically Modified Organisms," *Philippine Daily Inquirer,* October 16, 1999.

42. Quoted in Townsend, "Why Soya is a Hidden Destroyer."

43. Emails from Sue Dunwell and John Graham to Michael Dorsey, March 3 and March 6, 2000. Graham cited a February 1999 press release that attributed the findings to the fact that "our bodies are changing and food sensitivity is highly individual." "Are Vegetables Making You Ill?" York Nutritional Laboratory press release, February 1999.

44. Nestle, "Allergies to Transgenic Foods."

45. "Pusztai Interview," *GM-FREE,* vol. 1 no. 3, August/September 1999.

46. Arpad Pusztai, "Report of Project Coordinator on Data Produced at the Rowett Research Institute (RRI)," October 22, 1998, at http://www.rri.sari.ac.uk/gmo/ajp.htm.

47. Stanley W.B. Ewen and Arpad Pusztai, "Effect of diets containing genetically modified potatoes expressing *Galanthus nivalis* lectin on rat small intestine," *Lancet,* vol. 354, no. 9187, 16 October 1999, pp. 1353–4.

48. Mae-Wan Ho, Angela Ryan and Joe Cummins, "Cauliflower Mosaic Virus Promoter – A Recipe for Disaster," *Microbial Ecology in Health and Disease* 11, 1999, at http://www.scup.no/mehd/ho.

49. The reports were the May and Donaldson report (http://www.royalsoc.ac.uk/files/statfiles/document-29.pdf); the House of Commons Science and Technology Committee report, May 12, 1999; the Royal Society review (Noreen Murray et al., "Review of Data on Possible Toxicity of GM Potatoes," May 17, 1999); and the Nuffield Council on Bioethics' report (*Genetically Modified Crops: The Ethical and Social Issues,* London: Nuffield Foundation, May 1999).

50. Board of Science and Education, BMA, *The Impact of Genetic Modification on Agriculture, Food and Health,* London: British Medical Association, May 1999.

51. Richard Hindmarsh, "The Flawed 'Sustainable' Promise of Genetic Engineering," *Ecologist,* vol. 21, no. 5, September/October 1991, pp. 196–205.

52. Sheldon Krimsky and Roger Wrubel, "Engineering Crops for Herbicide Resist-

ance, *GeneWatch*, vol. 11, nos. 1–2, April 1998 (Boston: Council for Responsible Genetics), pp. 8–10.

53. Kenny Bruno, "Say It Ain't Soy, Monsanto," *Multinational Monitor*, vol. 18, nos. 1 & 2, January/February 1997.

54. Dawkins, *Gene Wars*, p. 31.

55. Bruno, "Say It Ain't Soy, Monsanto."

56. Caroline Cox, "Glyphosate, Part 1: Toxicology," *Journal of Pesticide Reform*, vol. 15, no. 3, Fall 1995 (Eugene, OR: Northwest Coalition for Alternatives to Pesticides), pp. 14–19.

57. A recent study associating glyphosate exposure with cancer is L. Hardell and M. Eriksson, "A case-control study of non-Hodgkin lymphoma and exposure to pesticides," *Cancer*, vol. 85, no. 6, 1999, pp. 1353–60.

58. Caroline Cox, "Glyphosate, Part 2: Human Exposure and Ecological Effects," *Journal of Pesticide Reform*, vol. 15, no. 4, Winter 1995, pp. 14–20.

59. Ibid.

60. Marc Lappé and Britt Bailey, *Against the Grain: Biotechnology and the Corporate Takeover of Our Food*, Monroe, ME: Common Courage Press, 1998.

61. Union of Concerned Scientists, "Foods on the Way to Market."

62. "Urge EPA to Deny Tolerance Enabling Use of Bromoxynil on Transgenic Cotton," *PANUPS*, April 1, 1997 (San Francisco: Pesticide Action Network).

63. "Transgenic Plants May Create New Viruses," *GenEthics News*, July/August 1996, p. 10.

64. Keith Schneider, "Study Finds Risk in Making Plants Viral Resistant," *New York Times*, March 11, 1994.

65. Dawkins, *Gene Wars*, p. 40.

66. Rebecca J. Goldburg and D. Douglas Hopkins, "Are Mouse Genes in Your Tomatoes?" *New York Newsday*, August 10, 1992.

67. "Research Notes, Genetic Crops," *Nutrition Week*, January 12, 1996 (Washington, DC: Community Nutrition Institute). Also see Chapter 6 of this volume.

68. Fox, *Superpigs and Wondercorn*, p. 21.

69. "Proposed Field Tests of Virus Engineered to Produce Scorpion Venom Toxin," *The Gene Exchange*, December 1994 (Washington, DC: Union of Concerned Scientists), p. 11; Cummins, Joe, "Genetic Experiments Threaten Southern Ontario," *Alive*, November 1996.

70. "Biobit: Edible Vaccines," *The Gene Exchange*, July 1995, p. 13.

71. Fox, *Superpigs and Wondercorn*, p. 13.

5

Safety First

BETH BURROWS

When all is said and done, the question of who bears the cost for a product that results in devastation and catastrophe becomes a choice of alternatives: do we prevent damage and catastrophe by requiring adequate research before a product is released or do we pay the survivors once the damage is done?

Dr Elaine Ingham[1]

WHAT A QUESTION

Was there ever a technology more exciting than genetic engineering? Fast, powerful, able to outpace evolution and leap species barriers, it seems to promise all we can possibly imagine. If we imbibe the hyperbole of its proponents, we can believe that genetic engineering will make deserts bloom, clean degraded soils and oceans, feed the world and end disease. It will make all our children tall and beautiful and, above all, above average. And if only we give the technology enough funding and trust, it can give us immortality in return.[2]

Genetic engineering offers us the stuff of dreams. And to question the goals of its proponents and practitioners, let alone to be suspicious of them or to suggest that they should put safety before fantasy, is to risk seeming mean-spirited, unintelligent, old-fashioned, and – worst of all – neo-Luddite. Those are the labels given to those of us who ask where this technology is leading and at what cost.

Genetic engineering is, like any technology, a social creation that reflects the interests and perceptions of its creators. And, for the most part, the creators of this technology come from or are employed by the corporate and (corporate-endowed) academic laboratories of the USA, Europe and Japan. Their technology is, as Brewster Kneen once pointed out to a panel of the National Agricultural Biotechnology Council, "a technology of domination and control over nature, people, and life itself … the deliberate selection of 'superior' genes, organisms, and people … and [the] effective [disposal] of those not considered desirable."[3]

For some time now, the proponents of genetic engineering have sought to release genetically engineered organisms into the environment and bring genetically engineered products into the marketplace. Some of those whose environments and markets were being "accessed" – when they were aware of what was happening – wanted to know the consequences of allowing genetic engineering into their homes, their fields, their bodies and their futures; they were concerned about the impact on their world. As with other new technologies, they expected that this one too would pose hazards and risks.

Scientists have talked about the potential hazards for years.[4] Professor Philip Regal, biologist and biosafety expert at the University of Minnesota, delineates five basic concerns in his *Brief History of Biotechnology Debates and Policies in the United States*:[5] genetic engineering could lead to global social, economic, and political dislocations and ethical enigmas; it could present serious risks to human health and the environment; its potential for military use could lead to designer diseases and an international biological arms race; it could spur new economic forces that might lead to loss of agricultural and genetic diversity; and it could enable the genetic engineering of humans, thus creating incalculable political, social and physical dangers. In view of such possibilities, scientists and others sought from the outset to avoid the hazards and minimize the risks posed by genetic engineering.[6] The search for such means, as well as the elusive goal of that search, are both called "biosafety."

WHAT A PROBLEM

Biosafety is hardly a popular topic of discussion. To those lulled into dreaming of a paradise perfected by a single new technology, the subject may sound boring, technical and, as one person put it, "too scientificky."[7] To those already concerned, "biosafety" may represent just one more hard-to-explain problem competing for the attention of an issue-fatigued public.[8] To politicians and even to many scientists, the subject of biosafety may be simply too risky to broach. Genetic engineering increasingly means agribusiness and pharmaceuticals, two industries already important as sources of funding for science, higher education and those who run for office. Talking "biosafety" can mean putting one's job and financial security at risk.[9]

Even diplomats charged by their governments to discuss biosafety baulk at doing so, perhaps because they are also charged to protect their countries' industrial interests. The discussions that took place during the biosafety protocol negotiations begun in 1995 under the aegis of the United Nations Convention on Biological Diversity were almost surreal in their avoidance of the topic.[10] At one of the meetings, for example, a country delegate said of biosafety: "Mr. Chair, some of us would prefer not to have any information at all." The chair was not alone in his perplexity when he answered, "Will the delegate kindly repeat what I can hardly believe my undoubtedly mistaken ears have heard him say?"

Whatever the difficulty in discussing biosafety, three broad questions persist. First, what are the risks generated by the creation and release of genetically engineered organisms and their products? Second, are the risks being adequately and reliably assessed and addressed? And third, are the risks generated acceptable risks and who decides what is "acceptable"?

Decisions about whether to allow the release of genetically engineered organisms are not simply technical decisions. Risk per se can be defined and described scientifically: in terms of the probability of a hazard occurring; the amount of damage produced should the hazard occur; and the distribution of damage among different segments of the population and the ecosystem. Nevertheless, what is acceptable risk remains a political issue, requiring public information and a public process for resolution. The question of biosafety is a political question. No matter how scientific-sounding the details, the answer to questions of biosafety – indeed the decision to ask biosafety questions in the first place – is always based on socio-economic considerations.

All over the world there have been demonstrations against products of genetic engineering, especially against the food products. Some of the demonstrations have been about environmental risks. But more often they have been about perceptions of how much risk is "acceptable." Activists in Germany, England and Ireland have ripped genetically engineered crops from the ground rather than risk the perceived environmental impacts of those crops coming to harvest. In India, farmers burned the test fields of Monsanto because the regional authorities and local people had not been consulted about whether to plant them.[11]

In the United States, where the Food and Drug Administration allowed genetically engineered food onto the shelves unlabeled, concerned consumers went into the streets to dump genetically engineered soybeans, genetically engineered maize, and milk from cows injected with recombinant Bovine Growth Hormone. Enough American eaters turned to organic food as a safe haven from genetic engineering and pesticides to push the growth rate of that niche market in the US to almost 20 per cent a year. And, in 1997, when the US Department of Agriculture (USDA) sought to include genetically engineered foods in the official definition of "organic," more than a quarter of a million consumers wrote to Washington to register their outrage. Even a tiny risk may be deemed unacceptable, and therefore not worth taking by a community that believes it has not been part of the decision-making.

Since the mid-1990s, engineered crops have been grown on millions of acres and engineered foods have been put – unlabeled – on millions of dinner plates. And yet, surprisingly little is known about the environmental and human health impacts of these new organisms and foods. It was more than a decade ago that Dr Margaret Mellon, now of the Union of Concerned Scientists, wrote that,

[The] primary risks to human health from engineered organisms will be indirect since few companies will develop for release organisms known to harm humans. However, it is possible for certain genetically engineering projects now on the drawing boards to increase the levels of toxins in foods and increase the number of disease-carrying organisms that are resistant to antibiotics.[12]

Mellon pointed out that the most significant risks of biotechnology would be to the environment, noting that engineered organisms can be pests that displace existing plants and animals, disrupt the functioning of ecosystems, reduce biological diversity, alter the composition of species, and threaten some species with extinction.

Amazingly, even though the potential hazards have been recognized for a long time, to this day there is little published information about the actual hazards.[13] Biologists Allison Snow and Pedro Moran Palma, writing in 1997 about the need to evaluate the long-term effects of commercially grown plants, note:

the hundreds of small-scale field tests that have been carried out to evaluate the performance of genetically engineered crops have not been designed to investigate the ecological risks associated with widespread commercialization.... [F]ield trial reports ... often include statements such as "no characteristics associated with weediness were detected" or "no effects were seen in nontarget organisms" when little attention was paid to these effects. Thus, the fact that nothing happened is not useful in evaluating ecological risks unless these questions are the focus of carefully designed long-term experiments.[14]

Further complicating the problem of assessment is the enigma of genetic engineers (and their regulators) declaring some genetically engineered organisms "no risk" before testing. Even where testing is carried out, there remains – at least in the US – the possibility that portions of the risk assessment will be declared "confidential business information," thereby making critical details unavailable to the public.[15] Is it any wonder that some of us are unsatisfied with the "results"?

Genetic engineering is the product of a technological world that may not be looking for the negative impacts of its creations.[16] Those taking the risks are not necessarily those who will face the hazards. Do we really expect a corporate system that has not adequately repaid the acknowledged victims of Bhopal to take into consideration the potential thousands and millions of uncatalogued inhabitants of soil and marine ecosystems?

What biosafety information does exist indicates that transfer of traits from genetically engineered crops to their weedy relatives can occur and can cause problems,[17] that allergens can unexpectedly be transferred to foods engineered for human consumption,[18] and that beneficial non-target insects can be harmed indirectly by products of genetic engineering,[19] and so can soil micro-organisms.[20] The fact that science detects these adverse effects does not argue against regulation but for more careful scrutiny.

Consider the case of *Klebsiella planticola* SDF20, a genetically engineered organism designed to produce ethanol from crop waste. The organism accomplished the task for which it was engineered and it also did a little more. As doctoral candidate Michael Holmes of Oregon State University discovered, it had unfortunate and unexpected side effects. In soils where it was present, plants died. In some soils, all the plants died. Pathogens flourished. Beneficial soil organisms died, including the ones whose presence is essential for the uptake of nutrients by plants. (Plants need the help of microorganisms to extract nutrients from the soil. Kill those microorganisms and you create a barren soil.) Further, the genetically engineered *Klebsiella* proved to be highly competitive, persisting in the soil, causing changes in ecosystem structure and functions, and having substantial long-term effects. All in all, the unexpected outcome of the presence of this particular genetically engineered micro-organism was highly adverse. So adverse that, had the engineered organism been released into the environment based on the judgment that "it worked" for the purpose designed, and had it persisted and multiplied in other agricultural soils, we might never have had crops again.[21]

Most of us don't worry about these matters. We assume that "experts," professionally charged with looking out for the general welfare, are minding the store. We expect that they are testing the products, regulating the industry, enforcing the law, and protecting us. Here again we may find ourselves disappointed. In the case of *Klebsiella planticola*, Holmes and his graduate advisor, Dr Elaine Ingham, were using what at the time were unusual laboratory techniques for examining the effects of organisms on soils and soil food webs: they were testing in real agricultural soils. Orthodox technique at the time required testing in sterile soils (a technique that Ingham later described as akin to testing the harmlessness of a mass murderer by putting him in an empty room, holding him there for two weeks, and then concluding that he was not a murderer because no one had died recently).

Depending on where we live, we may find that current orthodoxies are insufficient to the biosafety task. Those agencies that exist may be insufficiently empowered, funded, informed or advanced to do the job. In the United States, for example, a great many of the products of genetic engineering have been exempted from or never included in regulatory oversight. Further, there has been a persistent exchange of high-level personnel between the regulatory agencies and the genetic engineering industry.[22] Those two factors – deregulation and the revolving door – have helped create the enduring suspicion that, at least in the US, industry is very gently handled by regulators with little incentive to act for the common good and no wish to offend a future employer.

Industry, in general, wants only "enough" regulation to gain its products the public status of being "safe" but not so much regulation that product development and market penetration are impeded or that industry is made liable for the costs engendered by its own errors. In the face of calls for more regulation, industry has bitterly claimed that further regulation would impose a greater burden on genetic engineering than had ever been imposed on any technology in the past. This argument, of course, tacitly required those concerned with biosafety to dumb themselves down, to forget the mistakes made in assessing and controlling past (nuclear and chemical) technologies, to blind themselves to the revolving door between industry and the regulators, and to neglect the progress that science has demonstrated – although not necessarily exercised – in assessing ecological and human health impacts. (Of course, people would regulate new technologies more stringently than they did old technologies; we know more than we once did.)

WHAT A QUESTION

Complete biosafety may be beyond our scientific ability and political will to accomplish. It would require us to examine all the economic, social and ethical questions raised by the avoidance of hazard. Are we really ready to take into consideration the impacts of our own aspirations on other societies and other economies? Are we willing to sustain those whose only tradable commodity may be one we can manufacture more cheaply using genetic engineering?[23] Are we disposed to ask whether this technology really will help the marginal or further marginalize them? Are we ready to look at the effects of forgoing respect for those who may wish to know which of their foods are entirely halal or kosher or vegetable or cow? Do we care whether this technology feeds need or feeds greed as long as we can be in on the investment? Are we as eager to protect the health of others as we are to protect the health of our own economy? Do we dare to ask if the planet can bear one more too-late-and-too-little-assessed technology? Are we so ignorant of the past that we exempt ourselves from responsibility to the future? Clearly, these are painful questions. Unfortunately, the degree to which we achieve biosafety – or risk biodevastation – depends on us finding the right answers.

NOTES

1. "Biosafety Regulation: Why We Need It," an occasional paper of the Edmonds Institute, Edmonds, Washington, 1995.

2. All these promises (and more) have appeared in industry advertisements and published essays by genetic engineering proponents. One famous early advertisement of the 1980s featured a giant corn plant growing in the desert. The insistence that genetic engineering will feed the world persists in industry advertising to this day.

3. B. Kneen, "Constructing Food for Shareholder Value," in *Agricultural Biotechnology: New Products and New Partnerships*, NABC Reports 8, Ithaca, NY: National Agricultural Biotechnology Council, 1996.

4. The meeting at Asilomar in 1975 was probably the best-known early conference on biosafety, but scientific concerns about biosafety predate that conference. For details, see P.J. Regal, "A Brief History of Biotechnology Risk Debates and Policies in the United States," Edmonds, WA: Edmonds Institute, 1998.

5. Ibid., pp. 1–2.

6. Ibid., and sources therein.

7. The term "scientificky" was first used by the head of a large association of countrywomen when she explained to the author why the audience was so small for a scheduled lecture in Ireland.

8. In 1994 at a transboundary conference of activists from the US and Canada, nonagenarian Hazel Wolf joked about the problem of issue fatigue when she told a conference speaker who wanted to hold a special briefing about recombinant Bovine Growth Hormone, "You don't get invited to many parties, do you?"

9. See Regal, "A Brief History," and S. Krimsky and R. Wrubel, *Agricultural Biotechnology and the Environment: Science, Policy, and Social Issues*, Urbana, IL: University of Illinois Press, 1996.

10. In 1995, the treaty members of the Convention on Biological Diversity (CBD) called for an international "action on biosafety." Recognizing "the relatively short period of experience with releases" of genetically modified organisms, "the relatively small number of species and traits used, and the lack of experience in the range of environments, especially those in centres of origin and genetic diversity," the CBD members began negotiating an international biosafety protocol, which was ultimately agreed to by signatory countries in January of 2000. For details of their decision to proceed, see "A Call to Action. Decisions and Ministerial Statement from the Second Meeting of the Conference of the Parties to the Convention on Biological Diversity," United Nations Environment Program document UNEP/CBD/96/1. For a short article on the decision, see B. Burrows, "Regulating Biotechnology: Biosafety Protocol Update," *Global Pesticide Campaigner*, March 1996, pp. 14–15, 17. On the nearly successful US effort to scuttle the biosafety negotiations, see B. Burrows, "Resurrecting the Ugly American," *Food & Water Journal*, Spring 1999, pp. 32–5. For the final text of the Cartagena Protocol on biosafety, see http://www.biodiv.org.

11. See various reports in the Indian press for December 3, 1998, and for the subsequent and preceding dates; for example, "KRRS 'cremates' cotton in Bellary," *Times of India*, Bangalore edition, December 3, 1998.

12. M. Mellon, *Biotechnology and the Environment*, Washington, DC: National Wildlife Federation, 1988.

13. See, for example, the review article J.D. Doyle, G. Stotsky, G. McClung and C.W. Hendricks, "Effects of Genetically Engineered Microorganisms on Microbial Populations and Processes in Natural Habitats," *Advances in Applied Microbiology* 40, 1995, pp. 237–87.

14. A.A. Snow and P. M. Palma, "Commercialization of Transgenic Plants: Potential Ecological Risks," *BioScience*, vol. 47, no. 2, 1997, pp. 86–96.

15. See, for example, J. Rissler and M. Mellon, "Public Access to Biotechnology Applications," *Natural Resources & Environment*, vol. 4, no. 3, 1990, pp. 29–31, 54–58. See also B. Burrows, "Confidential Business Information (and the Cost of Doing Business as Usual)," *Third World Network Briefing Paper*, August 1998.

16. See Snow and Palma, "Commercialization of Transgenic Plants."

17. R. Jørgensen and B. Andersen, "Spontaneous hybridization between oilseed rape (Brassica napus) and weedy Brassica campestris," *American Journal of Botany* 75, 1995, pp. 519–27. See also J. Bergelson, C.B. Purrington and G. Wichmann, "Promiscuity in Transgenic Plants," *Nature* 365, 1998, p. 6697.

18. J.A. Nordlee, S.L. Taylor, J.A. Townsend, L.A. Thomas and R.K. Bush, "Identification of a brazil-nut allergen in transgenic soybeans," *New England Journal of Medicine* 334, 1996, pp. 688–92.

19. A.N.E. Birch, I.E. Geoghegan, M.E.N. Majerus and J. Allen, "Interactions between plant resistance genes, pest aphid populations and beneficial aphid predators," Scottish Crop Research Institute Annual Report, Dundee, 1996/7, pp. 68–72. See also A. Hilbeck, M. Baumgartner, P.M. Fried and F. Bigler, "Effects of transgenic Bacillus thuringiensis corn-fed prey on mortality and development time of immature Chrysoptera carnea," *Environmental Entomology* 27, 1998, pp. 480–87.

20. See, for example, M. Holmes, E. R. Ingham, J.D. Doyle and C.W. Hendricks, "Effects of Klebsiella planticola SDF20 on soil biota and wheat growth in sandy soil," *Applied Soil Ecology* 326, 1998, pp. 1–12.

21. Ibid.; and Elaine Ingham, personal communications.

22. B. Burrows, "Biotech Bedfellows," *Terrain*, Winter 1998, p. 13. Also see http://www.edmonds-institute.org.

23. "Substitute" crops can drive down commodity prices and lead to significant drops in export income, especially in places that export only one or two major crops. Countries in Africa already have suffered big losses due to laboratory-created vanilla and cocoa. See, for example, Vandana Shiva, "Design for Dispensibility," in V. Shiva, *Betting on Biodiversity: Why Genetic Engineering Will Not Feed the Hungry*, New Delhi: Research Foundation for Science, Technology and Ecology, 1998, pp. 35–40.

6

Ecological Consequences of
Genetic Engineering

RICARDA A. STEINBRECHER

He has gone to the very root, he says, of existence. He has deciphered the secrets. As to the persistence of matter, he insists he can alter the structure of molecules. At his hands, the molecules change, and changed and changing they enter his skin, hide in what he eats, secrete themselves in his tissue, alter the molecular structure of his body. He goes inside the heart of life, he says. He takes apart even the form of matter itself, he strips energy from mass, he splits what is whole, he takes this force for his own, he says. But what he has split does not stop coming apart. Fractures live in the air, invisible fractures come into his body, split his chromosomes, unravel the secrets of life in him....

Barely seen, soundlessly surrounding him, with hardly a breath of evidence, all he has burned, all he has mined from the ground, all he cast into the waters, all he has torn apart, comes back to him. He is haunted.

Susan Griffin, *Woman and Nature*[1]

We stand anew, like children in the Garden of Eden, on the threshold of creation. The secrets philosophers have been searching for over the centuries seem to have been revealed at last; we have found the philosopher's stone, the elixir of life, the alchemical formula, the key to Pandora's box. Yet, we have been here before. As Robert Oppenheimer stood dazzled in the blinding light of the first atomic explosion, he quoted an ancient Sanskrit text, saying "I am become Death, the Shatterer of worlds." Now, standing in the dark shroud of a man-made womb, *I am become Life, the Creator of worlds...*

There is an excitement surrounding genetic engineering which, spreading like a bush fire, has moved beyond the scientific community, capturing the minds of investors and corporations, stimulating the curious minds of the public with so many possible futures limited only by our imagination. The impossible has become possible. As a geneticist, I share the excitement of discovery with my scientific colleagues, and I am sure with many others who want to know everything inside out, want to understand how life works. But

I am deeply worried about the rapid pace of application of this new technology and the emergence of an overconfident science that seriously lacks self-questioning.

THE BIG PICTURE

The new genetics raises questions that relate as much to philosophy and spirituality as they do to science. This is nothing new; in the past, the relationship between these disciplines was much closer. Humankind has always been driven to explore the origins of life and its deeper meaning; this, after all, is part of being human. In this respect genetics revives that old relationship, dealing as it does with the essence, the "blueprint" of life. Up until quite recently, the science we inherited from our forebears, such as Descartes, gave us a highly mechanistic understanding of nature, which influenced our worldview deeply. Our world was broken down into its constituent components, each with its own function, all following laws that were considered to be constant and predictable.

Several years ago Fritjof Capra wrote *The Tao of Physics*, arguing that a "new physics" had stepped out beyond the boundaries of mechanistic science into an uncertain world, one which Cartesian-reared scientists are unfamiliar with and ill-equipped to describe or even understand, since the old rules of physics no longer apply. Genetics seems to be going in a similar direction. Uncertainty, randomness and the relationship of the tiniest particle, be it atom, gene or quantum, with the whole, be it cell, organ, organism or its surrounding ecology, is more familiar ground to mystics, priests and philosophers. As with the new physics, the most fundamental of the old laws no longer apply – even the barriers between species have been broken down.

We should be careful that we do not get carried away by our excitement. It is exactly the unpredictability, the uncertainty and our excitement about this new biology that makes it so dangerous. Science should not lose sight of its ethical ground rules and willingness to question itself.

As scientists we learn to deal with quantifiable facts which we name and classify. Thus we tame nature, make her ours. A rabbit is a rabbit is a rabbit, no matter how big, what colour, where found. The same is true for cells or genes. In their basic structure all genes are seemingly interchangeable. Within this system of naming and classifying we identify different genes, the gene for a blood clotting protein, muscle fibre protein, hormone, a colour pigment. We go further and assign traits – the gene for herbicide tolerance, insect resistance, drought resistance, frost resistance; and further – the gene for breast cancer, obesity, homosexuality, or even intelligence. The mechanistic approach of literally defining the boundaries of a gene and its function suggests that genes are little more than Lego building blocks, which can be cut out, moved around, placed accurately somewhere else, and will still continue to perform in the same ways as they did before.

The failure of the mechanistic approach is that it overlooks the relationships between components, and that the whole can be much more than the sum of its parts. Thus there are many questions which mechanistic science is not equipped to tackle. How can a whole being grow out of a single cell? How does a cell know which kind of cell it is supposed to be? How does it know when to produce what? How do genes control the functioning of our cells and therefore our bodies? Isn't there more to a gene and its product(s) than just the one function we might have assigned to it? No scientist will deny that we actually know very little about genes, how they work and how they interact with each other and how their traits are expressed in the organism. We have only just begun the journey towards understanding molecular genetics and have a long way to go.

Today we are running into serious problems treating genes as Lego blocks. We are dealing with a highly interactive living system, finely tuned by three billion years of nature's own research and development, a system which maintains itself in a dynamic equilibrium by a process called homeostasis. This means that on both the micro- and macro-levels, nature's systems are constantly communicating, checking, adjusting, whether we are concerned with the maintenance of the oxygen supply within a cell, or the maintenance of the Earth's atmosphere. Just as we should apply the principles of ecology to the way we work the land, so there is also a need to apply these principles to the dynamic systems that define and maintain the nature of each species.

Combinations of genes — and sometimes individual genes in an organism — control particular traits, such as the ability to learn language in humans, muscle development in a bird's wing, the capacity to turn the sun's light into energy in plants, and so on. All these traits, in one way or another, help the organism to survive and thrive. Over hundreds of years humans have bred livestock and crops to enhance certain traits. Until ten or fifteen years ago, these breeding improvements were limited to the traits available within species and closely related relatives. Creation and determining evolution were out of the reach of scientists. No laboratory manual could show how to cross a scorpion with a cauliflower or a fish with a strawberry. Now all this has changed. With the discovery and isolation of genes, along with the tools of genetic engineering, any gene can be transferred across to any species and endless possibilities of combinations have opened up.

Biotechnology promises a new agricultural revolution in which resources are no longer limited by the boundaries that define each species. Now we can control nature and redesign life according to our needs. The potential for these new and transferable traits seems to be vast: plants which can tolerate drought or frost, low-allergy rice varieties; grains which contain vitamin A, tolerate salinity, increase yields, resist diseases or pests, tolerate herbicides (weedkillers), increase nitrogen fixation or have a different fat composition. It is an endless catalogue, a laboratory wish list.

Most of these qualities depend on the proper interaction of many genes (multi-gene traits) and are not easily transferred by a single gene. There are a few traits that depend mainly on the presence of a single gene, such as the colour of a flower's petal, tolerance to specific herbicides or some forms of pest resistance. Though laboratory research might be able to determine that an isolated gene is a key factor in a very particular trait in the donor organism, it does not necessarily mean that this gene will behave in the same way and deliver the same traits in a receptor organism. Even single gene traits have their unexpected side effects; for example, petunia flowers engineered to produce red blooms inexplicably produced white blooms a generation or two later.[2] Salmon engineered to grow faster and bigger developed deformed heads, jaws and gill flaps, some deformities so severe that the fish could not see, breathe or feed.[3] Even simple versions of genetic engineering do not necessarily behave as expected.

It is generally understood that molecular biology and genetic engineering are still in their infancy. We have only just begun to understand how genes interact with each other and to what extent their products participate in the different metabolic and biochemical pathways of a particular organism.

Engineered versions of most of the world's major food, fibre and commodity crops have now been produced. By November of 1999 the US had approved thirty-eight different genetically modified (GM) food crops for sale: herbicide-resistant and high oleic acid canola (oilseed rape), chicory (radicchio), twelve different field corn varieties, plus popcorn and sweet corn, five types of cotton, flax, papaya, three types of potato, three types of soybean, two kinds of virus resistant squash, sugarbeet, four tomato varieties and a cherry tomato.[4] Other engineered crops under development are bananas, coffee, cucumbers, eucalyptus, fodder beet, pine, poplar, rice, sunflower, tobacco and wheat.

FIELD TRIALS OF GENETICALLY ENGINEERED CROPS

Companies that are producing genetically engineered crops carry out experiments to monitor the expression of the traits that have been transferred. Most of these experiments, known as field trials, are performed in the open environment, rather than in an enclosed area. Such field trials started in the mid-1980s in the United States with genetically engineered tobacco. Between 1986 and 1997 a total of 25,000 open field trials were conducted worldwide in forty-five countries. Seventy per cent of these trials took place in the US and Canada, followed by Europe, Latin America, Asia and a few in Africa (mainly South Africa). Sixty different crops were tested in these trials with genetically engineered characteristics belonging to ten different trait groups.[5] The most common trait in field trials in the US between 1987 and 1997 was herbicide tolerance (30 per cent), followed by insect resistance (24 per cent), product quality (e.g., changed fat content or delayed ripening) and viral resistance (10 per cent).

While companies would like the public to believe that these field trials are carried out to assess environmental impacts, the truth is that no meaningful environmental data are being collected in the vast majority of the trials. There is no obligation to monitor for gene flow, whether by cross-pollination or by means of horizontal gene transfer to unrelated species (see below). Effects on soil organisms and on the general insect and arthropod life within the tested field and the surrounding area are being ignored. The only questions being asked relate to the agronomic performance of the genetically modified plants, and the unwanted re-emergence of engineered plants in the following seasons, so-called volunteer plants.

Company spokespersons commonly state that no negative impacts on the environment have been observed – which is an obvious result considering that no one has been seriously looking for them. In fact, there have been only a few specially designed field trials to study gene flow, hybridization and environmental impact. The majority of these results have been alarming and will be presented below.

The fact that so many genetically engineered plants have been through field trials or given marketing consent does not represent a proof that these crops are safe. Neither does it mean that genetic engineering is an appropriate solution for the problems it claims to address.

HERBICIDE-TOLERANT CROPS (HTCs)

The use of broad-spectrum herbicides allows total clearance of almost any green plant from a specific area. The most commonly used broad-spectrum herbicides are based on either glyphosate (e.g., Monsanto's Roundup) or glufosinate ammonium (e.g., Basta, Ignite, Weedmaster and Liberty). Tolerance to broad-spectrum herbicides is the most common trait genetically engineered into crop plants. One gene or at most two new genes – usually taken from bacteria – are enough to make the plant produce the antidote to specific chemical weedkillers. Such herbicide-tolerance genes have been engineered into crops like oilseed rape, maize, chicory, lettuce, peas, rice, wheat, cotton, sugar and fodder beet, soybeans, trees, cauliflower, potatoes, cranberries and alfalfa.

The idea is that fields growing such genetically engineered crops can be sprayed with the specific broad-spectrum herbicide at any stage in the growing season to kill off all the weeds without harming the crop plants. This would in theory eliminate the need to mix different, more specific herbicides and hopefully reduce the amount of chemicals used. It is further claimed that this will reduce the necessity of ploughing, which leaves the earth bare and exposed to the wind and rain, often leading to soil erosion. This in turn could reduce the need to establish new agricultural land and thus contribute to preserving wilderness. While this concept sounds promising, it carries a plethora of problems that could turn against farmers, biological diversity, environment and future food supplies. There are many questions that should have been asked and properly investigated before these crops began to be grown on a commercial scale:

- Are they necessary? Is there a problem that needs to be addressed through altering plant genetics rather than improved management or other solutions?
- Is genetically engineered herbicide tolerance going to work: will all the weeds be killed?
- Are the engineered plants going to behave as promised or are there unpredictable side effects? Such side effects could include altered fertility or enhanced invasiveness (e.g. invading refuges of wild plants); production of toxins, allergens or altered nutrient composition; shut-down (silencing) or multiplication of genes; altered disease and pest resistance.
- Can cross-pollination with other crops or wild relatives be prevented or controlled?
- Will crop plants themselves turn into weeds?
- Will unwanted plants (weeds) become resistant? Are there going to be superweeds?
- What are the indirect effects of these crops? We need to consider effects from the extensive use of the broad-spectrum herbicide, as well as from the altered plants themselves, such as changing their own behaviour and their interaction with the environment, including: (a) arthropods, especially pollinators, beneficial insects and spiders; (b) birds, bats, frogs and mammals; (c) soil fertility and soil organisms; (d) allergies and chronic low-level toxicity to humans and farm animals; (e) water and water life; (f) plant life, especially the biological diversity of wild plants.

Contrary to good and responsible science, hardly any of the hazards and potential dangers have been properly investigated.

Modern intensive agriculture is quick to label plants other than the crop as "weeds" and insects other than known beneficial insects as "pests" – meaning something to be got rid of. No plant is intrinsically a weed – the designation depends on context and human values. While other approaches are practised in ecologically sustainable farming, the "extermination approach" is getting the loudest hearing. Before we engage in accelerating the arms race against weeds and pests, we should reconsider if a war waged on nature can produce a sustainable agriculture for future generations. The driving force behind herbicide-tolerant crops is the agrochemical industry.

If a plant species is repeatedly exposed to a specific weedkiller it can develop resistance to that herbicide – an evolutionary response more likely to occur in some plants than in others.[6] Over the last five years grassy weeds have become resistant to previously reliable weedkillers, and farmers in the UK are seeing blackgrass, wild oats, brome, couch grass and rye-grass grow out of control.[7] Neither glufosinate nor glyphosate is likely to escape the fate of other herbicides – they too will be outsmarted by adaptable weeds. An Australian farmer discovered in 1996 that rigid rye-grass, the most common weed in Australia, was no longer affected by Roundup. Researchers at Charles Sturt University in New South Wales confirmed that the rye-grass (*Lolium rigidum*) was tolerant to nearly five times the recommended spraying dose.[8] A year later, glyphosate-resistant goosegrass (*Eleusina indica*) was reported from an orchard in Malaysia and in 1998 glyphosate-resistant Italian rye-grass (*Lolium multiforum*) was found in California.

If agriculture embraces herbicide-tolerant crops, weed problems are likely to get worse. The widespread use of one or two herbicides can cause a shift in the weed population to one that is not easily controlled. Furthermore, equipping plants like canola, wheat, barley or forage grasses with herbicide resistance will soon result in the spread of this trait far and wide by cross-pollination. Indeed, in Europe oilseed rape (canola), for example, is already becoming a weed, invading hedgerows, field borders and road sides and persisting in fields it was grown on in previous years; this can only be exacerbated by the introduction of herbicide tolerance.

While herbicide resistance in weeds is a problem in agrochemical farming systems, it is a serious threat to biodiversity and wildlife. The wide use of broad-spectrum herbicides over vast areas of land will indiscriminately kill a whole range of other plants that are important constituents of natural eco-systems, and harm humans and animals as well. Broad-spectrum herbicides can kill wild crop plant varieties which farmers, especially in the South, depend on for further breeding. They eliminate wild plants from fields and their surroundings, which are hosts to many insects that are an important component to a functional ecosystem. Many birds depend on such insects as well as on the seeds of such wild plants. Larvae of moths and butterflies are often dependent on specific wild plants, which could become scarce as a result of broad-spectrum herbicides, turning these moths and butterflies into endangered species. Many cultures also depend on wild plants for medicinal purposes and for their insect-repellent qualities.

IS THE USE OF CHEMICAL HERBICIDES SAFE?

The effects of chemical herbicides are well documented. They reduce soil fertility, pollute water, deplete earthworms and beneficial soil microbes, and have varying short-term and long-term effects on human health.[9] While Monsanto has claimed that its glyphosate weedkiller Roundup is "environ-mentally friendly," "biodegradable" and "practically non-toxic" to mammals, birds and fish,[10] there is mounting evidence that glyphosate-based herbicides can be lethal to beneficial insects such as ladybirds (ladybugs) and lacewings (*Chrysoperla carnea*), which are predators of common agricultural pests such as aphids.

The biodegradability of herbicides and their effect on soil are dependent on the type of soil – whether it is clay-, loam- or sand-based. Soil fertility and the kinds of soil organisms present are important factors.[11] Soil is a crucial mix of nutrients and minerals, air pockets and roots, teeming with life that is interdependent, little studied and so often ignored. Though scientists know little about this soil-food web, it has become evident that soil-food web organisms are crucial to plant survival, disease resistance, forest regeneration and ecosystem stability.[12] This highlights the need to investigate the impacts of herbicides, engineered genes, gene products and any genetically engineered crop on the various soils and ecosystems worldwide before any claims for their local or global safety can even be considered.

Agreva (slated to be separated again from Aventis) is promoting its "Liberty Link System" of glufosinate-tolerant crop varieties with claims that the active ingredient glufosinate ammonium is harmless. However, the herbicide Liberty contains several additional chemical substances. Regulators in Germany gave Liberty a category III ranking (V = noxious), listing it among the more harmful herbicides. In eco-toxicological tests, Liberty readily kills spiders and predatory mites. In 1998 only 14 per cent of all registered herbicides required labelling as being harmful to beneficial organisms, among them all five glufosinate-containing products.

Glufosinate ammonium is marketed by Aventis/Agreva as a natural product, identical to a toxic substance (L-phosphinothricin) produced by soil bacteria as the active component of an antibiotic, Bialaphos. The company feeds the notion that these bacteria are common in soil, yet the toxin is in fact derived from a mutant strain of the soil bacterium *Streptomyces viridochromogenes*, isolated in Cameroon.[13] Except for two *streptomyces* strains in Japan, neither the antibiotic nor the toxin is found in common soil bacteria; on the contrary, many microorganisms die from glufosinate, even the non-mutated *Streptomyces viridochromogenes*. Canadian researchers examined twelve agricultural soils and found on average a 40 per cent reduction in bacteria after glufosinate treatment.[14] They also found that on average 20 per cent of soil fungi are killed by glufosinate, in some cases even 70 percent. Most disturbing, they found in a follow-up study that most plant-pathogenic fungi remain unharmed, while most beneficial fungi either die or lose their competitiveness.[15] The use of glufosinate on agricultural land thus carries the potential to cause a severe shift in soil organisms, favouring those which are harmful to plants and soil fertility.[16]

Both Monsanto and Aventis further claim that the bacterial genes they use to engineer herbicide-tolerant crops are derived from common soil bacteria and that consequently their products are common in the soil. Yet Aventis's tolerance gene is derived from the same mutant strain of the soil bacterium that was the blueprint for the herbicide toxin. The gene is not present in the non-mutant strain and efforts to find it in microbes isolated from German agricultural soils failed.[17] Thus we do not really know if they are safe.

What about the impacts on our food safety? Increased levels of herbicide residues in food are a real concern. Research is under way to assess to what extent HTCs encourage herbicide absorption by plant tissues. Studies on the synergistic effects of these chemicals in our food have also begun.

Genetically engineered cotton, flax and canola tolerant to other herbicides such as bromoxynil and sulfonylurea (e.g., chlorosulfuron) are already on the US and/or Canadian market, produced by Monsanto, Aventis and DuPont. Mutational breeding, rather than genetic engineering, has resulted in imidazolinone-tolerant plants, including canola and maize. The seeds are promoted globally as the Clearfield Crop System, designed by companies such as American Cyanamid, Pioneer, Garst, Mycogen and DeKalb.[18]

INSECTICIDE-PRODUCING CROPS

"Pest resistance" is the second most popular trait engineered into plants. This application is tailor-made for intensive monoculture farming. Monoculture farming, especially where whole regions are planted with the same crop such as cotton or maize, creates the perfect conditions for plant-feeding insects and other organisms to become pests. Insects are usually very specific about what they eat; they have evolved highly specialized diets which avoid competing with the feeding and breeding grounds of too many other insect species. Acres and acres of monoculture crops present specific insects with an ideal environment in which to flourish. Agribusiness interests describe warlike scenes in which armies of herbivores attack and destroy wide swathes of defenceless plants, unless they are vanquished with applications of chemical pesticides.

The occasional use of certain pesticides, especially biodegradable biopesticides, can control pests. Constant or repeated use, especially of the same chemical control agent, provokes insect populations to build up resistance to the chemical, even if it is a powerful compound such as DDT, lindane/cyclodiene, a pyrethroid or an organophosphate.[19]

The techno-fix solution offered by biotechnology is to engineer plants with genes from other species that produce bio-toxins. These genes will then be present in every plant cell and produce toxins in most if not all the tissues. The idea is that when insect larvae start feeding on any plant part, whether leaves, cobs or maybe even seeds, they will soon die. Indeed, creating defences against feeding insects or even higher animals is nothing new for plants. Over time plants have developed their own defences such as hairiness, thorniness or the production of toxins. The plant world produces an estimated ten thousand pesticidal endotoxins and other natural chemical defence substances.[20] However, the interaction between plants and insects is more a process of co-evolution than of extermination; producing defences, especially deadly defences, will always come at a cost.

Cunningly, insecticide-producing crops (IPCs) are called "insect protected crops" in the industry's public relations terminology. Such terminology is misleading in many ways. First, such genetically engineered crops are only "protected" from certain plant-feeding insect larvae. Second such "insect protection" is highly dependent on the amounts of toxin produced by the plant as well as on the insect's own ability to develop resistance against the plant-produced insecticide. Furthermore, it implies that using IPCs will reduce the presence of chemical insecticides and thus greatly benefit the environment – while in reality every IPC plant increases the amount of insecticide present in the environment.

The dangers of providing plants with insecticide-producing genes are manifold. They include the spread of insecticide production in wild relatives by cross-pollination and the creation of "superweeds," the killing of non-target organisms such as lacewings, ladybirds (ladybugs) and butterflies, the build-up of resistance to the insecticide in target organisms, the damaging effect of the gene products on soil organisms, allergenic or toxic side-effects on mammals, secondary effects on birds and frogs, susceptibility of the crop plant to other diseases, and altered behaviour of the crop plant. This list should be enough to cause alarm and to slow the development of insecticide-producing crops, yet IPCs are already commercially grown on a wide scale. Such crops include Bt cotton, maize and potatoes. Many more are grown in open trial sites, including forest and orchard trees.

Bt toxin-producing crops

The most commonly engineered insecticides are the bio-toxins from the bacterium *Bacillus thuringiensis* (Bt), which have been engineered into many staple crops. Discovered in 1911 by Ernst Berliner, *B. thuringiensis* was found to be toxic to flour moths, and later experiments showed its toxicity to the corn borer. Different bacterial varieties were shown to have different host specificities, including butterflies (*lepidopterus*), flies and mosquitoes (*dipterous*), different beetle species (*coleopterus*), or even soil nematodes.[21] After the Second World War, preparations of the bacterium were introduced as biological pesticides and have since been used by many farmers, especially organic farmers.

B. thuringiensis produces crystalline proteins during sporulation, often termed Bt endotoxins or delta-endotoxins. These endotoxins are actually protoxins, which will only turn toxic after ingestion by insects or their larvae, whose alkaline digestive system will dissolve and then digest the protoxin into its active form. Unlike the naturally occurring bacterial protoxin, the genetically engineered versions have been shortened and altered to be already soluble and active as soon as they are produced by the plant. This might sound like an achievement at first glance. Yet its ecological consequences promise to be disastrous.

First, the new Bt toxin can kill a wider host range than the intended target organisms, not least beneficial insects. Second, the Bt toxin is going to be present in most parts of the engineered plant at any time during its growth. Predators of plant-feeding insects can die of the Bt toxin when preying on larvae that previously fed on Bt crops. Swiss scientists have shown, for example, that lacewings (*Chrysoperla carnea*) will die when feeding on larvae of the European corn borer that ingested Bt toxin.[22] These effects are bound to reduce wildlife and agricultural biodiversity, but there is more to come. Not all larvae will be killed by the Bt toxin. Some are naturally unaffected by the toxin and others will build up resistance. Lacewings fed on unaffected larvae (in this case *Spodoptera littoralis*) died from the Bt toxin ingested by their prey.[23]

Furthermore, Bt plants produce pollen. If the Bt toxin is produced in the pollen, as in the case of Novartis's Bt maize, insects such as moths and butter-flies may die from ingesting the pollen. Initial laboratory experiments with monarch butterfly caterpillars (*Danaus plexippus*) at Cornell University found that they die from eating pollen from Bt maize.[24] Caterpillars of many butter-fly species are very particular about what they eat and are often already threatened with extermination by intensive agriculture and current land use patterns. Bt crops will add to this burden. Monarch butterfly caterpillars de-pend solely on milkweed species (*Asclepias*), commonly found in road ditches, grassy areas between fields and within fields of maize. Bt pollen landing on the leaves of those weed plants may be sufficient to kill some and weaken other caterpillars, possibly leading to long-term or cumulative effects on the fitness and survival of this species.

Researchers from Iowa State University reported that 19 per cent of monarch caterpillars died within forty-eight hours after feeding on milkweed plants growing within or near the edge of a Bt maize field.[25] As the pollen count drops rapidly further away from the field, some argue that there is little to worry about. Such voices ignore the long-term and chronic effect of low-level exposure on a whole population of butterflies. There is still very little independent information on pollen flow and on how Bt pollen might affect any kind of moth or butterfly, possibly driving whole populations towards extinction in areas of Bt agriculture and beyond.

Cross-pollination of wild relatives of Bt crops would give the progeny of those plants a clear advantage, as they have powerful defences against many insects. This could turn them – or even the agricultural crop itself – into invasive species or unprecedented weeds, threatening wildlife and biodiversity even further. Pollen that produces a genetically engineered protein, such as Bt toxin, can also prove disastrous to pollinators, such as honey bees, bumble-bees, moths and butterflies.[26] While this would impact on bee-keepers it would also affect wild plants, gardens and allotments, where a healthy popu-lation of pollinators is crucial for ensuring the survival of plant species.

Resistance build-up in pests

Many scientists are convinced that the trait of "pest-resistance" will be short-lived, as the conditions in fields with Bt crops are usually ideal for creating the selective pressure for Bt resistance in insects. Resistance would not only mean the end for genetically engineered Bt crops but also the loss of Bt bacterial preparations as a powerful biological pest control for organic and other farmers. This would create an even higher dependence on chemical pesticides. So does resistance to Bt toxin occur, and can it be avoided?

Resistance to endotoxins from *B. thuringiensis* has been studied seriously since the 1980s. It was observed in a lepidopteran insect, the moth *Plodia interpunctella*, a pest to grain and grain products.[27] It has also been found in the diamondback moth, *Plutella xylostella*, a pest to cruciferous crops, including the cabbage family.[28] When moth larvae were fed on cabbage leaves treated with *Bacillus thuringiensis* in an experiment, selection pressure led to an initial build-up of resistance a thousand times greater than the level in larvae which had not eaten the treated cabbage. Even fifteen generations later, none of which had consumed the Bt toxin, the resistance level was still around 170 times the level of the control populations.

As total bacterial preparations of *Bacillus thuringiensis* contain many different forms of Bt endotoxins (to date 133 different Bt endotoxins have been isolated from different strains in different combinations),[29] insects will not find it that easy to develop resistance. Yet even then, resistance has occurred as a result of overuse. Bt crops, on the contrary, are only engineered with one form of the endotoxin, allowing much faster build-up of resistance. Additionally, resistance to one form often imparts multiple cross-resistance to other Bt endotoxins.[30] Bt crops can thus trigger multiple resistance in insects, rendering a whole range of conventional Bt preparations useless.

Additional problems occur when the engineered crops don't behave according to plan. Monsanto's "NuCOTN" Bt cotton suffered serious pest infestations in 1996, when it was first grown in the southern region of the US, and needed spraying with chemical pesticides. The Bt cotton was meant to be resistant to cotton bollworm (*Helicoverpa zea*) and to tobacco budworm (*Heliothis virescens*). An unusually hot and dry summer, however, meant that neither plants nor pests behaved according to plan. It is a general rule that plants will alter their protein production under stress, such as drought or heat; such altered behaviour is not taken into account in risk assessments of genetically engineered crops. In the case of Monsanto's NuCOTN, the plants appeared to produce lower Bt toxin levels than under "normal" climatic conditions. The bollworms though seemed to thrive in hot and dry conditions. This combination caused severe damage to thousands of acres, resulting in lawsuits by the affected farmers.

Professor Fred Gould of North Carolina State University pointed out that even without the heatwave, field tests had shown that the genetically engineered cotton did not kill all the bollworms but just 80 per cent of them. Gould pointed out that "80 per cent mortality is exactly what researchers use when they want to breed resistant insects."[31] Thus growing Bt cotton provides the perfect breeding ground for Bt resistant insects. Gould therefore suggests that refuges of 20 per cent of the crop area are needed to slow down the build-up of resistance. Others say that refuges of 50 per cent or more are needed.[32] Refugia are areas of non-engineered crops grown alongside the Bt crop. They are not allowed to be sprayed with any pesticides, as they are meant to be "safe havens" for the insects, allowing them to breed and hopefully dilute any Bt resistance genes.

Until 1999, researchers believed that the inheritance of resistance to Bt toxins was a recessive trait. Yet researchers at Kansas State University found in laboratory experiments that the European corn borer passes resistance on as an incomplete dominant trait, which would spread resistance to a much higher degree than anticipated. The researchers conclude that "the usefulness of the high-dose/refuge strategy for resistance management in Bt maize may be diminished."[33] Further, researchers in New York and in Venezuela found that Bt corn secretes Bt toxin into soils, where it was found to bind to soil particles and persist in an active, lethal state for more than seven months.[34]

Other insecticides engineered into plants

Other common toxin genes used to produce IPCs are lectin genes, such as snowdrop lectin. They have been engineered into potatoes and other crops to stop aphids from feeding on the plants. Researchers believed the lectin to be benign to mammals, birds and other insects, but they soon uncovered alarming results. As potato plants producing the snowdrop toxin reduced the aphid population by only half, the presence of ladybirds and other natural predators would be crucial to prey on the remaining aphids. Yet when ladybirds (*Adalia bipuncta*) were fed on aphids (here *Myzus persicae*) that had previously been fed on lectin-producing potato, the lifespan of the female ladybirds was unexpectedly halved, and ladybird reproduction was seriously affected.[35]

In the context of current agricultural conditions this is even more serious. For example in the UK, climatic changes have seriously disrupted the breeding pattern of ladybirds. In addition, milder winters increase the survival rate of a parasitic wasp that predominately lays its eggs in female ladybirds; the wasp larvae eat the ladybird from the inside out. Lack of springtime frost means that the wasp is breeding three times a year rather than twice. As a result the ladybird population is in rapid decline.

Dr Arpad Pusztai, a leading specialist in lectins, discovered that rats fed on snowdrop lectin-producing potatoes suffered damage to the immune system, liver and gut.[36] Rats fed on potatoes with just the snowdrop lectin added showed no symptoms. His results raise a serious question about the use of such engineered lectins in foodstuffs, but also about genetic engineering itself, as the engineered lectin gene, with its cauliflower mosaic virus promoter, had apparently led to unexpected and harmful side effects.

Other toxin genes are being explored for use in plant genetic engineering, as evidenced by the wide range of patents that have been granted for such products. For example, the patent for a spider venom is held by AstraZeneca (US 5763568), for plants expressing scorpion venom by Agracetus/Monsanto (US 5177308), and for genes associated with snake venom by Biogen Inc. (US 5182260).

A further approach is to introduce proteinase (protease) inhibitor genes, which produce small peptides that inhibit digestive enzymes. A proteinase inhibitor gene from rice has been engineered into poplars, to create resistance to attacks from the beetle *Chrysomella tremulae*, which is a pest on young poplar plantations.[37] New Zealand researchers found that the lifespan of bees was significantly reduced when they were fed on pollen that contained a potato-derived protease inhibitor.[38]

Consequences and unanswered questions

Many additional problems with genetically engineered pesticides are widely overlooked. If the main pest of a particular crop is killed with engineered pesticides, for example, it will open the door for other insects to move in and become pests. As these pests might not be susceptible to the engineered toxin, only the spraying of insecticides would help. There have already been reports about insects moving in on crops (e.g., poplar trees) where they were previously not known as (serious) pests.[39]

Furthermore, the targeted pests could alter their behaviour and attack other crops. If the chemical signalling between plant-feeding insects and plants is broken down by genetic engineering, affecting plant–insect relationships and the evolution of resistance or other survival strategies, it may prove impossible to predict the consequences.[40] Also of major concern are secondary effects on birds, frogs, bats and other animals whose primary food source is insects. Agricultural wildlife has already been depleted by current intensive farming practices and will suffer further from the commercialization of genetically engineered crops that aim to rid the fields of weeds and insects alike.

DISEASE RESISTANCE

Disease resistance is the third arm of crop engineering, aiming to reduce crop losses due to outbreaks of diseases. Researchers are looking into creating resistance to fungi and bacteria, yet the main focus thus far is on resistance to viruses.

Viruses are minute packages of genetic information, contained in a protein capsule or coat. To infect a plant, viruses are dependent on a vector, such as an insect, nematode worm or fungus, or on injuries to the plant. The conventional method of preventing insect-mediated viral infection is to apply large amounts of pesticides to kill the insect vector. Yet genetic engineering seems to offer new solutions. For reasons not fully understood, plants can develop a kind of resistance or "immunity" to a particular virus, and often its close relatives, if they are engineered to contain proteins or gene sequences of that virus. For this reason biotechnologists seek to engineer plants with the genetic information of a specific plant virus, usually with the gene for its viral coat-protein. These "virus protected" crops, however, seem likely to cause more problems than they are designed to solve, potentially giving rise to new viruses and viral diseases with greater virulence and pathogenicity.

As part of their natural evolutionary process, viruses acquire genetic information from each other. This process is known as "viral recombination" and usually occurs when two similar viruses have infected the same cell simultaneously. Using the new genetic information, the recombined virus can strike differently – as was the case with a recent Ugandan cassava virus outbreak. In 1997 a new cassava disease swept across Uganda, causing starvation in regions where cassava is staple food. Viruses from two different regions had recombined when infecting the same plant, giving rise to an extremely severe form of cassava mosaic disease.[41]

Simultaneous infections are quite rare events under normal conditions. However, every cell of every genetically engineered "virus protected" plant across a whole field will contain viral genetic material, and so any viral infection can be considered a simultaneous infection. Experiments have confirmed that viruses do indeed recombine with engineered viral sequences in plants.[42] The potential of such newly recombined viruses to overcome the defences of related wild plants, or even be able to infect new host plants, is a serious concern. In laboratory experiments infecting viruses have also swapped their protein coat for that of another virus that had been engineered into a plant. In a genetically engineered *N. benthiama* plant, the new coat enabled a virus to travel between plants, carried by aphids.[43]

In spite of strong cause for concern, Monsanto's "NewLeaf PLUS" potato, engineered with leafroll virus genes, was grown in huge monocultures in North America beginning in 1997. Of equal concern are Asgrow's and Upjohn's virus-resistant squash lines, especially as squash has wild relatives in the US. Both engineered squash varieties are marketed by Seminis Vegetable Seeds. Upjohn's squash contains genes of the watermelon mosaic virus and zucchini yellow virus and Asgrow's squash is additionally resistant to cucumber mosaic virus. Research and field trials have also been conducted with tomatoes (tomato and tobacco mosaic virus), melons (cucumber mosaic and papaya ringspot virus), papayas (papaya ringspot virus), plums (plum pox potyvirus), and cassava.[44]

Engineering plants with viral genes is a seriously short-sighted approach to protecting crops from viral infection. Genetically engineered virus-resistant crops can quite clearly pose a severe long-term threat to food security and the environment. With huge fields of engineered crop plants containing viral sequences, and the spread of viral genes through cross-pollination to wild and weedy relatives, a major increase in recombined viruses can be predicted. This poses a serious threat to wild plants and the ecological balance. Weedy species that might have been kept in check in the past by viral attacks might now grow freely and displace other plants crucial to an ecosystem.

DROUGHT, SALT AND COLD TOLERANCE:
THE HAZARDS OF 'ADAPTIVE' CROPS

For the next generation of genetically engineered plants, researchers are looking at traits such as growth in shorter day lengths, increased nitrogen fixation, and heightened tolerance to drought, frost, salt, ozone stress, and toxic residues in soil and water. Most of these are "multi-gene" traits, and thus a more complex application to get to work, often requiring "gene stacking" – the sequential addition of engineered genes. While "simple" traits such as herbicide resistance or insecticide production already interfere with the plant's own internal biochemical pathways and gene regulation, creating unpredictable side effects, this is likely to be exacerbated for complex traits.

For example, researchers engineered aspen trees to increase their growth rate. They used the phytochrome A gene from oats, a gene known to be involved in the plant's response to day length. This altered the trees' perception of day length, and the engineered aspen were now able to continue growing with as little as six hours' daylight, compared to fifteen hours in ordinary aspen. Yet the side effect was that the engineered trees had lost their capacity for cold-adaptation.[45] This example shows the complex interaction of genes and traits, which are often hard to determine – especially within today's short time frame of development and marketing. If side effects *are* discovered, will they be countered through additional genes, causing yet more unpredicted side-effects?

Plants are commonly well adapted to specific climatic as well as soil conditions. For example, while some plants thrive on chalky soil and easily survive milder frosts, they might not flourish on salty soil or cope with droughts. Seeds often require specific conditions for germination, such as extreme heat from fire, which would be unsuitable for or even destroy the seeds of other plants which require other specific conditions, such as frost. These adaptations to environmental conditions set limits to where plants can or cannot grow, and give rise to intricate ecosystems in defined habitats.

An obvious ecological problem with engineering "adaptive" traits into plants is their newly acquired ability to flourish in environments other than their own. They can easily invade, disturb or even destroy whole ecosystems, displacing plant communities and reducing diversity and wildlife. Cross-pollination to wild and weedy relatives will, as with the traits described earlier, amplify the negative effects, making genetic engineering of plants a dangerous gamble with the integrity of our natural world. Engineering of plants for frost tolerance must be viewed in this context, particularly efforts to create soft fruits with improved freezing and thawing qualities. Researchers have introduced synthetic versions of the anti-freeze genes from winter flounder into potatoes, tobacco and tomatoes,[46] and anti-freeze genes from plants and insects are also being investigated for this purpose.

A less obvious hazard is the likelihood that such plants might alter soil composition. Every plant affects its surrounding soil, for example by taking certain nutrients, enhancing and interacting with certain soil organisms, replenishing soil through decomposition, or releasing certain substances into soil. This can permanently change soil and displace plants which used to grow there. This is most evident in nitrogen-fixing plants. In Hawaii, for example, specialized ecosystems are established on lava, which is too nutrient-poor to sustain most other types of life. A naturally occurring nitrogen-fixing tree, capable of growing on impoverished soils, was introduced. It invaded the lava, pumping it full of nutrients and consequently driving out the diverse and specialized plant communities which would normally grow there.[47] There is no reason to assume that genetically engineered nitrogen-fixing plants would not be capable of exerting the same negative impact. Their spread could lead to a similar, permanent loss of ecosystems and whole habitats.

What are the reasons for engineering "adaptive" crops? It is argued that the soil and climatic conditions in many regions of the world do not allow for the growth of high-yield crops such as hybrid wheat. Yet many local plants are adapted to these conditions, enhanced through breeding and selection by generations of farmers – primarily by women. Those crops might not lend themselves to intensive agriculture, yet intermediate and indigenous technologies such as companion planting are often a better solution to produce sufficient food than genetically engineering "adaptive" plants designed for cash-crop agribusiness. Nitrogen-fixing plants are hailed as the solution to decrease the use of fertilizers; yet rotational systems can effectively enrich soil with nitrogen without the added danger of creating new invasive species that can spread their genes far and wide. Salt tolerance is seen as a way to make use of salty, marshy land for agriculture. Aside from the problems of invasiveness and of displacing native species, salt-tolerant crops are intended as a sticking plaster, or band-aid, in regions where intensive farming and irrigation is resulting in the salinification of the soil.

QUALITY TRAITS: HOW TO KEEP THEM APART?

Changing or enhancing product quality is another fast developing arm of second generation crop engineering. Products reported to be in the pipeline include: hypoallergenic rice; vitamin A-enriched canola and rice; biodegradable plant-derived plastic compounds; altered oil content for chemical industry feedstocks, cosmetics and nutrition; pharmaceuticals and vaccines in potatoes, bananas, tobacco, maize, kiwi fruit and oranges; plants high in minerals, and those able to sequester lead and other toxic metals; high-fibre crops; low-lignin trees – the list is endless.

The tailor-making of plants for any purpose needs to be viewed with the same stringent catalogue of questions as any other genetically engineered crop, looking at their unpredictability, toxicity, allergenicity, environmental impacts, the effects on birds, insects, small mammals – to mention a few. Serious dimensions are cross-pollination and horizontal gene transfer. Farmers who save seeds for the next planting season might find, due to cross-pollination, that their future crops are contaminated with inedible and toxic compounds or with altered components that reduce their value for processing. It is also highly unlikely that seeds can be kept absolutely segregated for any length of time. Seeds drop during harvest and settle in the ground; they find escape routes during transport and even during processing. As genetic engineering does not normally come along with any visible or tangible alteration, such as changed colour or glowing in the dark, there is little chance of preventing gene pollution. The consequences for the food chain and the environment could be tremendous – and yet they would be irrevocable once the genes have started to spread.

Gene escape

The phrase "gene escape" conjures visions of genes plotting and scheming, trying any possible route and means to get away from where they have been placed by biotechnologists. Genes may act as if they have a mind of their own, but this has more to do with their interaction with other genes and their products. Genes are passed around constantly within a species and to closely related species, whether engineered or not. This happens in the form of pollen, seeds, shoots, suckers, tubers or runners, all of which will give rise to new plants nearby or far away.

Pollen is the one means through which plants will pass genetic information on to other plants, whether these are wild or weedy relatives or crop plants of the same kind. Bees are known to fly and carry pollen anywhere within a radius of approximately five kilometres from their hive; butterflies, beetles, bumblebees, aphids and flies are also pollinators, for which little information about flight and pollination patterns is known. Wind can also carry pollen for miles. Pollen from trees was found in the treeless Shetland Islands, having travelled for more than 250 kilometres from the nearest forest. Other pollen was found to have travelled more than 600 kilometres, originating from the Himalayas and ending up in the heart of the Rajasthan desert in northwestern India.[48]

Field tests with genetically engineered potatoes, using antibiotic resistance marker genes, were carried out to determine whether the engineered marker gene would be passed on to neighbouring ordinary potatoes of a different variety by cross-pollination. The ordinary potatoes were planted in patches around the genetically engineered plot and seeds were collected for testing. The researchers found that 35 per cent of the seeds of the ordinary potatoes, grown at a distance of 1,100 metres from the engineered crops, contained the engineered gene.[49] Agronomists in Alberta, Canada, have identified volunteer (weedy) strains of oilseed rape (canola) that had acquired simultaneous resistance to Roundup, Liberty and Cyanamid's "Pursuit" herbicide through cross-pollination.[50]

What about plants that are highly selfing (self-pollinating): can or will they cross-pollinate? Using *Arabidopsis thaliana* (Thale cress) as a model, researchers at the University of Chicago made a disturbing discovery. They engineered a strain with a herbicide tolerance gene from a mutant *A. thaliana* plant and compared its performance to the non-engineered herbicide-tolerant mutant of *A. thaliana*. They found that the engineered strain was twenty times more likely to cross-pollinate than the ordinary mutant, which is especially worrying as this gene (Crs1–1, chlorosulfuron tolerance) has been introduced into dozens of agricultural crops.[51]

If you wanted to design a means to scatter a gene far and wide throughout the environment and its species, the best way to do it would be to put it in a plant which has lots of wild relatives and pollen that travels well. Increasingly scientists agree that risk assessment should not focus on the likelihood of a gene escaping, but on the impact that this gene will have *when* it escapes. "Because we conclude that the escape of [GM] pollen is inevitable, we argue that the focus of risk analysis should be shifted towards the 'invasiveness' of [GM] plants and 'mitigation' of their impact on natural as well as agricultural systems."[52] All the negative impacts of genetic engineering are hugely enhanced through gene escape. The benefits proclaimed by industry are only designed for when the specific crop is grown, processed or consumed. Yet all the claimed benefits, such as pest resistance, herbicide tolerance or delayed ripening, dissipate when the engineered plant leaves the field – either in the form of seed or pollen – and enters the wider environment. From this moment on, only the negatives count.

Many agricultural crops can cross with wild or garden relatives and produce fertile offspring, such as oilseed rape (canola) with wild radish, wild turnip, hoary mustard and mustard greens, occasionally even with charlock.[53] While canola poses a great biodiversity problem in Europe and maybe North America, the problems are exacerbated in places like Central and South America, which are home to maize, squash, pumpkin, beans, cassava, cotton, tomato, potato and sweet potato. Cross-pollination other than to neighbouring crops is primarily a problem in the world's centres of origin and diversity, where wild relatives have evolved as part of the ecosystem, or where the crop itself can act as a weed. The contamination of the gene pool and the loss of important old and wild varieties is a general problem linked to intensive agriculture, yet introducing genetically engineered species into centres of diversity is going to heighten these problems. It is of interest, in this context, that the world's poorest nations account for 95.7 per cent of the world's genetic resources of the major agricultural crops. North America and Australia, on the other hand, are totally dependent on external sources of genetic resources for their major agricultural crops.[54]

Horizontal gene transfer

Can unrelated species such as bacteria, fungi or even mammals get hold of the "escaping gene" from a plant? While such horizontal gene transfer was initially vehemently denied by biotechnologists, the table has turned. Population geneticists are pointing out that gene transfer is part of the overall evolutionary process. For example, species of nematode worm that feed on plants acquired a gene needed for plant digestion from soil bacteria.[55]

A standard argument by biotech industry representatives is that if gene transfer happened we all should look orange or green by now, given the amounts of carrots and lettuce we have eaten during the course of human existence! This argument ignores some basic facts. Genes of higher organisms, whether plant or animal, are usually very large. While the actual information for producing a protein might be small, the actual gene is on average eight times the size. This is because the information sequence in higher organisms is frequently interrupted by large sections of DNA which carry no direct information at all. These sequences are called "intervening sequences"or "introns." This does not mean that these DNA sequences do not have a function – the opposite is probably the case – but we just don't know what it is yet.

No bacteria or fungi can cope with such enormous genes, nor would they be able to distinguish between information and interrupting sequences. Genetically engineered genes, however, are pure information; the interrupting sequences have been removed. This makes it possible for other organisms to pick up a whole gene and use it. Very little is known about the actual process of horizontal gene transfer and funding is scarce for this work. Some data suggest that similarities between the genetic sequences can enhance gene transfer. By engineering bacterial, viral, plant or animal genes into plants, we are introducing genetic similarities where they did not previously exist. This indeed might enhance the exchange of genetic material.[56]

For example, a fungus (*Aspergillus niger*) was found to pick up engineered genes when grown in co-culture with genetically engineered Swedish turnips (*Brassica napus*) and black mustard (*B. niger*).[57] Another experiment looked at horizontal gene transfer from sugar beet to soil bacteria (*Acinetobacter*). The engineered gene, conferring resistance to a number of antibiotics, was transferred to the bacteria when they were exposed to either purified whole beet DNA or liquefied samples of whole beet plants. In both cases the bacteria were found to have acquired antibiotic resistance.[58]

GENETICALLY ENGINEERED MICROORGANISMS

Despite a fundamental lack of knowledge of soil, serious attempts are being made to release genetically engineered microorganisms into open fields on a commercial scale. Once released, even on a small scale, there is no calling them back.

A German company genetically engineered the soil bacterium *Klebsiella planticola* to produce ethanol from agricultural residues in closed containers. Once the ethanol production was completed, the remaining sludge was intended to be spread on the fields as natural compost. In the United States the EPA added the engineered bacteria to duck food, fish food and Daphnia food with no ill effect on these organisms, so the EPA ruled that the bacteria could move on to open field trials. Yet there was unease, and Elaine Ingham from Oregon State University was asked for additional testing. In laboratory tests, her team tested the effect on wheat plants by adding the engineered bacteria to natural soil. The effect was stunning; the engineered bacteria killed the plants within seven days, and led to a shift in the spectrum of soil microorganisms towards the pathogenic end of the scale. Furthermore, the engineered *Klebsiella planticola* produced about seventeen parts per million alcohol, while most plant roots can only tolerate about one part per million alcohol. As *K. planticola* lives in the root system of all plants and in litter material all over the world, the engineered strain would have been able to spread globally, perhaps killing all terrestrial plants in its wake.[59]

There is research to engineer killer microbes against drug-producing plants, while commercial releases of engineered bacteria (*Sinorhizobium meliloti*) for enhanced nitrogen fixation in soil have already taken place in the US. Such applications of genetic engineering strike many ecologists as highly irresponsible, as we know next to nothing about the interaction of microorganisms, especially soil microorganisms, their movements and their boundaries.

UNPREDICTABILITY AND STRESS RESPONSE

The term "genetic engineering" gives the false impression of everything being under control. "Engineering" evokes "precision," everything thoroughly thought through and meticulously executed – like engineering a designer bridge over a wide river. Yet nothing could be further from the truth. Genetic engineering of crops, imprecision and unpredictability go hand in hand, whether it is unexpected and unrelated side effects (pleiotropic effects), gene silencing or positional effects. "Genetic gambling" or "randomeering" might be a more appropriate term.

Treating genes like Lego blocks fails to recognize their very nature. The following examples depict the lack of knowledge and anticipation; they highlight the fact that genetic engineering of plants and animals are early experiments of an infant science and not established science.

Gene silencing

Gene silencing is a phenomenon discovered only about ten years ago, and one that caused a stir among researchers when half of a sample of genetically engineered red petunias lost or changed their colour. The flowers were engineered with an extra copy of their own pigmentation gene and were expected to display a uniform red flower.[60]

Rice engineered with an antibiotic-resistance marker gene lost this resistance in most of the second generation when plants were grown in the field.[61] When analysed, researchers found that the engineered gene had been copied up or reduced to varying degrees in different plants. Alfalfa cultures, engineered with a herbicide-resistance gene and an antibiotic-resistance gene, lost the herbicide resistance to an increasing degree when grown in warmer temperatures.[62] Dutch researchers, working with genetically engineered oilseed rape made the disturbing observation that the inheritance patterns of the engineered trait do not follow the "Mendelian rule," but rather that gene silencing occurs to varying degrees in generations down the line, even after back-crossing with ordinary varieties.[63]

Nobody knows how exactly gene silencing of engineered genes works. It appears that several factors play an important role, such as sequence similarity between the engineered gene and the plant's own genes, multi-copies of genes, and the use of distinct bacterial or viral sequences. Yet other influences, including temperature and stress, leave any predictable outcome to be a mere guess or hopeful wish. The result, in any case, is that the engineered gene stops working, and quite often a gene of the plant as well if they share certain similarities. Gene silencing was also found to be reversible, thus making genetic engineering even more unpredictable.[64]

Stress response

Every plant is adapted to tolerate a certain amount of stress, whether to heat, drought, frost, flooding, insect attack, altered light conditions or grazing. If the stress is too extreme, the plant will die. Yet there are stages when the plant survives by altering its own metabolism. It can cast off leaves, slow down growth processes, grow smaller or fewer fruit, or produce more chemical defences, to mention a few. Fava beans that were naturally tolerant to the herbicide glyphosate still experienced stress in the presence of the herbicide and responded with a heightened level of plant estrogens,[65] which can mimic our human estrogens.[66] In fact this seems to be a typical stress response for many beans.

Monsanto blamed bad weather conditions when both their cotton varieties failed to perform properly. In 1997 nearly fifty growers in the southern United States filed complaints and sought compensation when their cotton crop produced malformed bolls and prematurely dropped their bolls. They also reported that the plants had stunted growth and deformed root systems. Monsanto first claimed that it had been unusually warm, which was disputed by the attorney representing the growers. Then they said that abnormally cool and wet weather had caused the plants to grow slowly, making them vulnerable to the first herbicide spraying. Others now suggest that fruit abortion is due to poor pollination, common to all of Monsanto's engineered cotton lines, the causes of which are as yet unclear.[67] In all scenarios, application of Roundup exacerbated the problems already at hand. This illustrates the unpredictability and the unexpected side effects of engineered crops when growing conditions impose stress.

If stress-sensitive genetically engineered plants cross-pollinate with wild relatives it is anybody's guess what might happen. The genes might spread as long as no severe stress occurs, contaminating the gene pool. When eventually these wild plants are exposed to stress, their survival could be threatened.

Unpredicted side effects

When researchers looked at "insect protected" cotton that produced genetically engineered Bt toxin, they found that the leaves' rotting characteristics were different from ordinary cotton.[68] The researchers concluded that this was not due to the presence of Bt toxin but rather to an unexpected side effect of the Bt gene and the experimental technique. Such effects could seriously impact soil composition and composting, especially in forest ecosystems with Bt trees.

A field of 30,000 petunias, engineered with a red pigment gene from maize (A1 gene with 35S promoter) and an antibiotic resistance gene not only extensively reverted to white through gene silencing but also showed numerous unrelated side effects.[69] These included extra growth (more leaves, shoots and flowers), a lowered fertility rate and enhanced resistance to diseases, all of which cannot be explained through the introduced traits of the genes.[70] The gene silencing in these petunias is now thought to have been triggered by prolonged exposure to direct sunlight.

Lignin is a strengthening substance essential to trees and their structure, but it is a problem for the pulp and paper industry, as it needs to be removed for quality paper. Consequently genetic engineering is being used to reduce lignin production in trees by interfering with the genes involved in lignin synthesis. Trees are not an annual crop but live for hundreds of years, exposed to many stresses such as frost, fire, drought, storm and insect attacks. No risk assessment can predict the interference that genetic engineering will have on the stress response and the ageing of trees. For example, a French research group engineered tobacco with reduced lignin production, yet this also led to a weakened level of viral resistance.[71] Genetically engineered trees are a serious threat to forest ecosystems generations down the line and any open release of engineered trees should be called into question.

Monsanto's Roundup Ready soybeans have come under attack in recent years for being prone to fungal attacks in certain regions in the United States. A research team from the University of Georgia looked into these accusations and reported in 1999 that under hot temperatures (heat stress) the stem of glyphosate-resistant soybeans turns brittle and splits. At normal growing temperatures of 25°C, Roundup Ready soybeans produce 20 per cent more lignin – a strengthening substance of the cell wall – than their conventional counterparts. This increased lignin content at an early stage seems to alter the plants' ability to cope with higher growing temperatures, whereupon the tissue tears and the stem splits. Once the stems crack they are vulnerable to fungal attack. The researchers link this unpredicted and seemingly unrelated side effect to the newly introduced gene itself, which, possibly in combination with the herbicide, interferes with the plant's own biochemical pathways and its response to summer heat.[72]

CONCLUSIONS

Genetically engineered crops are tailor-made for intensive Western-style agribusiness monocultures. This can be regarded as a sticking plaster approach to problems born out of intensive farming systems. Commercialization of these crops will lead us further away from sustainable food production and cannot be interpreted as "progress." Real progress would lie in using our knowledge in working with nature, preserving wildlife and replenishing soil, ridding the agricultural system of its dependency on chemical drugs.

Genetically engineered crops, as illustrated here, are full of unpredictabilities and unintended side effects. Even the scary tale of "superweeds" and "super-bugs," fantastic as it sounds, is a reality just waiting to dawn. Multiple resistance in plants and insects, altered behaviour in viruses and pests, the ability to grow in salty or nitrogen-poor soils, plants that defy insect and viral attacks – any of these combinations can legitimately be given the attribute of "super." Whether these plants or insects will turn into serious crop pests or overrun wild habitats and threaten the survival of species, we don't know. It is certain, though, that this technology will enable them to do exactly this.

The spread of engineered traits, particularly those that confer survival advantages to plants such as resistance to herbicides, insects and viral attack, is only a matter of time. Even traits which confer disadvantages as far so the plant is concerned – delayed or disabled ripening for example – will enter the gene pool of the crop relatives. As researcher Paul Hatchwell concluded, "in sensitive ecosystems, particularly where certain species are already threatened, large numbers of new introductions could make the difference between extinction and survival."[73]

Some proponents argue that if genetic engineering spells disaster, should it not already have happened? As nothing obvious has occurred, isn't that proof that it is safe? Anything that is instantly toxic or lethal would be recognized immediately, we are told; it would probably not leave the laboratory. Yet it took us over three decades to acknowledge that low-level radiation is dangerous and ultimately lethal. And still some scientists, governments and corporations insist that nuclear power plants and low-level radiation are nothing to worry about.

The risks, as outlined here, are simply too great, especially as there is no need to walk down the road towards genetically engineered crops. Safer and more promising options are available for farmers all around the globe to produce enough food in a sustainable manner and supply their communities with local and diverse food.[74] Our generation carries the responsibility for future generations, who do not yet have a voice. As arguably the most fantastically inventive, curious, creative, inspirational creatures on this planet, it is well within our capabilities to come up with a better solution to our food production than genetic "randomeering," which is an affront to science and to our creative potential. We still should learn all there is to learn about plants, environmental interactions and producing healthy food. But we might also have to learn that turning food into a globally profitable business for corporate shareholders runs counter to what food and food security are all about.

NOTES

I would like to thank my friends and colleagues Douda Bensasson, Hartmut Meyer, Viola Sampson and Rowan Tilly for their support and expertise.

1. London: The Women's Press, 1984, p. 134.

2. P. Meyer et al., "Endogenous and environmental factors influence 35S promoter methylation of a maize A1 gene construct in transgenic petunia and its colour phenotype," *Molecular and General Genetics* 231, 1992, pp. 345–52.

3. R.H. Devlin, T.Y. Yesaki, C.A. Biagi and E.M. Donaldson, "Extraordinary Salmon Growth," *Nature* 371, 1994, pp. 209–10.

4. http://www.ucsusa.org/.

5. Clive James, *Global Status of Transgenic Crops in 1997*, International Service for the Acquisition of Agri-biotech Applications (ISAAA), Brief no. 5, 1997.

6. For herbicide-resistant "weeds," see http://weedscience.com/.

7. A. Blake, "Rise in Number of Grass Weeds Causes Concern," *Farmers Weekly*, July 23, 1999, p. 54.

8. J. Gressel, "Fewer constraints than proclaimed to the evolution of glyphosate resistant weeds," *Resistant Pest Management*, vol. 8, no. 2, 1996, pp. 2–5; and J. Pratley et al., "Resistance to glyphosate in Lolium rigidum: I. Bioevaluation," *Weed Science*, vol. 47, no. 4, 1999, pp. 405–411.

9. For example, S.A. Briggs, Rachel Carson Council, *Basic Guide to Pesticides: Their Characteristics and Hazards*, London and Washington DC: Taylor & Francis, 1992. See also L. Hardell and M. Eriksson, "A case-control study of non-Hodgkin lymphoma and exposure to pesticides," *Cancer*, vol. 85, no. 6, 1999, pp. 1353–60.

10. After complaints from New York State authorities and the Attorney General's ruling that Monsanto's advertisements were misleading, the company agreed in November 1996 to change its advertisements in the state and to pay $50,000 costs. In 1998 a similar case occurred in Sweden.

11. Soil that has been degraded by intensive farming practices is unable to degrade Roundup properly, for example. Elaine R. Ingham, personal communication

12. Earthworms, nematodes and burrowing insects make up just a small portion of the subterranean population; many millions of microscopic bacteria and fungi (including mycorhiza) are central to nutrient cycling and the health and fertility of soil. See http://www.soilfoodweb.com.

13. E. Strauch et al., "Cloning of a phosphinothricin N-acetyltransferase gene from Streptomyces viridochormogenes Tu494 and its expression in *Streptomyces lividans* and *Escherichia coli*," *Gene* 63, 1988, pp. 65–74.

14. I. Ahmad and D. Malloch, "Interaction of soil microflora with the bioherbicide phosphinothricin," *Agriculture, Ecosystems, and Environment* 54, 1995, pp. 165–74.

15. I. Ahmad, J. Bissett and D. Malloch, "Effect of phosphinothricin on nitrogen metabolism of *Trichoderma* species and its implications for their control of phytopathogenic fungi," *Pesticide Biochemistry and Physiology* 53, 1995, pp. 49–59; I. Ahmad, J. Bissett and D. Malloch, "Influence of the bioherbicide phosphinothricin on interactions between phytopathogens and their antagonists," *Canadian Journal of Botany* 73, 1995, pp. 1750–60.

16. H. Meyer and V. Wolters, "Ecological effects of the use of broad-spectrum herbicides in herbicide-resistant crops," *Proceedings of the Society for Ecology* 28, 1998, pp. 337–344.

17. Strauch et al., "Cloning," and H. Meyer, personal communication.

18. Herbicides of the imidizolanone family include brand names such as Pursuit, Raptor, Lightning, Scepter, Steel, Arsenal and OnDuty.

19. C.P. Srivastava, "Insecticide resistance in *Helicoverpa armigera* in India," *Resistant Pest Management* 7, 1995, pp. 4–5; R.H. French-Constant, "The molecular and population genetics of cyclodiene insecticide resistance," *Insect Biochemistry and Molecular Biology* 24, 1994, pp. 335–45; R.L. Metcalf, "Insect resistance to insecticides," *Pesticide Science* 26, 1989, pp. 333–358.

20. D. Pimentel, "Herbivore population feeding pressure on plant hosts: feedback evolution and host conservation," *Oikos* 53, 1988, pp. 289–302.

21. Reviewed in B. Tappeser, "The difference between conventional *Bacillus thuringiensis* strains and transgenic insect resistant plants – Possible reasons for rapid resistance development and susceptibility of non-target organisms," Freiburg: Institute for Applied Ecology, 1997.

22. A. Hilbeck et al., "Effects of transgenic *Bacillus thuringiensis* corn-fed prey on mortality and development time of immature *Chrysoperla carnea* (*Neuroptera: Chrysopidae*)," *Environmental Entomology*, vol. 27, no. 2, 1998, pp. 480–87; A. Hilbeck et al., "Toxicity of *Bacillus thuringiensis* Cry1Ab toxin to the predator *Chrysoperla carnea* (*Neuroptera: Chrysopidae*)," *Environmental Entomology* vol. 27, no. 5, 1998, pp. 1255–63; A. A. Hilbeck, "Prey-mediated effects of Cry1Ab toxin and protoxin and Cry2A protoxin on the predator *Chrysoperla carnea*," *Entomologia Experimentalis Et Applicata* vol. 91, no. 2, 1999, pp. 305–16.

23. Ibid.

24. J. Losey et al., "Transgenic Pollen Harms Monarch Larvae," *Nature* 399, 1999, p. 6733.

25. L. Hansen and J. Obrycki, "Non-target effects of Bt-corn pollen on the Monarch butterfly (Lepidoptera: Danaidae)," Abstract, North Central Branch Meeting of the Entomological Society of America, March 1999.

26. For example, L.A. Malone et al., "In vivo responses of honey bee midgut proteases to two protease inhibitors from potato," *Journal of Insect Physiology*, vol. 44, no. 2, 1998, pp. 141–7.

27. W.H. McGaughey, *Science* 229, 1985, pp. 193–5.

28. J.D. Tang et al., "Stable resistance to *Bacillus thuringiensis* in *Plutella xylostella*," *Resistant Pest Management* 7, 1995, pp. 8–9.

29. N. Crickmore et al., "Revision of the nomenclature for the *Bacillus thuringiensis* pesticidal crystal proteins," *Microbiology and Molecular Biology Reviews*, vol. 62, no. 3, 1998, pp. 807–14. See also Tappeser, "The difference," 21.

30. B.E. Tabashnik et al., "One gene in diamondback moth confers resistance to four *Bacillus thuringiensis* toxins," *Proceedings of the National Academy of Science, USA* 94, 1997, pp. 1640–44.

31. F. Gould, cited in J.L. Fox, "Bt Cotton Infestations Renew Resistance Concerns," *Nature Biotechnology* 14, 1996, p. 1070.

32. M. Mellon and J. Rissler, eds, *Now or Never: Serious New Plans to Save a Natural Pest Control*, Cambridge, MA: Union of Concerned Scientists, 1998.

33. F. Huang et al., "Inheritance of resistance to *Bacillus thuringiensis* toxin (Dipel ES) in the European corn borer," *Science*, vol. 284 no. 5416, 1999, pp. 965–7; see also B. Tappeser, *Von Restrisiken, Risikoresten und Risikobereitschaft. Dokumentation einer Ringvorlesung*, Graz 2000.

34. D. Saxena, S. Flores and G. Stotzky, "Insecticidal toxin in root exudates from Bt corn," *Nature* 402, 1999, p. 480.

35. A.N.E. Birch et al., SCRI Annual Report 1996/97 pp. 70–73.

36. S.W.B. Ewen and A. Pusztai, "Effect of diets containing genetically modified potatoes expressing *Galanthus nivalis* lectin on rat small intestine," *Lancet*, vol. 354, no. 9187, 1999, pp. 1353–4.

37. J.C. Leple et al., "Toxicity to chrysomela-tremulae (coleoptera, chrysomelidae) of transgenic poplars expressing a cysteine proteinase-inhibitor," *Molecular Breeding* 1(4), 1995, pp. 319–328.

38. Malone et al., "In vivo responses of honey bee midgut proteases," pp. 141–7.

39. D. Ewald and Y. Han, "Freisetzungsversuche mit transgenen Pappeln in China," UBA Symposium "Freisetzung transgener Gehölze – Stand, Probleme, Perspektiven" September 20–21, 1999, Humboldt University of Berlin; M. Bernhardt, F. Thomas and B. Tappeser, "Gentechnik und biologischer Pflanzenschutz, Analyse und Bewertung gentechnischer Ansätze in der biologischen Schädlingsbekämpfung," Row 73, Ecological Institute e.V., Freiburg, 1991.

40. K.F. Raffa, "Genetic engineering of trees to enhance resistance to insects – Evaluating the risks of biotype evolution and secondary pest outbreak," *Bioscience* vol. 39, no. 8, 1989, pp. 524–34.

41. X. Zhou et al., "Evidence that DNA-A of a geminivirus associated with severe cassava mosaic disease in Uganda has arisen by interspecific recombination," *Journal of General Virology* 78, 1997, pp. 2101–11.

42. A. Greene and R.F. Allison, "Recombination between viral RNA and transgenic plant transcripts," *Science* 263, 1994, pp. 1423–5.

43. H. Lecoq et al., "Aphid transmission of a non-aphid transmissible strain of zucchini yellow potyvirus from transgenic plants expressing the capsid protein of plum pox potyvirus," *Molecular Plant-Microbe Interactions* 6, 1993, p. 403.

44. A virus-resistant cassava was recently developed through traditional, selective breeding with naturally resistant plants.

45. J.E. Olsen et al., "Ectopic expression of *oat* phytochrome A in hybrid *aspen* changes critical daylength for growth and prevents cold acclimatization," *Plant Journal*, vol. 12, no. 6, 1997, pp. 1339–50.

46. J.G. Wallis et al., "Expression of a synthetic protein in potato reduces electrolyte release at freezing temperatures," *Plant Molecular Biology* 35, 1997, pp. 323–30; R. Hightower et al., "Expression of antifreeze proteins in transgenic plants," *Plant Molecular Biology*, vol. 17, no. 5, 1991, pp. 1013–21.

47. M.J. Crawley, *Plant Ecology*, Oxford and Cambridge MA: Blackwell Science, 1997, pp. 595–632.

48. J.B. Tyldesley, "Long-range transmission of tree pollen to Shetland," *New Physiologist* 72, 1993, pp. 175–90, 691–7; G. Singh et al., "Pollen-rain from vegetation of North-West India," *New Physiologist* 72, 1993, pp. 191–206.

49. I. Skogsmyr, "Gene dispersal from transgenic potatoes to conspecifics: A field trial," *Theoretical and Applied Genetics* 88, 1994, pp. 770–74.

50. Mary MacArthur, "Triple-Resistant Canola Weeds Found in Alberta," *Western Producer*, February 10, 2000, at http://www.producer.com/articles/20000210/news/20000210news01.html.

51. J. Bergelson, et al, "Promiscuity in transgenic plants," *Nature* 395, 1998, p. 25.

52. P. Kareiva et al., "Studying and managing the risk of cross-fertilization between transgenic crops and wild relatives," *Molecular Ecology* 3, 1994, pp. 15–21.

53. R.B. Jorgensen, "Regional patterns of gene flow and its consequences for GM oilseed rape," presented at Gene Flow and Agriculture – Relevance for Transgenic Crops; University of Keele, Staffordshire, 1999.

54. C. Juma, in *The Gene Hunters*, London: Zed Books, 1989, p. 21.

55. Y.T. Yan et al., "Genomic organization of four beta-1,4–endoglucanase genes in plant-parasitic cyst nematodes and its evolutionary implications," *Gene* 220, 1998, pp. 61–70.

56. K.M. Nielsen et al., "Horizontal gene transfer from transgenic plants to terrestial bacteria – a rare event?" *FEMS Microbiology Reviews* 22, 1998, pp. 79–103.

57. T. Hoffmann, C. Golz and O. Schieder, "Foreign DNA-sequences are received by a wild-type strain of *Aspergillus niger* after co-culture with transgenic higher-plants," *Current Genetics* vol. 27, no. 1, 1994, pp. 70–76.

58. F. Gebhard and K. Smalla, "Transformation of Acinetobacter sp strain BD413 by transgenic sugar beet DNA," *Applied Environmental Microbiology*, vol. 64, no. 4, 1998, pp. 1550–54.

59. M.T. Holmes, E.R. Ingham, J.D. Doyle and C.W. Hendricks, "Effects of *Klebsiella planticola* SDF20 on soil biota and wheat growth in sandy soil," *Applied Soil Ecology* 326, 1998, pp. 1–12; and Elaine Ingham, personal communication.

60. C. Napoli et al., "Introduction of a chimeric chalcone synthase gene into petunia results in a reversible co-suppression of homologous genes *in trans.*," *Plant Cell* 2, 1990, pp. 279–89.

61. W. Schuh et al., "The phenotype characterisation of R2 generation transgenic rice plants under field conditions," *Plant Science* 89, 1993, pp. 69–79.

62. C. Walter et al., "High frequency, heat treatment-induced inactivation of the phosphinothricin resistance gene in transgenic single cell suspension cultures of *Medicago sativa*," *Molecular and General Genetics* 235, 1992, pp. 189–96.

63. P.L.J. Metz et al., "Occasional loss of expression of phosphinothricin tolerance in sexual offspring of transgenic oilseed rape (*Brassica napus L.*)," *Euphytica*, vol. 98, no. 3, 1997, pp. 189–96.

64. O. Mittelsten Scheid et al., "Release of epigenic gene silencing by trans-acting mutations in *Arabidopsis*," *Proceedings of the National Academy of Sciences, USA* 95, 1998, pp. 632–7.

65. H. Sandermann and E. Wellmann, in *Biosafety*, Bonn: German Ministry of Research and Technology, 1988, pp. 285–92.

66. For example, K.D.R. Setchell et al., "Exposure of infants to phyto-estrogens from soy-based infant formula," *Lancet* 350, 1997, pp. 23–7.

67. Keith L. Edmisten, *Concerns with Roundup Ready Cotton*, North Carolina Co-operative Extension Service, 1999.

68. K.K. Donegan et al., "Changes in levels, species and DNA fingerprints of soil-microorganisms associated with cotton expressing the *bacillus-thuringiensis* var kurstaki endotoxin," *Applied Soil Ecology*, vol. 2, no. 2, 1995, pp. 111–24.

69. H. Meyer et al., "Endogenous and environmental factors influence 35S promoter methylation of a maize A1 gene construct in transgenic petunia and its colour phenotype," *Molecular and General Genetics* 231, 1992, pp. 345–52.

70. B. Tappeser, *Gutachten zu der wissenschaftlichen Zielsetzung und dem wissenschaftlichen Sinn der Freisetzungsexperiments mit transgenen Petunien*, Freiburg: Institute for Applied Ecology, 1991.

71. S. Maury et al. (IBMP du CNRS, France), "Transgenic tobaccos with depressed OMTs have reduced lignin content, modified vascular organisation and weakened anti-viral resistance," Presentation at "Forest Biotechnology '99," Keble College, University of Oxford, July 11–16, 1999.

72. J.M. Gertz, W.K. Vencill and N.S. Hill, "Tolerance of transgenic soybean (*Glycine max*) to heat stress," paper given at the 1999 Brighton Conference of the British Crop Protection Council. Also Andy Coghlan, "Monsanto's Modified Soya Beans are Cracking Up in the Heat – Splitting Headache," *New Scientist*, November 20, 1999.

73. P. Hatchwell, "Opening Pandora's Box: The Risks of Releasing Genetically Engineered Organisms," *The Ecologist*, vol. 19, no. 4, 1989, pp. 130–36.

74. For example, recent research in Kenya established that a number of common grasses will help fend off common pests of maize (corn). These grasses either work as repellents or traps for the stem borer or as weed competitors; their use in farming systems is being extended into Uganda, Tanzania and Ethiopia. See Fred Pearce, "Farmer's Friend," *New Scientist*, October 24, 1998, p. 25.

7

Biotechnology to the Rescue?
Ten Reasons Why Biotechnology is
Incompatible with Sustainable Agriculture

JACK KLOPPENBURG, JR AND BETH BURROWS

Proponents of biotechnology like to portray genetic engineering as both "natural" and inevitable. By doing so, these industrial conquistadors effectively plant biotechnology's flag on the future, claiming it as their own and denying the space for other visions to be articulated. Yet there is nothing inevitable about biotechnology, and to see it as such is to succumb to a determinism that is disempowering to many who might otherwise oppose the transformations that capital is so busy engineering. Whether biotechnology is widely adopted or not will depend on choices made by people in all walks of life, not least the choice of whether or not to resist its application and further development.

However, even among some whose knowledge of history renders them sceptical of corporate claims as to the benefits of biotechnology, there is a reluctance to throw the biotechnology baby out with the corporate bathwater. Might there not be *some* truth, for example, in Monsanto's assertion that "sustainable agriculture is possible only with biotechnology and imaginative chemistry?"[1] Could judicious use of biotechnology's powerful capabilities by people of good will not help facilitate a transition to sustainable agriculture? Following are ten reasons why it cannot.

I. THE COMMODITY FORM

Commodification is the central tendency in contemporary society. The principle on which businesses operate is that everything should be bought and sold, that nothing should be beyond the reach of the market. Neither knowledge nor the production of knowledge have escaped this commodification. By definition, commodities are available to those who can purchase them, and, since purchasing power is a function of the distribution of wealth and income, business interests are the ones now in a position to dominate the purchase of knowledge and thus to control its development and deployment.

In the United States, some 75 per cent of all research and development is now undertaken by private industry. Worldwide, companies such as Monsanto, Novartis, DuPont, Aventis, AstraZeneca and Unilever have the almost exclusive capacity to develop and introduce biotechnologies. Given the increasing commodification of science and technology and the reality of where purchasing power lies, it is naive to expect a new tool such as biotechnology ever to fall outside of corporate control.

2. YOU CAN'T TAKE THIS BABY OUT OF THE BATHWATER

Technological utopianism, even among progressive people, manifests itself in the hope that, if removed from its corporate entanglement, biotechnology could be used for sustainable ends. But the baby of modern biotechnology is not so easily separated from the corporate bathwater. The new genetic technologies are being developed in corporate and academic laboratories in the industrial, commoditized, capitalist North to serve industrial, commoditized and capitalist ends. It may be possible to use genetic engineering technologies for other ends, but they will best serve those purposes for which they were originally and expressly designed.

To put a different face on these technologies, Monsanto, USAID and other organizations have already started distributing surplus biotechnology (noncritical techniques and knowledge) just as foreign aid programmes were used to distribute surplus US wheat in the 1950s and 1960s. As they did with "Green Revolution" technologies, the Rockefeller and Ford Foundations are financing the development of biotechnology expertise in the network of international agricultural research centres associated with the CGIAR system and in selected developing nations. But such technical "aid," like the wheat in earlier years, will bind recipients more closely to an expanding global market for these technologies and to a commitment to a particular approach to the production of knowledge.[2]

3. AGAINST TECHNOLOGICAL DETERMINISM

Which elements of nature are explored and how nature is interpreted are determined largely by those who have the power to fund research. How such new knowledge or technologies are deployed depends in large measure on social rather than technical parameters.

In their advertising, annual reports and public presentations, companies researching biotechnology continue to speak of technical miracles as both necessary and sufficient solutions to all the problems we face.[3] But ecological and social sustainability follow from social arrangements, not from the technologies developed.[4] A focus on technologies as causal agents is a deflection from the hard but necessary task of social transformation, which is precisely what corporations intend.

4. STRUCTURAL LIMITS TO CORPORATE RESEARCH

Companies need to sell products and therefore need to develop saleable tech-nologies as quickly as they can. Private research money is invested in those areas of research that are commodifiable: they are not invested in those areas that cannot be quickly or easily packaged and sold or that are antagonistic to commodification because they are practices rather than discrete technologies (crop rotations for insect control, for instance). Industry will not develop "management technologies" (practices which involve manipulation of multiple variables in an agroecosystem and are therefore systemic) because they cannot be easily sold.

Thus industry tends to develop technical products which redress symptoms rather than solve underlying problems. For instance, against the depredations of the Colorado beetle, Monsanto offers not better methods of crop rotation but potato plants engineered to produce bacterial toxins in their tissues which will kill the beetle. Monsanto has constructed the problem as the potato beetle, not as potato monoculture. Existing systems of agricultural production are left intact while the root causes of unsustainability are left untouched.

5. THE ROAD NOT TAKEN

In the United States as elsewhere, there are all manner of traditional and visionary farmers at the margins and in the interstices between technological convention and scientific orthodoxy – Amish, Mennonites, Native Americans, organic farmers, perennial polyculturists, low-input producers, seed savers and horse farmers – who continually produce and reproduce a landscape of alter-native agricultural possibilities. Despite an agricultural policy environment actively hostile to their interests, many have managed not just to survive but to thrive. Would we not have a more sustainable agriculture if the $500 million Monsanto spent in developing plants resistant to its Roundup herbi-cide had been spent on gaining a better understanding of rotational grazing of dairy herds or the structure of Amish farming?

Little effort has been made to understand the farming systems developed by those who, by their very survival outside conventional agriculture, have demonstrated useful and workable alternatives. The University of Wisconsin, for example, raised $27 million for a new Biotechnology Center. Meanwhile, it houses both its Center for Integrated Agricultural Systems (a phrase that intimates sustainability without offending corporate sensibilities) and its Agri-cultural Technology and Family Farm Institute in a tiny, remodelled furnace building on the university campus.

6. OUR COMPROMISED UNIVERSITIES

If companies cannot or will not carry out alternative research agendas using biotechnology, could publicly funded agricultural universities do so? Univer-sities have long been the servants of agribusiness, and industry's influence on the direction of university research is growing rather than diminishing. Companies engaged in biotechnology have reached into universities in an unprecedented fashion, establishing a wide variety of arrangements to draw knowledge out of academia and into the corporate labs. Popular mechanisms include consulting, participation of university researchers on corporate scientific boards, university/industrial consortia, and direct, sometimes very large, industry contracts for research to be undertaken by universities. And the companies do not make the grants out of the goodness of their corporate hearts. Hoechst (now Aventis), for instance, expected something in return for the $70 million it provided to Harvard; Monsanto expects something out of the $62 million it has given to Washington University. What the companies get is exclusive licences, patent rights, placement of corporate scientists as observers in university labs and early access to information.

Such arrangements present several challenges to academic and scientific integrity: conflicts of interest may arise; the free flow of information may be constrained; and the public may lose a pool of experts who might have been relied upon for a disinterested analysis of scientific issues and options, since the best and the brightest scientists are often most closely associated with industry. Public science, because of its close connections to industry, is constrained in its capacity to provide alternatives to industrial modes of developing biotechnologies. It has become complementary to, rather than competitive with, corporate biotechnology.

7. BETTING AGAINST THE FUTURE

Corporate and academic biotechnologists argue that they are or can be reasonably certain of the environmental effects. Indeed they have succeeded in lifting nearly all restrictions on the release of genetically engineered organisms into the environment, largely on the grounds that gene transfer is nothing more than what nature itself does. Yet techniques such as recombinant DNA transfer and protoplast fusion are qualitatively different from sexual techniques of reproduction.

Given biotechnology's power to create wholly new organisms that could not be produced without the intervention of a scientist, and a corporately funded one at that, it represents a qualitatively new threat to the environment from the possible escape and proliferation of recombinant organisms, from unintended gene transfers, or simply from mutation or interaction which no one foresaw or intended. The degree of the threat is unknown, but it is possibly very large and certainly, given our track record with exotic species (for example, kudzu and zebra mussels introduced into the US and rabbits released for hunting in Australia), the risks are far from zero.[5]

That there is still a relative lack of ecological data to gainsay or confirm claims about the environmental effects of genetically engineered organisms reflects more the paucity of funding for ecological studies than it does the robustness of claims of low potential risk. Apparent lack of evidence of hazards should not be taken to imply safety but to reflect lack of testing for such hazards.[6] The very difficulty that biologists have often had in getting transgenes to function is evidence of how imperfect their understanding is not only of the genome itself but also of genome–environment interaction.

Health and environmental problems associated with biotechnology will doubtless increase relatively slowly over time but it will be difficult, if not impossible, to partition the effect of genetic engineering in causing the problem from other variables. Each of those variables acting in isolation may be relatively benign. An increase in cancers, for example, might not be attributable just to atrazine or just to a genetically engineered potato. But individual benign components may have adverse synergistic effects. Why add to the chemical and biological burden our bodies are already bearing?

8. AGAINST REDUCTIONISM

It is also possible to oppose biotechnology on epistemological grounds. What is at issue is not simply the nature of biotechnology, but the nature of Western science.[7] In this critique, science is regarded not as a neutral tool, but as an undertaking stamped with the character of the society which produced it. Our society has been characterized by domination and control (of women, of classes, of ethnic or cultural groups, of other species and of natural resources) and the frames of reference we use to analyse and interpret the physical world reflect that character.

Hence, even if biotechnology were to come under broader social guidance, simple redirection of objectives might not be enough to develop sustainable technologies in agriculture. Agricultural scientists are, after all, working within a dominant epistemological paradigm commonly described as reductionist. That is, the methodological principle of scientific research is to break a problem down into manageable pieces and look for those elements which provide the greatest amount of control over the phenomenon of interest.

For example, in trying to increase yields in corn, agricultural scientists in effect reduced a cornfield to corn genes and bags of nitrogen. The development of genetically uniform hybrid corn, needing high applications of inorganic fertilizers, produced high yields, and also created the fertilizer industry and the seed corn industry. Because they have inorganic fertilizer, most corn farmers no longer worry about soil fertility and thus, instead of planting rotations of leguminous crops to maintain fertility, plant corn continuously. With apparently unlimited fertility, they plant the plants closer together, for which they need to get rid of weeds and thus use herbicide. When insects and weeds become resistant to the herbicide, farmers use more and more pesticides, which flow into the groundwater. Because nothing but corn is grown, soil begins to wash away; in addition, the pesticides kill off microorganisms in the soil and organic additions are reduced. With the use of heavy mechanical equipment, the soil is compacted, compounding erosion problems. Costs continue to rise because the equipment runs on oil and the chemicals are petroleum based. Farmers end up paying more and more for inputs that eventually degrade and destroy the environment into which they are being ploughed.

Such effects of reductionistic and linear thinking take a while to emerge. By the time they do and are recognized as such, a powerful set of interests has built up to sell the various piecemeal solutions to the different problems and thus sustain the reductionist system interests which want to carry on selling their solutions regardless.

Biotechnology merely slots into this approach rather than challenging it. Private industry thrives on the reduction of complex systems to component parts, a simplification which creates space for perpetual rounds of product development. The very nature of modern biotechnology, operating as it does at the molecular and cellular levels, makes the reductionistic, piecemeal route especially tempting.

9. BIOTECHNOLOGY DISTANCES US FROM OTHER SPECIES

Biotechnologists are not limiting their activities to plants and microorganisms. Scorpions, salmon, pigs and cattle are all slated for "improvements" of various sorts. In what has been dubbed "pharming," sheep, goats and cows are being genetically engineered to produce human proteins in their milk. The intention is to make animal milk a more perfect substitute for human milk. This will doubtless be marketed as better than human since it will be subject to rigorous quality control and will not contain the toxins now found in all humans.[8]

Such a wholesale loss of respect for the integrity of other species is endorsed even by many conservation biologists captivated by the "price it to save it" approach. In writing of "unmined riches," E.O. Wilson describes what he calls "The New Environmentalism": "The race is on to develop methods, to draw more income from the wildlands without killing them, and so to give the invisible hand of free market economics a green thumb."[9] All species come to be viewed as little more than sets of genetic codes.[10] The UN Convention on Biological Diversity comes to be little more than a means to legitimate a market for genes.[11]

10. BIOTECHNOLOGY DISTANCES US FROM EACH OTHER

If we regard other species as mere commodities, can we fail to see each other or parts of each other as commodities as well? Corporate and academic biotechnologists have begun to focus on human genetic information as a raw material. The leading journal *Science* reports on one of many programmes of bioprospecting that target our own species: "Geneticists want to collect DNA from such groups as the Arewete. Just 130 members of this tribe remain on the Xingu River in Brazil."[12] Reading that, one wonders what kind of sensibility would rather have the genes than the people? If species, including our own, are treated as commodities, our treatment of each other will increasingly come to reflect differential values in the market.

LOOKING ELSEWHERE

For these reasons, and in the absence of dramatic social, political and economic changes, we should not expect companies, universities, scientists and technocrats to use biotechnology to develop a sustainable, regenerative, low-input or diversified agriculture. There is little reason to believe that rBGH, herbicide-resistant corn or cloned Douglas firs assist a move towards sustainable agriculture. Biotechnology for the foreseeable future will continue to be dominated by and respond principally to the needs of industry.

Technology and social power are not separable. The way knowledge is developed and the uses to which knowledge is put reflect the distribution of social power. The development of modern biotechnology is being shaped by a narrow range of private interests which considerably influence the agendas of our public universities. There are no reasons to believe the "planetary patriots" of Monsanto who tell us otherwise, or to support the technical applications they are promulgating. We should look elsewhere for genuine approaches towards sustainable agriculture.

NOTES

1. H.A. Schneiderman and W.D. Carpenter, "Planetary Patriotism: Sustainable Agriculture for the Future," *Environmental Science and Technology*, vol. 24, no. 4, April, 1990, p. 472.

2. For discussion of whether indigenous farmers receive appropriate compensation for patented materials derived from seed they have been induced to store in international germplasm collections, see Rural Advancement Foundation International, *Conserving Indigenous Knowledge: Integrating Two Systems of Innovation*, commissioned by the United Nations Development Programme, 1995. See also D.F. Mackenzie, "Battle for the World Seed Banks," *New Scientist*, July 2, 1994, p. 4.

3. See J. Kloppenburg, Jr., "Sustainable Agriculture and the New Biotechnologies," *Science as Culture*, vol. 2, no. 13, p. 4, 1991, pp. 482–506; D.L. Kleinman and J. Kloppenburg, Jr., "Taking the Discursive High Ground: Monsanto and the Biotechnology Controversy," *Sociological Forum*, vol. 6, no. 3, September 1991; *From Field to Plate* and *Biotechnology: Agriculture and Your Food*, Saskatoon, SK: AgWest Biotech.

4. See the essays in M.R. Smith and L. Marx (eds.), *Does Technology Drive History?*, Cambridge, MA: MIT Press, 1994.

5. For the latest problem, see P. Shenon, "In Vietnam, the Snails Dine Regally," *New York Times*, May 17, 1995, p. A9. See also E. Ingham, *Biosafety Regulation: Why We Need It*, Edmonds, WA: The Edmonds Institute, 1995.

6. See J. Doyle et al., "Effects of genetically engineered microorganisms on microbial population and processes in natural habitats," *Advances in Applied Microbiology* 40, 1995, pp. 237ff.

7. A different criticism of biotechnology as outmoded reductionist science has been made by Mae-Wan Ho, Brian Goodwin and others. Much of what we have labelled "science," they might term "technology" parading as science. For their criticism of biotechnology, see, for example, M.-W. Ho "Genetic Engineering: Hope or Hoax?"

Third World Resurgence, no. 53/54, pp. 28ff.

8. See M. Crouch, "Like Mother Used to Make," *The Women's Review of Books*, vol. 12. no. 5, February, 1995, pp. 31–3.

9. E.O. Wilson, *The Diversity of Life*, New York: W.W. Norton, 1992. p. 283.

10. See, for example, "From Wolf to Poodle: Taming and Changing Animals," *Your World, Biotechnology & You*, vol. 5, no. 1, 1995, developed by the Pennsylvania Biotechnology Association "for seventh and eighth grade classroom use."

11. Note, for example, the argument on ratification made by US Secretary of the Interior Bruce Babbitt, Secretary of State Warren Christopher, and Secretary of Agriculture Mike Espy in an August 16, 1994, letter and Accompanying Memorandum of Record to congressional leader: "US ratification of the Convention benefits US agriculture by ... safeguarding US access to agricultural genetic resources, and encouraging conservation of such resources in other countries.... The US depends on access to foreign germplasm for plant breeding programs of such key crops as corn, wheat, soybeans, potatoes, cotton, and most vegetables.... By becoming a party to the Biodiversity Convention, the US will ensure continued access to genetic resources."

12. L. Roberts, "A Genetic Survey of Vanishing Peoples," *Science* 252, June 21, 1991, pp. 1614–17.

8

From Native Forest
to Frankenforest

ORIN LANGELLE

While genetically engineered food crops have held centre stage in the debate over genetically modified organisms (GMOs), the genetic engineering of trees has mostly remained outside the global spotlight. But while little is being said publicly, much is being done, as the back door for this biotechnology is wide open. It is propped in place by the timber and chemical industries, with even the petroleum and auto industries taking part. In the near future there is cause to speculate that all industrial forestry plantations might include genetically engineered trees. Genetically altered trees, combined with economic incentives for plantation forestry – including international agreements that offer carbon dioxide credits to polluting corporations – could prove to be a recipe for environmental disaster.

The timber industry has plans to grow hundreds of thousands of hectares of fast-growing monoculture plantations that are herbicide and pest resistant, and contain lower levels of the complex polymer lignin for easier pulping. All of this is designed to feed the insatiable appetite of the pulp and paper industry and to produce more uniform traits for lumber production. The industry touts the business of "designer trees" as one that will promote forestry health and prove beneficial to the economy.

The World Rainforest Movement (WRM) reported in December of 1999:

Field trials of GM tree species have expanded in different regions of the world. Countries with confirmed trials at present are: Australia, Belgium, Canada, Chile, Finland, France, Germany, Indonesia, Italy, Japan, New Zealand, Portugal, Spain, UK, US and Uruguay. In 1998, there were 44 new trials and, in the last three years, the number of trial tree species doubled. Since that year there have been 116 confirmed GM tree trials in 17 countries, using 24 tree species, 75 per cent of which being timber-producing species. The situation is especially dangerous in Southern countries, where there is often little or no regulation regarding the setting up of such trials. They are often driven by the private sector, and notably by those multinationals that wish to invest in genetically modified organisms (GMO) but are restricted by regulations in Northern countries.[1]

Although this brave new world of engineered trees may seem promising to some, "Frankenforestry" has the potential for catastrophic effects on ecosystems, local communities and future generations. Too many questions remain unanswered with regard to long-term safety and biodiversity issues, ethical considerations, and the all-important question of who will control the land that these plantations will be located on. Moreover, these trees defy natural ecological patterns, and the technology used to create them is entirely unproven.

THE RISKS OF GENETICALLY ENGINEERED TREES

In July of 1999, London's *Daily Telegraph* reported, "Terminator trees, genetically engineered never to flower, could ensure a silent spring in the forests of the future. Such trees will grow faster than before, but will be devoid of the bees, butterflies, moths, birds and squirrels which will depend on pollen, seed and nectar."[2] This may seem like fiction today, but the facts suggest that this scenario could very well be the future of commercial forestry.

The World Wide Fund for Nature (WWF), for example, has warned of the dangers of genetically engineered trees which can cross-pollinate with native trees over a distance of 400 miles and are being grown in field trials with little knowledge of what the final result will produce. Francis Sullivan, director of programs for WWF in the UK, said, "The genie of genetically modified super-trees is already out of the bottle. We must make sure it does not get out of control otherwise such trees could run riot through the forests of the world without us knowing what are the consequences."[3] "It's a bit like rabbits in Australia," Sullivan continued, "they just take over and cleaning up afterwards is extremely expensive."[4] Unfortunately, that expense may be too costly for the inhabitants of earth.

The environmental safety, ethical and moral questions raised by altered trees, and the possible effects these engineered organisms may have on humans and other species, are of utmost importance. Europe's ASEED (Action for Solidarity, Equality, Environment and Development), in an article reprinted from the Soil Association's Responsible Forestry Programme, reports that there are various classes of environmental risks.[5] These are partly extrapolated from our experience with genetically engineered food crops, combined with our understanding of forest ecosystems:

- Released GMOs could disrupt natural ecosystems, especially if a particular variety of engineered trees comes to dominate an ecosystem.
- Pesticide-secreting trees could promote pest resistance to natural biological controls, or to chemicals that are commonly used to control pests (creating 'super pests' for which there are no known controls).

- Tree crops engineered to be immune to viral diseases could, through re-combination and mutation, create new viruses more pathogenic than the original strains.
- Genes engineered into trees may flow to wild relatives, modifying or elimi-nating the original varieties. The capacity for altered genetic material from flowering trees to infiltrate the natural environment is undisputed and indeed seen as inevitable, a phenomenon which has sparked considerable effort by researchers to locate genes that will enable the creation of sterile trees.
- Trees engineered for pesticide or insecticide tolerance may result in in-creasing application of such chemicals.
- Ecological disruptions from engineered trees may lead to reductions in biological diversity.
- Genetic modifications may prove to be either highly unstable or unexpected-ly persistent. There is already much evidence for the inherent unpredictabil-ity of genetic modifications, which can carry through to future generations.

THE PLAYERS: USUAL, UNUSUAL AND THE ROLE OF CARBON CREDITS

In April of 1999, a landmark forest biotechnology venture was announced, one that did attract some international attention. International Paper, the largest private landholder in the United States, which operates in thirty-one countries and exports to 130 nations, announced plans to join with New Zealand's Fletcher Challenge Forests, the New York-based Westvaco Corpo-ration (both timber and pulp companies), and the biotechnology giant Monsanto to form Arborgen, a forestry biotechnology joint venture to produce and market tree seedlings.[6] The companies announced their intent to contract with New Zealand's Genesis Research and Development Corporation Ltd., a biotechnology research company, to provide genomics research and access to its intellectual property.

Other seemingly odd players in the genetically engineered tree business are oil behemoths British Petroleum and Shell Oil.[7] Both are leading players in the funding of research into and development of engineered trees in the hope of obtaining carbon emission credits. The Kyoto Protocol (an agreement reached at the Third Conference of the Parties to the Framework Convention on Climate Change in December 1997) was originally designed to cut emissions of greenhouse gasses that lead to global warming. But, as a pre-condition for agreeing to reduce its emissions, the United States insisted on a system of internationally tradable carbon dioxide credits. The Kyoto Proto-col's Article 12 has been widely interpreted to mean that planting trees to offset carbon emissions will count towards emission credits.[8]

Thus many industrialized countries (and their transnational corporations) hope to plant trees instead of cutting emissions. England's well respected CornerHouse research group reports that "some Northern governments, led by the US, are pressing hard for countries or companies with enough money to be allowed to avoid emissions reductions … by investing in forestry projects which purportedly 'sequester' or 'store' carbon."[9] Corner House quotes a report by Daphne Wysham of the *San Jose Mercury News* that by 1998 climate meetings had come to resemble "trade shows" in which, "instead of focusing on global warming, attendees jostled to get a piece of the lucrative emerging market: trading in pollution credits."[10] Companies hope that investments in the genetic engineering of trees will lead to faster-growing varieties that may ultimately gain them more credits to offset their increasing carbon dioxide emissions.

Not to be outdone by the oil industry, Japan's leading auto maker, Toyota, announced early in 1998 that it was moving into the biotechnology business, specifically forestation and domestic animal feed. Since 1990, Toyota has been doing biotechnology research and has developed varieties of genetically engineered trees that they say will be used to plant areas damaged by slash-and-burn agriculture; they also say they plan to plant in deserts to help sequester carbon.[11] Toyota's trees are "designed" to remove large amounts of CO_2 produced by cars and other sources.[12]

PLANTATIONS

In order for genetically engineered trees to be profitable, which is their sole reason for existence, they must have a medium. That medium is industrial plantations. The WRM reports, "Plantations are especially attractive in that they promise to be able to furnish exceptionally uniform raw material more quickly than natural forests and on a smaller land base."[13] Currently, industrial plantations account for between 15 and 30 per cent of the world's supply of pulpwood, an amount that is bound to rise. "Already by 1990, 95 per cent of Chile's industrial wood production, 93 per cent of New Zealand's and 60 per cent of Brazil's came from plantations. By the year 2030, Indonesia plans to increase the share its plantations take in production of industrial wood from the present 20 per cent to 80 per cent," the WRM further states.[14] In New Zealand alone, 1.6 million hectares (or approximately 4 million acres) are in industrial plantations and are growing by an estimated 100,000 hectares per year.[15] In the past twenty-five years, the international demand for wood has grown 36 percent.[16] This has put pressure on industrial plantations to increase production, which leads to larger tracts of land being taken over by industrial plantations, as well as the proposed use of genetically engineered trees.

Global trade in forest products grew from US$80 billion in 1985 to US$152 billion in 1995, and the focus of trade has shifted from logs to processed products (plywood, paper, packaging, etc.).[17] By 2007, pulp exports from Latin America are expected to grow by more than 70 percent.[18] Clearly, the earth's remaining native forests are being destroyed at an alarming rate. Tropical forests are lost at a rate of 13.7 million hectares annually.[19] From 1980 to 1995, approximately 200 million hectares of tropical forest were lost, and in the Brazilian Amazon almost 50 million hectares of rainforest were destroyed between 1979 and 1999. Only some 8 per cent of the world's forests are protected from commercial extraction.[20]

With the decimation of the earth's remaining native forests, the timber industry has been planting large-scale monocultural plantations aimed at the production of low-cost products for export. Plantations are not forests. They are a monoculture crop and have drastic consequences for the people who live nearby and for the remaining remnants of the once intact ecosystem. Water, wildlife, plants and people's livelihoods suffer as agrochemical use increases and people are deprived of most of the available agricultural land. In tropical countries all plantations have resulted in the destruction of native forests, and their proliferation is supported by international lending agencies and university-based research.[21] According to the June 1998 Montevideo Declaration, singed by NGO representatives from fourteen countries:

> Money badly needed to support the development of local livelihood security ... is directed into forestry research supporting the use of fertilizers, herbicides, pesticides, biotechnology, cloning and a Green Revolution-like package of techniques which has proven detrimental to local environments and livelihoods.[22]

SOUTHEASTERN MEXICO: A CASE STUDY

ACERCA – Action for Community and Ecology in the Regions of Central America – through its campaign office in Burlington, Vermont, has launched an ongoing investigation regarding the North American Free Trade Agreement (NAFTA) and its effects on indigenous peoples and the land of southeastern Mexico. ACERCA went to southeastern Mexico in April of 1999 to investigate the incursion of eucalyptus plantations and the multinational companies behind them. One of the most important results of the investigation was the discovery that genetically engineered trees are sprouting up at an alarming rate.

The power and weight of the timber industry's influence in southeastern Mexico is evidenced by the weakening of Article 27 of the Mexican constitution and the introduction of other new laws to open up Mexico for timber exploitation. Article 27, which emerged from the Mexican Revolution of 1910, promised agrarian reform and land redistribution to peasant communities. In part, the reform of Article 27 to permit the privatization of communal land holdings, as required by NAFTA,[23] forced the indigenous people of Chiapas to declare war on the government. The indigenous Zapatista uprising began on January 1, 1994, the day NAFTA went into effect.

The Mexican government also promoted a 1992 forestry law that allowed commercial tree plantations and a 1997 revision of the law that literally granted big timber's wishes. The law implemented a series of proposals made by Edward Krobacker, International Paper's forestry division vice-president, allowing big timber to acquire land parcels of unlimited size.[24] Mexican activists have charged that multinational corporations rent the land for a few years from the campesinos to grow their plantations and then give it back to them after the land is ruined.[25]

The World Rainforest Movement's 1999 report, *Tree Plantations: Impacts and Struggles*, highlighted the increased demand for forest products in Mexico:

> The increased activities of the "maquiladora" industry (installed within Mexico and based on imported inputs and external export markets) have resulted in an enormous deficit in packaging papers – which are currently being imported from the US and Canada – used in the necessary packaging of the industrial goods for the supply of external markets. Responding to pressures from the country's industrial sector, the Mexican government is now paving the way for the promotion of large scale pulpwood plantations to provide industry with raw material to produce cheap pulp and paper to fill in that gap.[26]

"The characteristic most desirable for chip and pulp production is rapid growth," says Mick Petrie, former coordinator of the Genetically Engineered Tree Campaign at the Native Forest Network's Eastern North American Resource Center. He continues:

> Trees that are genetically engineered to grow to harvestable size in 3–5 years require huge amounts of water and nutrients. Rapid cycles of growth and removal will quickly lead to abandonment of land when soils created over millennia are exhausted. Fiber produced in this way is primarily made into unnecessary packaging for consumables and virgin paper for which ecological alternatives are already available.[27]

Plantations of eucalyptus and African palm are in production in southeastern Mexico, particularly in Chiapas, Tabasco, Veracruz and the Isthmus of Tehuantepec.[28] Fast-growing eucalyptus is especially well known for its ecologically destructive properties and leaves a virtual desert in its path.[29] Mexican journalist Jaime Aviles writes in *La Jornada*, "Eucalyptus is the perfect neoliberal tree. It's fast growing, kills everything near it, and makes a lot of money for a few people."[30]

Gerardo Gonzales of Foro para el Desarrollo Sustentable (Forum for Sustainable Development) in Chiapas reports that Grupo Pulsar, a company based in the northern Mexican city of Monterrey, has a research centre in Tapachula, Chiapas.[31] The research centre serves as NAFTA's humid-tropics research laboratory for agricultural biotechnology. This facility is involved in the genetic engineering of trees, as well as the development of new altered strains of vegetables. Grupo Pulsar, a Mexican multinational corporation, is working on new genetic strains of vegetables and other crops. The company is headed by Alfonso Romo, whose agribusiness subsidiaries control 40 per cent of the vegetable seed market in North America.[32] Romo's Empresas La Moderna has engaged an estimated 800 partners to farm melons, chili, papayas, bamboo and eucalyptus trees.[33]

Grupo Pulsar has announced plans to invest $300 million in order to create 300,000 hectares of commercial tree plantations in southern Mexico, according to Lloyd's Mexican Economic Report. The company claims that this will be sufficient to double Mexico's timber production. According to Lloyd's:

> Most of the plantings will be of tropical eucalyptus trees native to Australia and Indonesia, which have been genetically improved in Brazil. From 10-centimeter high seedlings, the trees take about six years to grow to a harvestable height of 35 meters with a diameter of 25 centimeters.[34]

Planfosur (Forest Plantations of the South) is a venture of the Texas-based multinational, Temple-Inland Forest Products International, Inc. Temple-Inland is the fifth largest private landowner in the southern US and the ninth largest in the US. Temple-Inland additionally holds land in Puerto Rico, Argentina and Chile. Bobby Inman, former CIA director, is on the board of directors.[35] According to a Planfosur spokesperson, the company has planted 21,000 hectares of eucalyptus in Tabasco and Veracruz, some of which is genetically engineered.[36] Planfosur's own literature says:

> The mission of the Planfosur partners is to plant, grow and harvest the fast-growing eucalyptus species on a total of 21,000 hectares on a continuing basis.... As forestry becomes a major industry in Mexico, larger projects will be considered ... includ[ing] the establishment of additional tree farms and the construction and operation of a pulpmill somewhere near the tree farms.[37]

Avelino B. Villa Salas, the Planfosur spokesman, said that the purpose of the plantations is to provide wood chips to be transported to where the market is best for paper pulp and particle board. Some would be used in Mexico to replace the one third of all packaging that is now imported. At present 4.5 million seedlings are grown per year in Planfosur's Los Choapas nursery in Tabasco, many genetically engineered. The company claims that it intends to plant genetically engineered eucalyptus species entirely after the third rotation.[38] Villa Salas said that Monsanto is supplying agrochemicals (specifically FAENA, a glyphosate herbicide comparable to Roundup) for these plantations and that Temple-Inland, Inc. has just signed on with Monsanto for additional genetic research.

Villa Salas and Jaime Cruz, who is responsible for the protection of the Planfosur's plantations, took the ACERCA delegation to visit plantation sites in Tabasco and Veracruz. Some sites reportedly contained genetically engineered eucalyptus tests, but most were monocultures of conventional first-rotation eucalyptus. One of the ACERCA participants, Lupito Flores, the Development Director of the Kettle Range Conservation Group in Spokane, Washington, stated while touring one of the sites:

> I climbed to the top of the fire lookout to talk with a Mexican worker who stood there shirtless, on duty all day without cover, an umbrella, or even a hat to shade him from the sweltering heat. I looked out over miles and miles of neatly planted rows of eucalyptus trees, rolling off into the distance in all directions as far as the eye could see; my heart sank to think that soon thousands of these trees will be genetically engineered.[39]

Flores continued, "Villa Salas told us they use seeds with genes from New Zealand, Australia, Southeast Asia and Germany and plan to use genes from Zimbabwe, Venezuela, Brazil, and Colombia. Very shortly we will be seeing the bitter fruit of Planfosur's genetic research."

RESEARCH ON GENETICALLY ENGINEERED TREES

Genetic engineering of trees is a new discipline, and trees have a very different biology to most crop plants. Research is costly and evaluating engineered trees is difficult and time-consuming due to the extended life cycles of tree species, even though genetic manipulation can speed up the process. This research has historically lagged behind research on the engineering of food crops, but times are changing.[40]

"A publicly traded firm named ForBio, which is based in Brisbane, Australia, aims to revolutionize forestry, with the help of the latest techniques of 'mapping' a tree's DNA," *Forbes Global Business and Finance* reported in August 1998.[41] Bill Henderson, the company's co-founder and chief executive says, "We're looking at changing the whole economics of plantation crops. We can identify which genes are controlling performance, give ourselves a road map and we just shorten the growing times dramatically."[42] *Forbes* states that ForBio's yield improvements through genetic selection "have been startling." ForBio and Monsanto have entered a joint venture, Monfori Nusantara, and are conducting field trials in Indonesia. In May of 1997 Monfori Nusantara planted over fifty different clones of hybrid eucalyptus that have been growing up to a metre a month, doubling the normal rate of growth. The eucalyptus are expected to be harvested for wood chips in four to five years. ForBio is also active in the US, with a facility in Boise, Idaho, a major agribusiness centre. "More and more we're being driven to dealing with the major multinationals," Henderson says. "Companies like Monsanto want to get their insect- and herbicide-resistant genes into forestry and the quickest and most effective way is a tie-up with us."[43]

Many universities have teamed up with the private sector for research. One such effort, Oregon State University's Tree Genetic Engineering Research Cooperative (which has a dozen members including timber giants Fort James and Potlatch), has been cloning trees that tolerate glyphosate, the active ingredient in Monsanto's Roundup.[44] These "Roundup Ready" trees allow widespread applications of the herbicide, which will not hurt the trees but will kill off all native vegetation after the application, greatly reducing biodiversity.

Another area of research is incorporating Bt toxin (from *Bacillus thuringiensis*), a biological pesticide, into trees. The Canadian Forest Service began pioneering Bt research in the 1950s. The CFS boasts that it continues to lead the world in Bt tree research.[45] The implications of increased Bt use in genetically engineered products have been highlighted by evidence that pollen produced from genetically engineered Bt corn can kill monarch butterflies.[46]

Several companies are also developing trees that are easier to pulp. Trees are engineered to produce a weakened form of lignin, the natural strengthening agent that must be broken down in the pulping process by chemicals.[47] The British biotechnology company AstraZeneca, now part of Syngenta, has developed technology that genetically modifies lignin in paper pulp trees.[48] The company, which was responsible for the first test planting of genetically engineered poplar trees in Britain,[49] is collaborating with Shell Oil and Nippon Paper Industries to insert its genes into trees such as eucalyptus and poplar.[50] The industry claims that trees engineered to make lignin easier to remove from the cellulose will make paper manufacture less energy- and chemical-intensive. What is actually needed, though, is a reduction in paper and packaging consumption, not ways to make it easier to produce. The Native Forest Network's Mick Petrie says,

> Industry claims that genetic altering of trees to lower lignins will decrease toxic emissions during pulping. Alternative technologies such as closed-loop and oxygen delignification, which would eliminate toxic releases altogether, already exist, but the pulp and paper industry has vigorously resisted upgrading aging mills to stop pollution. Clearly the industry's motive for pursuing genetically altered trees is profit-based and not because of concern for the environment.[51]

Sunflower Consulting's Billy Stern adds,

> Attempting to engineer trees with weakened lignin is yet another example of backwards thinking by the pulp and paper industry. Enough low lignin fiber already exists, in the form of agricultural residue byproducts from wheat, rice, barley, corn and other food crops, to satisfy all of the world's paper needs. Research money would be better spent looking at producing stronger paper from shorter fibers, and from mixed fiber blends using kenaf, hemp and a small percentage of wood pulp.[52]

In order for genetically engineered trees not to cross-pollinate with other species, buds need to be hand-picked before they flower, or they must be made sterile. Much research is under way to harness Terminator-like technologies to make genetically engineered trees sterile and thus genetically isolate them from wild populations. However, it is quite possible, if not probable, that mistakes will be made which will have detrimental effects on many wild species of trees, plants and potentially animals (see Chapter 1 in this volume).

Other research is concentrating on creating trees that can produce higher quality fibre for saw logs. ForBio's co-founder and head of research, biochemist Bob Teasdale, says, "Everything you can think of is under genetic control. We can change the number of branches, the angle they grow at and even their diameter."[53] This genetic control governs the size and number of knots and affects the quality of wood. Will multinational timber corporations someday want to produce trees that are genetically engineered to have rectangular instead of round trunks, making a two-by-four much easier to mill?

One of the next challenges for the Doctors of Frankenforesty is the mass production of their creations. Until recently a tedious and time-consuming method of replicating seedlings via tissue culture was done by hand. To speed up this process ForBio has built the world's first tissue-culture robot for automated plant production.[54]

Jack Eaton, plant manager with the Potlatch Corporation, is optimistic about the use of transgenic plant materials in operational plantations, emphasizing, "It's no longer science fiction."[55] Westvaco states in its brochure *Ecosystem-Based Forestry*, that its primary goal is "to maximize fiber production through the use of genetically improved tree species, plantation forestry and the appropriate use of fertilizers and herbicides."[56] John Pait from The Timber Company (which tracks stock for Georgia Pacific) said that Georgia Pacific generates 12 tons of genetically "improved" seedlings a year, some of which appear to be GMOs.[57]

Dominated by researchers from forestry schools, the US Forest Service and big timber companies, the US National Academy of Sciences held a workshop on "Capacity in Forestry Research" in July of 1999. In attendance were a handful of environmentalists from the Dogwood Alliance, Wilderness Society, and other groups. Biotechnology was identified as one of the top research priorities between now and 2020. There was a great deal of discussion at this conference about genetic engineering and the genomics of commercial timber species, according to Danna Smith, Executive Director of the Dogwood Alliance, a coalition of sixty organizations in fifteen states across the southeastern United States that is working to protect the region's forests from the expansion of industrial forestry. "It was mentioned more than once at the workshop that funding for research in biotechnology related to forest productivity was not nearly at the levels that exists in agriculture, and that funding in forest biotechnology is considered critical," Smith reports.[58]

Research and development are moving particularly fast in the southeastern United States. Smith points to US Forest Service estimates that approximately 40 per cent of the coastal pine forest in the southeast has already been converted to pine plantations and that at present rates 70 per cent of the natural pine forest in the region will be converted to plantations by the year 2040. Hardwood forests are also being converted to pine plantations, but how much conversion is taking place in the hardwood region of the South is not known. "Replacing natural forests with rows of pine trees has devastating ecological impacts; the use of genetically engineered trees in pine plantations is like rubbing salt in an already serious wound," Smith explains.[59] Currently, a large fraction of the tens of millions of pine seedlings planted annually in the region are genetically improved,[60] mostly through the use of conventional breeding, advanced tissue culture techniques and cloning.

But the timber industry is being encouraged by forestry leaders in North Carolina and throughout the country to use genetic engineering to grow more wood on less land and produce hardier trees with better traits. North Carolina, an acknowledged leader in other areas of biotechnology, is moving full steam ahead with engineered tree research. "This … is the time to merge trees and biotechnology in ways we haven't done before," said Robert B.

Jordan III, president of Jordan Lumber and former North Carolina lieutenant governor in the fall of 1999.[61]

North Carolina State University's biotechnology program has been awarded a \$4.4 million grant from the National Science Foundation to lead a study of the genetics of the loblolly pine, in the hope of creating better, stronger wood.[62] The project leader, Forestry Professor Ronald Sederoff, said that the grant marks the first time a public-sector study of genetically engineered trees has received the funding and respect it deserves,[63] and announced that he would coordinate the sequencing of the loblolly pine tree's genetic material.[64]

The North Carolina Biotechnology Center has created an Advisory Committee on Forest Biotechnology, chaired by former lieutenant governor Jordan, that consists of twenty-eight forest industry representatives, government officials and university researchers.[65] Robert Kellison, committee vice-chairman and director of forest technology and forest products for the Champion International Corporation, said, "The industry is just absolutely convinced we're going to have to have biotechnology to be competitive."[66]

STEPPING INTO THE FUTURE

Although local communities have rallied against forestry plantations in many areas of the world, there has been little concerted effort to oppose the introduction of genetically engineered trees on those plantations. Greenpeace and other environmental groups have been opposing the introduction of GMOs into the food chain, but they and other large environmental groups have only begun to address the worldwide introduction of engineered trees and are just beginning to research the subject. However, underground direct action activists have started activities that are bound to escalate.

In July of 1999, a dramatic direct action occurred at AstraZeneca's research site near London, during which over a hundred genetically altered trees were destroyed in a dawn raid. An anonymous communiqué denounced the research, stating "those who are manipulating the DNA of trees, using a very powerful but dimly understood technology, show contempt for our planet and the life it supports, including human life. They respect only profit for themselves and their shareholders."[67]

It did not take long before this style of direct action hit the North American continent. On October 27, 1999, a group calling themselves Reclaim the Genes destroyed five hundred trees and seedlings at Silvagen, Inc.'s test plots near the University of British Columbia in Vancouver. Damage was estimated at \$250,000 and five years of research was reportedly lost.[68] Four days later, the Genetix Goblins sabotaged a thousand assorted conifer saplings at Western Forest Products (WestFor) in Saanich, British Columbia, claiming that WestFor was involved in genetic engineering.[69]

There has been controversy regarding the WestFor action in Saanich. Kermit Ritland, professor in the Department of Forest Sciences, University of British Columbia, says that the trees destroyed "were produced by traditional tree breeding techniques (phenotypic selection). There were no trees that have genes 'inserted' into them, for example, they don't have anything like the 'Roundup Ready' genes that Monsanto has inserted into soybeans and poplars." Ritland continues that "it's tragic that innocent researchers are being caught in this crossfire."[70] Melissa Sorensen of the Eugene, Oregon, based Bioengineering Action Network, responds,

> As industrial agriculture has shown, it's a slippery slope from genetic "improvement" to genetic "engineering". Tree breeding and genetic engineering are just ways of speeding up production and cutting costs. It might not seem as horrific as a tree that contains flounder genes, or glows Christmas lights – but even trees "improved" in a laboratory to grow faster and more uniform than the more conventionally bred plantation species will speed up the degradation of forest biodiversity.[71]

More actions followed before the end of 1999. On November 18, hundreds of fruit trees at Okanagan Biotechnology Inc. in Summerland, British Columbia, were destroyed by "The WTO Welcome Committee." A week later, two institutions near Seattle practising genetic engineering of trees were sabotaged. Genetically altered poplars, saplings and a greenhouse were destroyed.[72]

With the earth's forests on the decline and industrial plantations on the rise, there is every reason to believe that the future of plantations will be in seedlings of genetically engineered trees. Globally, forests as we know them may become no more than isolated, ecologically fragile preserves, with an almost science fiction-like presence.

If global trade agreements and neoliberal policies continue to flourish, corporations will be in a position to maximize the economic potential of genetically engineered trees and, consequently, both people and the natural world will suffer. Schemes like the Kyoto Protocol, which establishes a trading system to buy and sell carbon emissions, will only lead to further environmental degradation, as corporations like Shell and BP delve further into the forestry business and economic incentives encourage transforming native forests into genetically engineered plantations. As genetically engineered trees come out of the greenhouse and into industrial plantations, people and their communities need to be organized, outraged and discuss appropriate actions.

"Genetically engineering trees that grow faster and contain more fiber will not solve the industry's problem of overconsumption and unsustainable logging," explains Danna Smith of the Dogwood Alliance. "In fact it exacerbates the problem by creating the illusion that the industry can somehow 'manufacture' nature. The move towards genetically engineered trees is yet another example of how the timber industry avoids addressing the issues related to unsustainable logging by identifying short-sighted solutions that have potentially devastating long-term consequences."[73]

NOTES

1. "Call for Global Moratorium on Genetically Engineered Trees," *World Rainforest Movement Bulletin* 29, December 1999.

2. Oliver Tickell and Charles Clover, "Trees that Never Flower Herald a Silent Spring," *Daily Telegraph*, July 17, 1999.

3. Paul Brown, "Forests in Danger from GM Super-tree Says WWF. Field trials, including five in UK, 'not properly controlled'," *Guardian*, November 10, 1999.

4. Australian Associated Press/Reuters, "Conservation Groups Call for Worldwide GM Forestry Ban," November 9, 1999.

5. A SEED Europe, "The Issues Surrounding GMOs in Forestry," in *Europe's Forests, A Campaign Guide*, January 1999, pp. 58–9.

6. PR Newswire, April 6, 1999, http://www.prnewswire.com. Monsanto has since withdrawn from this venture.

7. Hugh Warwick, "Frankenstein Forests," *Green Direct Magazine*, http://www.greendirect.co.uk/trees.htm.

8. Hugh Warwick, "The Next GM Threat: Frankenstein Forests," *Ecologist*, vol. 29, no. 4, July 1999, pp. 250–51. This insistence led to a breakdown in the Kyoto Process in November 2000.

9. Larry Lohmann, *The Dyson Effect: Carbon "Offset" Forestry and the Privatisation of the Atmosphere*, CornerHouse Briefing no. 15, July 1999, Sturminster Newton: Corner House, p. 2.

10. Daphne Wysham, "Profiting from Pollution," *San Jose Mercury News*, 22 November 1998, quoted in ibid., p. 3.

11. "Toyota to Enter Biotechnology Business," *News on Genetic Engineering*, February 13, 1998, at http://www.netlink.de/gen/Zeitung/1998/home.html.

12. "Sequestered Carbon Dioxide," *Green Energies Newsletter*, August 9, 1998, at http://www.nrglink.com/archives/nrgs898.html#nrgs8998.1.

13. Ricardo Carrere and Larry Lohmann, "Pulping the South – The Expansion of Commercial Tree Plantations," World Rainforest Movement Plantations Campaign, http://www.wrm.org.uy/english/plantations/material/tree.htm.

14. Ibid.

15. Michael Grace, quoted in "Plant Biotechnology, Final Report and Proceedings from the Talking Technology Forums," August 22–24, 1996 and May 8, 1999, p. 30, Consumers' Institute of New Zealand, http://www.consumer.org.nz/tech/updated_report.pdf.

16. Press Release from the 1997 IRRDB (International Rubber Research and Development Board) Annual Report.

17. World Resources Institute, "Forest Trade Facts," *Forest Notes*, November 1999, p. 6.

18. Ibid.

19. *Global Futures Bulletin* 83, Institute for Global Futures Research (Australia), June 14, 1999, email communication from George Marshall, forwarded by Patrick Reinsborough, Rainforest Action Network.

20. Ibid.

21. Ibid. p. 16.

22. "The Montevideo Declaration, June 1998, Montevideo, Uruguay," in Lohmann, *The Dyson Effect*, p. 18; full version at http://igc.apc.org/wrm.

23. Dan La Botz, *Democracy in Mexico*, Boston: South End Press, 1995, p. 24.

24. John Ross, "Big Pulp vs. Zapatistas, Cellulose Dreams in Southeastern Mexico," *Dark Night Field Notes* 14, December 1998, p. 53.

25. Gustavo Castro, CIEPAC (Center for Investigation of Economic and Political Action), interview April 19, 1999, San Cristóbal, Chiapas, Mexico.

26. Alejandro Villamar, "Datos de la 'version mexicana' de la estrategia global de la industria maderera–papelera internacional bajo el TLCAN," April 1998, quoted in World Rainforest Movement, *Tree Plantations: Impacts and Struggles*, February 1999, p. 121.

27. Mick Petrie, Native Forest Network Eastern North American Resource Center, Burlington, Vermont, telephone interview, February 7, 2000.

28. "Eucalyptus, Neoliberalism and NAFTA in Southeastern Mexico," ACERCA report, June, 1999, http://www.acerca.org/eucalyptus_Mexico1.html.

29. Melissa Burch, "Lacandona: The Zapatistas and Rainforest of Chiapas, Mexico," ACERCA Green Paper 1, p. 7, January 1999.

30. Ibid.

31. Gerardo Gonzalez, Foro para el Desarrollo Sustenable (Forum for Sustainable Development), interview April 20, 1999, San Cristóbal, Chiapas, Mexico.

32. Jonathan Friedland and Scott Kilman, "As Geneticists Develop an Appetite for Greens, Mr. Romo Flourishes," *Wall Street Journal*, January 28, 1999, p. A1.

33. James F, Smith, "Biotech Farmers in Chiapas Lead Peaceful Agricultural Revolution," *Los Angeles Times*, July 26, 1998, p. D1.

34. Lloyd's Mexican Economic Report, June 1999, "Trees to Cut Down…" http://www.mexconnect.com/MEX/lloyds/llydeco0699.html#6.

35. Ross, "Big Pulp vs. Zapatistas," p. 52.

36. Avelino B.Villa Salas, Planfosur spokesman, ACERCA interview in Villahermosa, Tabasco, Mexico, April 23, 1999.

37. Planfosur document, "Information about Planfosur S. De R.L. De C.," obtained April 23, 1999, Villahermosa, Tabasco, Mexico.

38. Villa Salas, interview.

39. Lupito Flores, Kettle Range Conservation Group, Spokane, WA, email communication, January 12, 2000.

40. Based on Press Release from the 1997 IRRDB (International Rubber Research and Development Board) Annual Report.

41. Bob Johnstone, "Supertrees," *Forbes Global Business & Finance*, August 10, 1998, p. 52.

42. Ibid.

43. Ibid. p. 53.

44. *ASTI Connections*, vol. 8, no. 2, 1996, p. 5 (Corvallis, OR: Advanced Science and Technology Institute).

45. Canadian Forest Service, "Biotechnology at the CFS," http://nrcan.gc.ca/cfs/proj/sci-tech/prog/biotech/pest_e.html (updated January 5, 1999).

46. John Carey, "Imperiled Monarchs Alter the Biotech Landscape," *Business Week*, June 7, 1999, p. 36; John Losey et al., "Transgenic Pollen Harms Monarch Larvae," *Nature* 399, 20 May 1999, p. 214; see also Brian Tokar, "Biotech Corn and the Deaths of Butterflies," *Food & Water Journal*, Summer 1999, pp. 20–23.

47. Nick Nuttall, "'Paper' Trees Will Cut Pollution," *The Times*, June 8, 1999.

48. *Intellectual Property and Biodiversity News*, vol. 4, no. 4, March 15, 1995 (Minneapolis: Institute for Agriculture and Trade Policy).

49. Marie Woolf, "Biotech and Paper Companies Team Up to Alter Trees Genetically," *Independent on Sunday*, May 16, 1999, posted to IGC electronic conference wall.events, May 25, 1999.

50. *Intellectual Property and Biodiversity News*, vol. 4, no. 4, March 15, 1995.

51. Petrie, telephone interview, February 7, 2000.

52. Billy Stern, Sunflower Consulting, email communication, January 11, 2000.

53. Johnstone, "Supertrees," pp. 52–3.

54. Ibid. p. 53.

55. *ASTI Connections*, vol. 8, no. 2, 1996, p. 5.

56. Danna Smith, Executive Director, Dogwood Alliance, email communication, November 8, 1999.

57. Ibid.

58. Ibid.

59. Ibid.

60. Temple-Inland Forest Co., *A Standard for Today, A Commitment for Tomorrow*, p. 1.

61. PR Newswire, November 2, 1999, http://www.prnewswire.com.

62. Smith, email communication, November 8, 1999, quoting *Raleigh News and Observer* article of September 27, 1999.

63. Ibid.

64. PR Newswire, November 2, 1999, http://www.prnewswire.com.

65. Ibid.

66. Ibid.

67. Genetic Engineering Network Media Release (quoting anonymous communiqué), July 12, 1999.

68. Bioengineering Action Network (BAN), Eugene, Oregon, email communication from Annie Oakley, January 21, 2000.

69. Ibid.

70. Kermit Ritland, response to web page posting http://vancouver.tao.ca/~ban/1099bctreesresease.htm, "More doubts about GE tree actions," email forwarded by Brian Tokar, November 8, 1999.

71. Melissa Sorenson, Bioengineering Action Network (BAN), Eugene, Oregon, email communication from Annie Oakley, January 21, 2000.

72. Bioengineering Action Network (BAN), Eugene, Oregon, email communication from Annie Oakley, January 21, 2000.

73. Smith, email communication, November 8, 1999.

PART II

Medical Genetics,
Science and Human Rights

Beyond today's heated debates over genetically engineered food and the environment, the biotechnology industry's boldest claim for the future is that it will thoroughly revolutionize medical care. Advocates of cloning, genetic screening, "gene therapy" and other new technologies promise cures for intractable genetic diseases, drugs precisely tailored to an individual's genetic makeup, earlier detection of chronic health problems, and more. We read of drugs and other useful substances produced by genetically engineered sheep and goats in their milk, and of animal organs to be made suitable for human transplants. Parents are promised that doctors will soon be able to cure genetic diseases before a child is born, and that someday they may be able to choose the genetic makeup of their offspring.

Each new development in genetic medicine, whether real or purely speculative, is granted headline status, as biotechnologists promise to satisfy everyone's longing for a healthy, disease-free future. This can be difficult to criticize, as nearly everyone has seen loved ones suffer from conditions that may someday be curable, and parents will always have the highest aspirations for their children. But just as the biotechnology industry's claims about the benefits of genetic engineering in agriculture have proven largely fraudulent, developments in biotech medicine pose a host of new problems that have barely begun to be addressed. As the chapters in this section will reveal, today's biotech medicine is largely a story of exaggerated promises, overlooked risks, exorbitant costs, and research funds diverted from far more sustainable and equitable alternatives.

The first applications of genetic engineering in medicine relied on the same technology that brought us recombinant Bovine Growth Hormone: the insertion of structural genes for simple proteins and peptides into the DNA of bacteria. The engineered bacteria reproduce rapidly, and the desired substance is then extracted from the bacterial suspension. The first product of this technology to cause serious problems was L-tryptophan, a single amino acid, which is involved in the synthesis of the neurotransmitter serotonin in human nerve cells.

Tryptophan supplements were commonly used in the 1980s as a remedy for insomnia. A Japanese company, Showa Denko, developed a way to alter bacteria genetically to increase tryptophan production by duplicating regulatory genes and adding an additional enzyme from another strain of bacterium. In 1989, more than five thousand people contracted a previously unknown, debilitating blood disease, eosonophilia myalgia syndrome (EMS), and twenty-seven died from the condition. In several American states, virtually all the reported cases were linked to Showa Denko's brand of tryptophan.[1] Chemical analyses demonstrated the presence of a number of substances in Showa Denko's product that are not normally found in commercial preparations of tryptophan, including very unusual pairings (dimers) of tryptophan molecules.[2] The company claimed that the problem was simply one of improper puri-fication, but its production facilities were dismantled before a thorough investigation could be completed.

One of the most widely used products of genetically altered bacteria is synthetic human insulin, marketed as Humulin. Before the mid-1980s, people suffering from diabetes relied on daily injections of insulin extracted from cattle and pigs in slaughterhouses; the genetically engineered alternative was initially promoted as a more humane option, and a safer alternative for those who are allergic to animal insulin. Soon, synthetic human insulin became the treatment of choice and began to be prescribed by doctors worldwide. According to a suppressed 1993 report commissioned by the British Diabetics Association, however, the medical consequences are quite troubling. The report found that 10 per cent of diabetics using the new "human" insulin were suffering frequent, wildly uncontrollable hypoglycaemic comas, and up to 20 per cent had a strong preference for animal-derived insulin due to problems ranging from memory loss and an inability to concentrate to troubling person-ality changes.[3] Heavy promotion of the synthetic human insulin by companies such as Eli Lilly and the Danish biotechnology company Novo Nordisk has made it difficult for patients to switch back to animal-derived varieties.

As scientists learn to identify genetic markers correlated with particular medical conditions, companies have begun developing genetic tests for con-ditions ranging from sickle-cell anaemia to spina bifida to breast cancer. While the identification of genetic markers rarely leads to cures or treatments, employers and insurance companies have not hesitated to use this information to discriminate against those now labelled genetically "defective." Probably the best known example is sickle-cell trait, a condition that scientists have under-stood in rather well defined biochemical terms since the early 1970s. It is one of the rare conditions that can be attributed to a single-gene mutation. One in ten African-American babies is born with at least one copy of the sickle cell gene, but only one in five hundred – those who inherit the sickle-cell trait from both their parents – actually has sickle-cell anaemia. While no one has yet devised a "cure" for this condition – which may benefit those living in tropical regions by imparting resistance to certain forms of malaria – there are numerous documented cases of insurance and workplace discrimination. A DuPont chemical plant in New Jersey was cited for restricting the activities of those who tested positive for sickle-cell trait, and it took a lawsuit to overturn a US Air Force Academy policy of excluding sickle-cell carriers.[4]

In 1997, the US National Institute of Environmental Health Sciences announced a $60 million Environmental Genome Project, purportedly to study the relationship between genes and the environment. A review of the project's documents, however, reveals little interest in investigating the environmental *causes* of disease; rather, it is dedicated to surveying the population distribution of genetic markers that may be related to environmental illness. Focused narrowly on the analysis of risk, the project seeks "to understand the nature, significance, and distribution of susceptible genotypes in the general population."[5] The focus is entirely on mapping individual susceptibilities to environmental toxins, rather than studying the biological effects of these toxins and working to eliminate them from the environment. It is yet another instance of blaming the victims. Given the tacit acceptance of genetic discrimination in the US, and the genetic reductionism that pervades nearly all areas of medical research today, it appears likely that groups seeking redress from environmental contamination will increasingly be blamed for their individual sensitivities to environmental stresses.

Indeed, the rise of genetic engineering and other biotechnologies has severely distorted the research agenda in nearly every area of the biological sciences. Whether one is studying cancer, environmental illness, the structures of the nervous system, or problems in evolution and development, there is a pervasive and systematic bias towards research methods that involve the identification and cloning of genes. In some areas of research, this is entirely appropriate, given the ability of genetic methods to reveal important information about the molecular structures underlying some biological processes. But in the decades since the first gene was isolated and cloned, the hegemony of genetic techniques in biology has grown far beyond those well-defined areas. In today's world of high-tech medicine, the problem is nearly always believed to lie in one's genes.

The entrenchment of this reductionist, gene-centred thinking in the biological sciences has been vastly enabled by new developments in biotechnology; however, the origins of the manipulative world-view of biotechnology can be traced to the very beginnings of science. Francis Bacon, who is credited with codifying the "scientific method," wrote in 1624 of a "New Atlantis," an imagined future society run by an elite of scientist-priests who would "enlarge … the bounds of the human empire" and "improve" nature so as better to satisfy particular human ends. Bacon predicted efforts to alter the life cycles of edible plants, to "make commixtures and copulations" of diverse species and create entirely new ones.[6]

Nearly three centuries later, the Franco-American biologist Jacques Loeb

began to advocate a utilitarian "technology of living substance," a new biological engineering that would "take up the problems of transforming species
and creating life."[7] The instrumentalism of Loeb and his scientific descendants
was in considerable contrast to the outlook of many early molecular biologists, whose profound curiosity about the living world was partly inspired by
the pioneering quantum physicist Edwin Schrödinger. In his path-breaking
1943 lectures, published under the title, *What is Life?*, Schrödinger wrote that
"we must be prepared to find [living matter] working in a manner that
cannot be reduced to the ordinary laws of physics."[8] Today's biotechnologists,
firmly in the camp of a utilitarian, reductionist world-view, would likely
mischaracterize this quest for a more holistic biology as a mere throwback to
nineteenth-century vitalism.[9]

The outlook of the biotechnology industry is rooted in a unique social
and economic reality as well as a reductionist philosophy. The past two decades
have seen an unprecedented rise in corporate funding of basic biological
research. Biotechnology's "university–industrial complex," as described well
over a decade ago by Martin Kenney, is a world dominated by private funding
of university laboratories, research contracts aimed squarely at patents and
product development, and a new elite of entrepreneurial professors who travel
freely between the laboratory and the corporate boardroom.[10] Academic freedom has been overrun by a profit-driven obsession with secrecy, and yester-
day's academic colleagues have become today's commercial competitors.

This sea change in the practice of scientific research went largely unnoticed until very recently, but a few observers have documented disturbing
trends. The first systematic survey of academic–corporate ties in biotechnology
was in 1991, when Sheldon Krimsky and his colleagues surveyed faculty
members in the biological sciences at prestigious research universities and
found that between 10 and 30 per cent had direct ties to biotechnology
companies.[11] The figure was higher for members of the prestigious National
Academy of Sciences. A 1998 study reported that over 40 per cent of scientists surveyed had received gifts from biotechnology companies, and that a
third had ties to companies that expected to review their academic papers
prior to publication.[12] An investigation by *New York Times* reporter Sheryl Gay
Stolberg, in the aftermath of the scandalous death of an 18-year-old gene
therapy patient at the University of Pennsylvania, found that today, "academic
scientists who lack industry ties have become as rare as giant pandas in the
wild."[13]

While these ties are often claimed to benefit patients by speeding up the
commercialization of new scientific discoveries, the evidence suggests otherwise. Researchers at RAFI (Rural Advancement Foundation International)
have shown how the biotechnology industry contributes to the rising costs of
medical care by narrowing the scope of research, and by procuring patents for
a wide array of scientific procedures. "Companies need biotech patent monopolies … in order to consolidate their control over medical research," reports
RAFI's Hope Shand,[14] and the concomitant increase in the costs of medical

research is widely acknowledged as the main driving force behind the unprecedented wave of pharmaceutical industry mergers in recent years. Researchers often complain about the publication delays, licensing fees and other constraints that corporate funding and biotech patents impose on research practices, but are generally unwilling to sacrifice the lucrative consultancies and stock options that often come with industry ties.

The chapters in this section examine a vast array of recent developments in medical biotechnology, and seek to expose many of the industry's extravagant claims for the future. Marcy Darnovsky analyses current debates over "gene therapy," a euphemism for the genetic manipulation of human beings, and reveals the interests that underlie today's aggressive advocacy of genetic enhancement and "designer babies." Barbara Katz Rothman dissects the contemporary discourse around cancer, and examines the deeper consequences of viewing disease as merely something "in our genes." Sarah Sexton challenges the claimed medical benefits of animal and human cloning, and situates the cloning debate in the context of today's increasingly inequitable health care system. David King then elaborates further on the disturbing eugenic implications of developments in genetic medicine, and shows how today's emerging "laissez faire eugenics" reinforces social inequities and further marginalizes the disabled, while trying to hide behind myths of "free choice."

Alix Fano's chapter scrutinizes research into xenotransplantation – the attempt to alter the organs of animals so they can be transplanted into humans – and reveals disturbing implications for the welfare of animals and the priorities of medical research. Zoë Meleo-Erwin describes the rise of new reproductive technologies, and shows how the increasing commercialization of reproduction especially victimizes women. Finally, Mitchel Cohen explains how research seeking a genetic basis for violence specifically targets African-American youth, perpetuating long-discredited myths of racial inferiority. Each of these chapters highlights the subtle interplay of scientific and social factors, and helps reveal the disturbing political and ideological underside of today's "brave new world" of biotech medicine.

NOTES

1. Philip Raphals, "Does Medical Mystery Threaten Biotech?" *Science*, November 2, 1990, p. 619, and "New Epidemic Linked to Unidentified L-tryptophan Agent," *Medical Post* (Montreal), August 7, 1990.

2. Arthur N. Mayeno and Gerald J. Gleich, "Eosinophilia-myalgia syndrome and tryptophan production: a cautionary tale," *Tibtech* 12, September 1994, pp. 346–52; Robert H. Hill, Jr, et al., "Contaminants in L-tryptophan associated with Eosinophilia Myalgia Syndrome," *Archives of Environmental Contamination and Toxicology* 25, 1993, pp. 134–42.

3. Paul Brown, "Diabetics Not Told of Insulin Risk," *Guardian*, March 9, 1999, at http://www.swissdiabetes.ch/~fis2/englvers/englnews.htm.

4. Brian Tokar, "The False Promise of Biotechnology," *Z Magazine*, February 1992,

pp. 27–32, Ruth Hubbard and Elijah Wald, *Exploding the Gene Myth*, Boston MA: Beacon Press, 1997, pp. 33–5. For ongoing coverage, see the monthly newsletter *GeneWatch*, published by the Council for Responsible Genetics in Cambridge, Massachusetts.

5. "Understanding Gene–Environment Interactions," *Environmental Health Perspectives*, vol. 105, no. 6, June 1997, at http://ehpnet1.niehs.nih.gov/docs/1997/105–6/niehsnews.html; see also http://www.niehs.nih.gov/envgenom/.

6. Francis Bacon, "New Atlantis," in Frederic R. White, ed., *Famous Utopias of the Renaissance*, New York: Hendricks House, 1955, esp. pp. 240–43.

7. Quoted in Philip J. Pauly, *Controlling Life: Jacques Loeb and the Engineering Ideal in Biology*, New York: Oxford University Press, 1987, pp. 51, 177.

8. Edwin Schrödinger, *What is Life?*, Cambridge: Cambridge University Press, 1962, p. 76.

9. On non-reductionist alternatives, see Stuart A. Newman, "Idealist Biology," *Perspectives in Biology and Medicine*, vol. 31, no. 3, Spring 1988, pp. 353–68, and Brian Goodwin, *How the Leopard Changed its Spots: The Evolution of Complexity*, New York: Scribner's Sons, 1994.

10. Martin Kenney, *Biotechnology: The University-Industrial Complex*, New Haven: Yale University Press, 1986. Many of the issues raised by Kenney are updated in Eyal Press and Jennifer Washburn, "The Kept University," *Atlantic Monthly*, vol. 285, no. 3, March 2000, pp. 39–54.

11. Sheldon Krimsky, James G. Ennis and Robert Weissman, "Academic-Corporate Ties in Biotechnology: A Quantitative Study," *Science, Technology & Human Values*, vol. 16, no. 3, Summer 1991, pp. 275–87. The figure for National Academy members was 37 per cent.

12. Described in Sheryl Gay Stolberg, "Gifts to Science Researchers Have Strings, Study Finds," *New York Times*, April 1, 1998, p. A17.

13. Sheryl Gay Stolberg, "Financial Ties in Biomedicine Get Close Look," *New York Times*, February 20, 2000, p. A1.

14. Quoted in Pat Mooney, "Biotech Patents Distort, Discourage Innovation and Increase Costs for Dubious Drugs," *Third World Resurgence* 84, August 1997, p. 24.

9

The Case against Designer Babies:
The Politics of Genetic Enhancement

MARCY DARNOVSKY

Life is entering a new phase in its history. We are seizing control of our own evolution.

Gregory Stock[1]

No one really has the guts to say it.... If we could make better human beings by knowing how to add genes, why shouldn't we do it?

James Watson[2]

The technology to produce human clones – genetic copies of existing people – is within reach of today's researchers and fertility doctors. By the time you, or perhaps the children you know, decide to have babies, cloning may well be an option. Designer babies – children whose genes have been permanently altered to "enhance" them physically or behaviourally, and who will pass their modified genes on to their own children – represent a far more difficult technical manoeuvre. Some influential geneticists are convinced that designer-baby technology, too, will be available within the next few decades. Together with a small but increasingly assertive bunch of biotech entrepreneurs, bio-ethicists, social theorists and journalists, these scientists are avidly promoting a world in which affluent parents are as likely to arrange genetic enhancements for their children as to send them to private school. Their vision is explicitly political as well as technical: they fully support an onslaught of consumer-driven eugenic engineering, anticipating and accepting that designer babies and their further enhanced progeny will come to constitute a powerful genetic aristocracy.

This loose alliance of designer-baby promoters believes that new genetic and reproductive technologies are both inevitable and a boon to humanity. They exuberantly describe near-term genetic manipulations – within a generation – that may increase resistance to diseases, "optimize" height and weight, and boost intelligence. Further off, but within the lifetimes of today's children, they foresee the ability to adjust personality, design new body forms, extend life expectancy, and endow hyper-intelligence. Some confidently predict splicing traits from other species into human children. In late 1999, for example, an *ABC Nightline* special on cloning speculated that genetic engineers would learn to design children with "night vision from an owl" and "supersensitive hearing cloned from a dog."[3]

Gregory Stock, a central figure in the campaign for designer babies that was launched in 1998, goes even further. Genetic enhancement technology, he says, "promises ... eventually to transform our very beings as ever more significant genetic changes are introduced into our genomes. This technology will force us to re-examine even the very notion of what it means to be human [as] we become subject to the same process of conscious design that has so dramatically altered the world around us."[4]

To anyone concerned about environmental degradation, sceptical about the unalloyed virtues of technoscience, dubious about the all-determining power of DNA, or unsure that elite scientists should be the ones to decide our future, Stock's words read more as threat than promise. His enthusiasm bespeaks a wilful blindness to the unintended (though not unforeseeable) consequences of genetic enhancement technology. On the other hand, his brash confidence, and that of others eager to engineer a post-human future, does serve to clarify what is at stake.

The genetic and reproductive technologies that put designer babies on the near horizon open the door to "seizing control of human evolution." In the world that lies on the far side of that portal, parent-consumers will flock to fertility clinics to pick and choose offspring options from a list of traits said to be determined by DNA. As proponents of this "post-human" future openly acknowledge, since most designer children will be born to the well-off, the new technologies will significantly exacerbate socio-economic inequality.

How plausible are such scenarios? In this era of wildly overblown claims about the power of genes to determine everything from sexual orientation to thrill-seeking to homelessness, scepticism would seem the better part of futuristic prudence.[5] But this is also an era of dizzyingly rapid technoscientific change, and it would be foolish to dismiss the possibility that scientists will achieve enough mastery over the human genome to wreak enormous damage – biologically, culturally and politically.

Because human beings are far more than the product of genes – since DNA is one of many factors in human development – the feats of genetic manipulation will turn out to be much more modest than what the designer-baby advocates predict. Nonetheless, their political vision is dangerously powerful. Marketing the ability to specify our children's appearance and abilities encourages a grotesque consumerist mentality towards children and all human life. Fostering the notion that only a "perfect baby" is worthy of life threatens our solidarity with and support for people with disabilities, and perpetuates standards of perfection set by a market system that caters to political, economic and cultural elites. Channelling hopes for human betterment into preoccupation with genetic fixes shrinks our already withered commitments to improving social conditions and enriching cultural and community life. Promoting a future of genetically engineered inequality legitimizes the vast existing injustices that are socially arranged and enforced.

VARIETIES OF GENETIC ENGINEERING: CRUCIAL DISTINCTIONS

The only kind of genetic engineering currently practised on human beings is experimental, and involves efforts to fix the genes of somatic (or body) cells in patients with relatively rare health problems that reflect the functions of single genes. In about five hundred clinical trials since the early 1990s, doctors have tried to introduce genetic modifications to patients' lungs, nerves, muscles, and other tissues.

The 1999 death of an 18-year-old in one of these somatic gene therapy trials resulted in revelations of almost seven hundred "serious adverse effects" that researchers and doctors had somehow failed to report to the proper regulatory authorities.[6] But well before the safety of somatic gene therapy had been so tragically and starkly called into question, its effectiveness had begun to appear dubious. Even pioneer genetic engineer W. French Anderson has voiced doubts, saying bluntly in 1998, "There is no evidence of a gene therapy protocol that helps in any disease situation."[7] At this point, "gene therapy" would more accurately be called "genetic experiments on human subjects."

Whether they work well or not at all, somatic gene manipulations affect only the individuals who undergo them. Designer-baby technology, more widely and formally known as "germline engineering" or "germline enhancement," targets embryos at the earliest stage of their development. Such genetic alterations would be replicated in every cell in the body of a child born after a germline procedure. Since the germ cells (reproductive cells) would also be affected, the child would pass any modified or added genes to future generations. The engineered DNA could thus become a permanent part of the human genetic legacy.

Many genetic scientists say they favour human germline engineering because it could allow people who carry serious genetic mutations to avoid passing them to their children. But this argument is disingenuous, since these scientists know full well that safer and simpler options are already available. Instead of attempting to engineer the human germline, by tampering with the genes of embryos created by in vitro fertilization, technicians can examine those embryos and transfer only unaffected ones to the body of the mother – a technique known as pre-implantation screening. Alternatively, people who carry genetic defects and prefer to avoid IVF can use prenatal screening, with the option of abortion if necessary. Others who want a child will choose to adopt.

Prenatal and pre-implantation screening are themselves technologies that can easily be abused. Sex-selection abortions are already common in some parts of the world, especially India and China. Many disability rights activists point out that prenatal and pre-implantation screening constitute a form of eugenics, and fear that their widespread availability would encourage intolerance for anyone perceived as having a gene-associated "imperfection." They oppose human germline engineering, but do not accept prenatal or pre-implantation screening as an alternatve to it. Other opponents of human germline engineering feel that these procedures are justifiable for the prevention of serious gene-associated disease.[8]

The bottom line, in any case, is that the medical justifications for human germline engineering are strained, while the medical, ethical and political risks it poses are profound. Fortunately, the distinction between somatic and germline engineering is a clear technical demarcation. It provides a clear stopping point on a very slippery slope.

THE DESIGNER-BABY VISION: IS IT EUGENICS?

Many advocates of human germline engineering are abandoning their untenable claims about its medical benefits, and openly acknowledging that their real interest is in genetic enhancement. Yet most of them energetically reject the characterization of their project as "eugenics." Bioethicist Arthur Caplan, for example, asserts that "it is simply a confusion to equate eugenics with any discussion of germline therapy."[9] There are, of course, significant differences between eugenic choices made voluntarily by parents, and eugenic mandates enforced by state-coerced sterilization (as in the United States in the early and mid-twentieth century) or state-sponsored genocide, as in Nazi Germany. As biotech critic Jeremy Rifkin puts it, "The old eugenics was steeped in political ideology and motivated by fear and hate. The new eugenics is being spurred by market forces and consumer desire."[10]

Richard Hayes, one of the first people to call attention to the accelerating push for germline engineering in the late 1990s, has coined the phrase "techno-eugenics" to describe this new phase. In "The Threat of the New Human Techno-Eugenics," Hayes writes:

> The single most portentous technological threshold in all of human history is close upon us: the ability of humans to deliberately modify the genes that get passed to our children.... This vision of the human future celebrates ... a world in which parents select their children's genes, literally from a catalogue, to give them an edge in the quest for "success." It celebrates nothing less than the end of our common humanity, as we segregate into separate genetic castes and eventually separate species.[11]

Alarmist? Overwrought? Listen to the words of Lee Silver, one of the key players in the pro-germline engineering camp. Silver is a molecular geneticist and developmental biologist at Princeton University, and an unabashed promoter of consumer-driven "reprogenetic" technologies. After a few centuries of these practices, he believes, humanity will bifurcate into genetic Übermenschen and Untermenschen – and not long thereafter into different species.[12] Here is Silver's prediction for the year 2350:

> The GenRich – who account for 10 percent of the American population – all carry synthetic genes. Genes that were created in the laboratory.... The GenRich are a modern-day hereditary class of genetic aristocrats.... All aspects of the economy, the media, the entertainment industry, and the knowledge industry are controlled by members of the GenRich class.[13]

How do the other 90 per cent live? Silver is quite blunt on this point as well: "Naturals work as low-paid service providers or as laborers."

That rich and poor already live in biologically disparate worlds can be argued on the basis of any number of statistical measures: life expectancy, infant mortality, access to health care. Of course, medical resources and social priorities could be assigned to narrowing those gaps. But if the promoters of designer babies and human clones have their way, precious medical talent and funds will be devoted instead to a technically dubious project whose success will be measured by the extent to which it can inscribe inequality onto the human genome. The human genetic technologies will then serve to legitimize and expand injustice, to create a kind of inequity new in human history, and to make obsolete even rhetorical gestures towards equality. Silver writes:

> There is still some intermarriage as well as sexual intermingling between a few GenRich individuals and Naturals. But, ... as time passes, the mixing of the classes will become less and less frequent for reasons of both environment and genetics....
> If the accumulation of genetic knowledge and advances in genetic enhancement technology continue ... the GenRich class and the Natural class will become the GenRich humans and the Natural humans – entirely separate species with no ability to cross-breed, and with as much romantic interest in each other as a current human would have for a chimpanzee.

Silver understands that such scenarios are disconcerting. He counsels realism: that is, he celebrates freedom of the market and perpetuates the myth that private choices have no public consequences:

> Anyone who accepts the right of affluent parents to provide their children with an expensive private school education cannot use "unfairness" as a reason for rejecting the use of reprogenetic technologies.... There is no doubt about it ... whether we like it or not, the global marketplace will reign supreme.

When I first read Silver's book, I imagined that these sorts of bizarre prognostications must be the musings of an academic researcher indulging in mad-scientist mode. I soon learned differently. They are not ravings from the margins of modern science, but emanations from its prestigious and respected core. Silver vividly and accurately represents the techno-eugenic vision, a horrifyingly grandiose ideology shared by a disturbing number of Nobel laureate scientists and other influential professionals.

To some scientists, the lure of the technical ability to "enhance" human DNA is tantamount to its inevitability. Like many other techno-scientific elites, eugenic engineers are deeply offended by suggestions that the direction or pace of their work should be subject to social controls. They are motivated by a technocratic utopianism similar to the impulse that has always motivated supporters of eugenics: the urge to engineer human "perfection."

Proponents of germline enhancement have appropriated the phrase "brave new world," stripping it of the cautionary irony with which Aldous Huxley imbued it. Lee Silver, for example, used the phrase in the subtitle of *Remaking Eden*, which of course is itself a utopian reference. The exuberant techno-utopianism of the eugenic engineers is especially seductive in an era when the values and visions of the democratic left are widely perceived as faded and tattered.

The techno-eugenicists believe that human nature is written in our DNA and can be rewritten by genetic engineers in the service of a better and brighter future. They insist that scientific research and development should be controlled by researchers themselves, working in a "free market" with minimal interference from regulators and certainly without democratic oversight, and that reproductive decisions should be subject to the logic of consumer culture and commodity production. In their techno-eugenic future, human beings will be engineered by "life sciences" corporations that own all the bits and pieces of the human genome; well-heeled consumers will compete for genetic superiority; and all of us will learn to tolerate unprecedented inequality between the genetically "advantaged" and "disadvantaged."

THE CAMPAIGN FOR HUMAN GERMLINE ENGINEERING

Explicit public advocacy of human germline engineering was triggered by the public-relations disaster that followed the surprise announcement, in early 1997, that agricultural scientists in Scotland had cloned a sheep. The birth of "Dolly," as she was dubbed by media-savvy researchers, did in fact represent a technical breakthrough: this was the first mammal to be cloned from an adult rather than from an embryonic cell. But Dolly was also a cultural and ideological watershed. Surprise about the existence of a cloned sheep was dwarfed by disquiet over the prospect of a cloned person. Polls registered overwhelmingly negative public sentiment about human cloning: more than 90 per cent of Americans opposed it. From the Dolly episode, designer-baby crusaders took the lesson that they should start now to prepare the public for human germline engineering. This lesson was reinforced by the widespread hostility towards genetically modified food that had emerged as people in many countries suddenly discovered how much of it they were eating.

The techno-eugenicists lost no time in jump-starting their campaign. Their kick-off event, organized by Gregory Stock and neurobiologist John Campbell, was a symposium called "Engineering the Human Germline."[14] Its goal, Stock explained, was to make germline engineering "acceptable" to the public. Held in March 1998 at the University of California, Los Angeles (UCLA), the symposium attracted nearly a thousand people. James Watson, Lee Silver, W. French Anderson, former *Science* editor Daniel Koshland, and other scientific luminaries proclaimed the virtues and inevitability of germline enhancement of the human species. The *New York Times*, the *Washington Post* and other major newspapers echoed these conclusions in front-page coverage. "The question is not if, but when and how," the conference report stated.

Just a few months later, W. French Anderson – who won a reputation for strategic brilliance and confrontational gamesmanship while spearheading the late 1980s' push for somatic genetic engineering – made a move designed to "force the debate" on germline engineering. Anderson formally submitted to the Recombinant DNA Advisory Committee (RAC) – the committee of the National Institutes of Health (NIH) that oversees new uses of genetic engineering – a proposal asking for permission to try gene therapy on fetuses that had been shown by prenatal tests to have a condition called adenosine deaminase (ADA) deficiency, a fatal childhood disease. Anderson called the experimental procedure he was proposing somatic therapy, but freely acknowledged that it was likely "inadvertently" to modify the developing germ cells of the treated fetuses.[15]

Anderson's move and its significance were duly noted by the popular media. But most of the coverage hewed closely to the agenda of the germline engineers. Within months, both *Newsweek* (November 8, 1998) and *Time* (January 11, 1999) magazines had run articles entitled "Designer Babies." Nearly every article included the requisite queasy quotation from a bioethicist, but drastically underreported both the social vision that underlies germline engineering and the range of arguments against it. The media treated Anderson's proposal as a spectacular but nonetheless by-the-book example of scientific progress. In fact, it was an object lesson in science-as-public-relations, a gambit carefully designed to provide the NIH with a narrow procedural cover for giving its go-ahead, and to trump all doubts by evoking the real and terrible suffering of (the very few) ADA-deficient children.

Since Dolly, the UCLA symposium and Anderson's proposal, designer babies and human clones have indeed become more thinkable, and far more frequently discussed in the US popular media. Favourable and prominent coverage has appeared on network television, in national news and popular science magazines, and in books for young readers.[16] Even Francis Fukuyama, the pundit renowned for proclaiming the "end of history" in the aftermath of the fall of the Soviet Union, has chimed in. Fukuyama now believes that "biotechnology will be able to accomplish what the radical ideologies of the past, with their unbelievably crude techniques, were unable to accomplish: to bring about a new type of human being." Soon, he writes, "we will have definitively finished human History because we will have abolished human beings as such. And then, a new posthuman history will begin."[17]

Despite this spate of publicity, few people realize that an orchestrated campaign for public acceptance of human clones and designer babies is under way. Nor are they aware how tightly these technologies are bound to a grandiose political vision of market-based eugenics and genetically engineered inequality.

The only forum in which human cloning and germline manipulation have been debated in a sustained way is in the literature of academic bioethics. The bioethicists have devoted a great deal of thought to benefits and harms to parents who want cloned or enhanced children, to the children who are cloned or enhanced, and to the researchers who want to produce them. They tend to give less consideration to the political and social dimensions. Thus the bioethics debate is strong on issues of safety, informed consent, reproductive rights, individual identity, family relationships, the right to an unaltered genetic endowment (or to an improved one), and the right to "play God" or "tamper with Nature." It is weak on questions of ownership and control of genetic and reproductive research and technologies; the commodification of children, women, reproduction, and the human genome; the exacerbation of social and economic inequality; and the ideological consequences of the techno-eugenic vision.

As sociologist Barbara Katz Rothman puts it, the job of bioethicists is to mediate "between biomedical science and research on the one hand, and the concerns of the public on the other. But with all the power and the big money in the hands of science, bioethics becomes a translator, sometimes an apologist, sometimes an enabler, of scientific 'progress.'"[18]

An effective opposition to the techno-eugenic vision will have to challenge both the allegiances of the bioethicists and their political analysis. But that is just one task among many for critics of a designer-baby future.

DESIGNER-BABY TECHNOLOGY: THE STATE OF THE "ART"

Some of the new genetic and reproductive technologies are already on offer in fertility clinics and doctors' offices. Others are being tested in petri dishes or animals in university, government or corporate labs. Plans to study and develop far more powerful techniques are being made; some of these may ultimately turn out to be projections of scientists' informed but wishful or frenzied imaginations.

As far as is known, no one has tried to transfer a genetically altered human embryo into a woman's womb. Genetic manipulation of animals, plants and bacteria, however, is routine. Transgenic organisms have foreign genes not only transferred into their cells, but stably incorporated into their germlines: they bequeath their new or manipulated DNA to all subsequent generations.

Germline engineering in non-human animals has become fairly common. Genetic researchers have produced animal chimeras, including hybrid animals like the "geep" whose cells contain and express the genes of both a goat and a sheep. Since 1982 they have been engineering larger-than-normal mice by splicing in human genes that produce human growth hormone. In 1999 researchers at Princeton created a strain of mice said to have enhanced memories. Other genetic engineers have learned to inject human DNA into the fertilized eggs of cows, sheep, goats and pigs, creating animals that secrete human insulin, human blood-clotting proteins, and human enzymes in their milk.

The success rate of such procedures, however, remains very low. Most genetic engineering in animals takes advantage of the ability of certain viruses to insert themselves into a host organism's DNA. Scientists modify these "viral vectors" by removing the genetic sequences that cause disease, and adding the genetic sequences they want to insert into the host cell. But the chromosomal location at which the viral vector lands is essentially random. An inserted gene can introduce all sorts of errors into the genome, causing unpredictable problems for the transgenic animal. Depending on whether or where the new gene is incorporated, it can fail to express the proteins it is intended to produce, or disrupt normal cell functioning.

Biologists are far from being able to predict how such manipulations will interact with other genes, or with cellular or systemic mechanisms. Adding or changing multiple genes, necessary for the ambitious kinds of engineering that the techno-eugenicists envision, will doubtless be far more difficult and unpredictable. Today, researchers manipulate hundreds or thousands of embryos before producing a single viable animal that exhibits the intended genetic change. Many of the "mistakes" either fail to develop at all, develop abnormally and die before birth, or are born with mild or serious disabilities.[19]

A decade ago, when researchers were struggling to get governmental approval for somatic procedures, they offered repeated assurances that they wouldn't even consider germline modifications until somatic therapies were shown to be safe and effective. Now that somatic procedures have proved ineffective, conventional wisdom among geneticists is that working on early embryos is the way to go. University of Utah geneticist Mario Capecchi, who first produced "knock-out" mice with targeted genes inactivated, stated in 1998 that "germline therapy is ... actually much simpler than somatic gene therapy."[20]

Formidable technical obstacles remain, but some designer-baby advocates are convinced that solutions are in view. Many pin their hopes on the development of human artificial chromosomes (HACs). These synthetic "microchromosomes" are "assembled from individually isolated components of naturally occurring chromosomes." Geneticists believe that HACs may overcome the delivery problems and the limited capacity of viral vectors, allowing them both to avoid interfering with natural chromosomes, and to put into one package the multiple genes that would be needed to influence most traits.[21]

Researchers have already synthesized and inserted artificial chromosomes into mice; in October 1999, Chromos Molecular Systems announced that mice thus engineered had passed the artificial chromosomes to their offspring.[22] Researchers have also put HACs into human cells outside the body. Some have remained stable for six months, faithfully replicating with the rest of the chromosomes when the cultured cells divide.

Introducing an additional chromosome would likely be disastrous for a developing embryo. Even if such manipulations were "successful" in the short term, their implications could be enormous. Would a person with extra chromosomes be able to have children with someone whose DNA hadn't been similarly modified? Would the children and grandchildren of artificial-chromosome parents eventually develop problems, or change their minds about the alterations in body and behaviour that the artificial chromosomes produced?

Proponents of germline engineering have countered concerns about the inheritance of germline manipulations by future generations by proposing various technical fixes. Conceding that heritability "would be an undesirable property for the germline modifications envisioned today," Gregory Stock describes a proposal by Mario Capecchi for "non-heritable germline therapy." Capecchi believes that germline engineers could devise an external signal that would deactivate or dismantle the artificial chromosomes in a genetically engineered person's germ cells, so that the modifications would not be passed on to the next generation.

The reasoning and rhetoric in Stock's description of this plan are instructive. "By the time recipients of even the best engineered chromosome are ready to have children," he wrote, "it will be twenty or thirty years after they them-selves were conceived. Their once state-of-the-art artificial chromosome will be hopelessly out-of-date, and they'll want to give their child the latest gene cassettes and artificial chromosomes. It's not so different from upgraded soft-ware; they'd want the new release."[23]

HUMAN CLONING: PREREQUISITE FOR DESIGNER BABIES

The ability to clone human embryos is a technical prerequisite for the effective production of designer babies. Reproductive cloning – creating a cloned child from an existing person – would not be technically necessary. But it would constitute an important cultural and ideological breakthrough toward a designer-baby world. The techno-eugenicists recognize this point; as usual, they are blunt in explaining it. "[H]uman cloning is most significant as a symbol," Stock writes. "Whether or not human cloning is banned will have little impact ... because biotechnology is racing ahead on a broad front."[24]

Stock's assertion is, in effect, a brush-off of the cloning brouhaha that has engulfed biotechnologists. The wave of repugnance that greeted the Dolly-inspired prospect of human clones prompted repeated assurances from many scientific quarters that no responsible researcher or fertility doctor would attempt to clone a child "at this time." Biotech companies that clone stem cells taken from human embryos for their research in areas such as trans-plantation, cancer and longevity have staked out this position quite forcefully. Fearful of restrictive regulations, they take great pains to draw a sharp line between reproductive cloning and cloning that would not result in the creation of a human being.

Not a few scientists and bioethicists, however, openly support human cloning as a reproductive method. To be sure, they distance themselves from the
likes of Richard Seed, the Chicago physicist-turned-biologist who announced
with great fanfare in January 1998 that he would soon open a cloning clinic.
The "responsible" advocates of human reproductive cloning characterize Seed
as a renegade publicity-seeker, and acknowledge that human cloning must be
proven safe before it is used. They patiently explain that worries about tyrants
replicating themselves or mass-producing super-soldiers are exaggerated.

Many people have come to realize that these horror scenarios are indeed
beside the main point. They understand now, if they didn't before Dolly, that
a cloned child would not be an exact copy of the person who supplied the
original genome. Almost by definition, the clone would be born into a
different world, raised differently, surrounded by different people. She would
not even be biologically identical to the person from whom she was cloned
because of differences in the uterine environment in which she was gestated,
random developmental variations, differences in mitochondrial DNA, and
other biological deviations. Setting straight this "confusion over cloning," as
Richard Lewontin terms it, has provided a valuable corrective to simplistic
varieties of genetic reductionism.[25]

Advocates of reproductive cloning argue that it would "cure" some difficult
cases of infertility, provide an option for people who might pass an undesired
gene to their children, allow gay and lesbian couples to have genetically
related children, and allow people without partners to have children with no
"foreign" genes. In fact, other less ethically problematic reproductive
procedures are available for all these situations. What reproductive cloning
does provide is a way to have a child who is genetically related to you, but
to no one else, an unusual requirement at best.

Before Dolly, most scientists – even those working in the fields of genetics,
embryology and developmental biology – had believed that producing a clone
from an adult was impossible. Repeated experimental efforts had convinced
them that once cells have specialized (as had the mammary cell in the Dolly
example; as has almost any somatic cell beyond the earliest embryonic stages),
they no longer have the potential to "de-differentiate" and give rise to a
complete organism. This long-presumed inability to clone animals from adults
was a major barrier to the commercial production of genetically engineered
animals, which is why Wilmut and his colleagues had been working on it.
Somatic nuclear transfer, the method used to clone Dolly, allows the mass
production of valuable genetically designed animals: entire herds can be cloned
from the rare transgenic successes that scientists manage to create.

In humans, of course, the goal is quite different. The demand that the
techno-eugenicists are trying to stimulate and satisfy is for the "perfect baby"
– the genetically engineered child designed for gourmet parental sensibilities.
Accordingly, the genetic and reproductive techniques appropriate for human
germline engineering would diverge a bit from those used in livestock. Researchers would start with an embryo a week or so old. At this stage, human

embryos contain several dozen "stem cells," none of which has begun to differentiate or specialize, each theoretically capable of developing into a complete person. Breaking the embryo apart, the researchers would put each stem cell into a Petri dish. There they would attempt DNA manipulations, adding, deleting or modifying genes, or inserting artificial chromosomes. Researchers would then coax each manipulated stem cell to divide without differentiation, yielding colonies of identically modified embryonic stem cells. Removing a few cells from each colony, they would test to determine which had been "successfully" altered, though in fact there would be no reliable way to measure "success" at this stage. The researchers would then fuse one of the engineered embryonic stem cells with an enucleated human egg, creating a new single-celled zygote. If all went according to plan, this zygote would begin to divide, differentiate and develop into a human embryo. Implanted into a woman's womb, it would grow into a genetically engineered child – a designer baby.

WHO OWNS HUMAN CLONING?

The distinction between the procedures for germline engineering and human reproductive cloning can be summarized like this: for a cloned baby, use the Dolly technique – take a cell from the person you want to clone, then transfer the nucleus to an emptied-out egg; to make a designer baby, take a cell from an early embryo, tweak, and then transfer the nucleus.

As it happens, patents for pieces of both procedures are held by the Geron Corporation, a small California biotech company. Based on work done by its in-house scientists, Geron owns exclusive rights to the use of human embryonic stem cells. It also owns a piece of somatic nuclear transfer since its purchase in May 1999 of Roslin Bio-Med, the for-profit company established by the not-for-profit Roslin Institute (home of Dolly) and British venture capitalists. Geron says it will use these patented technologies to develop new transplantation methods, its stated core business interest.

So who now controls the technology to clone human beings? This question was almost never asked in all the hoopla surrounding Dolly. It was left to the Rural Advancement Foundation International (RAFI), a Canadian-based non-governmental organization, to determine that the Roslin Institute's original patent on the procedure covers all animals – including humans. RAFI found that, "while some other biotech companies … have worded their patents to specifically exclude humans, the Roslin patents deliberately make no such exclusion."[26] Spokespeople at the Roslin Institute told RAFI that they have no commercial interest in human cloning, and no moral tolerance of it. They said that the inclusion of humans in their patents "would ensure that nobody else could lay claim to human cloning." But, as RAFI argued at the time, the sincerity of the Roslin Institute's intentions is not the issue. "The ethics and fate of human cloning is not a matter to be entrusted to the Roslin Institute," said RAFI's Hope Shand. When the Roslin Institute spun off Roslin Bio-Med, it wrote a licensing agreement that explicitly excluded human reproductive cloning. Presumably, then, Geron's rights to the Roslin Institute's cloning technology are also limited by that license: it can clone animal embryos and fully formed animals; it can clone human embryos; but it cannot clone people.

How close are we to the day when we wake up to headlines announcing the birth of a cloned human? Informed answers range from "very far off" to "any day." It took Wilmut and his team 277 tries to produce a live cloned sheep [27] Since then, researchers in Honolulu working with mice and a slightly different technique have reduced that ratio somewhat. Other biotechnologists, vying to perfect cloning by somatic nuclear transfer, have found a variety of serious medical problems in the cloned animals and the mothers pregnant with them. Some of the mothers have died. In some species a "large offspring syndrome" has been noted. Other clones have been born with placental and umbilical cord abnormalities, severe immunological deficiencies, anaemia, organ deformities, or retarded development.[28]

In addition, researchers have yet to figure out whether clones will be "born old." Reports that Dolly's chromosomes are older than she is — they look to be about the age of the sheep from whom she was cloned — were widely published. But this may not prove to be an insurmountable obstacle, for several reasons. First, though her chromosomes are prematurely aged, Dolly appears normal. Second, Geron is working on a technique that they believe will prevent the ageing of chromosomes in humans as well as other animals. The implications for human longevity and health are unknown.

MAKING THE CASE AGAINST DESIGNER BABIES

Arguments about safety and risk will be important in building an opposition to human cloning and designer babies. If current practices in the fertility and biotech businesses are any indication, it won't be hard to argue convincingly that researchers and practitioners are moving ahead recklessly, putting commercial and competitive considerations ahead of human well-being.

The technologies to produce designer babies and human clones may turn out to be inherently unsafe — as is, for example, nuclear power. Or they may not. It is certain that a few years of tests in animals cannot definitively demonstrate safety. It is also certain that calls for assessing risks will be used to argue for more research and funding, rather than as reasons to forgo techno-eugenic goals. In any case, safety concerns are not the only reason to oppose germline engineering and human cloning. The strongest and most honest case against designer babies is an ethical and political one.

The policy arena is currently very much in flux. In the United States, neither federal law nor policy currently forbids germline engineering. The RAC, which would have to grant permission for federally funded work involving germline procedures (but which has no authority over other research), says only that it won't consider germline proposals "at this time."

The policy situation on human cloning in the US is similar: no federal law actually forbids it, though federal moneys cannot be used. Bill Clinton issued an executive order declaring a five-year moratorium on human reproductive cloning in 1997. In 1998, Republican senators introduced legislation that would have prohibited both reproductive and embryo cloning. Democrats who wanted to exempt embryo cloning for research purposes introduced a bill that would have banned only reproductive human cloning. The Republican proposal was defeated and the Democratic bill never came to a vote. The FDA claims that it has authority to regulate human cloning, but it is required by law to evaluate only questions of safety and effectiveness, and is expressly forbidden from considering ethical or social concerns.

Many other countries are far more restrictive. Germany's Embryo Protection Act of 1990 makes human cloning and germline engineering criminal acts. A number of other European countries forbid cloning and germline engineering indirectly by outlawing non-therapeutic research on human embryos. As of 1998, twenty-two European countries have signed a Council of Europe bioethics convention that includes similar restrictions. The World Health Organization supports a global ban on human cloning, though its recently released draft ethical guidelines on human germline engineering stop short of recommending a permanent ban. The UNESCO Declaration on the Human Genome, which has been adopted by the United Nations General Assembly, calls for global bans on both human cloning and germline engineering.

If the legal and regulatory framework covering designer-baby technologies is best described as uneven, activist engagement with these technologies can only be called – at least until very recently – low-level. But there have been some important exceptions. Nearly twenty years ago, Jeremy Rifkin's Foundation on Economic Trends organized a coalition of fifty-eight scientists and religious leaders to call for a worldwide ban on human germline experiments. Their "Theological Letter Concerning the Moral Arguments," presented to the US Congress on June 8, 1983, stated: "Genetic engineering of the human germline represents a fundamental threat to the preservation of the human species as we know it, and should be opposed with the same courage and conviction as we now oppose the threat of nuclear extinction."[29]

The religious coalition re-emerged and expanded to two hundred members in 1995, issuing a statement opposing patents on animals and human materials. Though this statement did not explicitly mention germline engineering, many of the coalition members expressed concern that the US policy of granting broad and vague patents on human tissues, genes, and cell lines could bolster economic incentives for germline "enhancement."[30]

The most consistent organizational presence in the opposition to human germline manipulation has been the Council for Responsible Genetics (CRG). A Cambridge-based group composed mostly of biologists, attorneys, social scientists, public health advocates, women's health and disability rights activists, and physicians, CRG has played a key role in defining and promoting a public-interest agenda for biotechnology since its founding in 1983. CRG's Human Genetics Committee issued a position paper calling for a ban on human germline manipulation in 1992. In 1998, after learning of Anderson's

RAC proposal, CRG sent out an alert; the RAC then received seventy pages of mostly negative public comments.[31]

Anderson's proposal also sparked the formation by David King and others in Britain of the Campaign Against Human Genetic Engineering. Their mission is to widen the debate about human genetics, now "dominated by scientists, doctors, and a narrow group of academic ethicists," and "to bring both genetic research and its application under democratic control."[32]

In the US, organizing against techno-eugenics has begun to emerge. In August 1999, activists in San Diego staged a protest outside a meeting of the California Advisory Committee on Human Cloning, complete with signs, banners and Frankenstein masks. A group in the San Francisco Bay Area has held a series of public meetings and workshops, established an on-line newsletter, and taken steps towards forming a public-interest organization.[33]

The political challenge for those opposed to a world of designer babies is daunting. Success will require widespread recognition that eugenic engineering would both trample human rights and undermine inviolable commitments to social and economic equality. Eventually, a permanent global ban on human germline manipulation and reproductive cloning will be necessary. With that safeguard in place, opponents of techno-eugenics could actively support genetic technologies that actually may help alleviate human suffering, if they are developed in a manner consistent with safety, social justice and democracy.

A global ban, of course, is a long-term goal. One of the first steps towards it is to craft a critical stance toward techno-eugenics that is politically effective, intellectually honest and morally rich. We will need to articulate a position that simultaneously opposes genetic determinism and genetic reductionism, and recognizes the expanding powers of biological manipulation. We will have to make it clear that we are neither anti-science nor anti-progress, and that our opposition to eugenic engineering is fully compatible with our commitment to reproductive rights. We will need to challenge technoscientific hubris, elitism, and the sense of inevitability promoted by advocates of human germline engineering. We will have to establish that social control of technology is a human right and a democratic principle.

In our opposition to a future of eugenically engineered inequality, we must take care not to downplay or obscure the inequalities that plague the present. We will have to confront the disturbingly comfortable fit between the techno-eugenic vision and key features of contemporary culture: consumerism run wild, hyper-competitive individualism, ever-increasing privatization of public goods, and passivity in the face of technoscientific "progress." Arguments against a designer-baby future that are based on qualms about "playing God" will not resonate with the secular. Objections to "tampering with Nature" will not persuade those who understand "nature" as a social category. It won't be easy to navigate the slippery slopes and fuzzy distinctions that characterize the new human genetic and reproductive technologies. But now that the techno-eugenicists are on the move, we urgently need a smart, broad-based campaign to protect what can be called, with chilling new meaning, a "human future."

NOTES

1. Gregory Stock is director of the UCLA Program on Science, Technology, and Society. He organized the March 1998 symposium "Engineering the Human Germline," at which he made this comment. For a summary report of the symposium, see http://www.ess.ucla.edu:80/huge.

2. James Watson is co-discoverer of DNA structure (for which he received a Nobel prize) and former director of the Human Genome Project. He made this comment at the "Engineering the Human Germline" symposium.

3. From the transcript of the programme, which aired on August 19, 1999. Available from ABC News.

4. Gregory Stock, "The Prospects for Human Germline Engineering," January 29, 1999, http://www.heise.de/tp/english/inhalt/co/2621/1.htm.

5. A "gay gene" was announced in 1993 by a research team at the National Cancer Institute led by Dr Dean Hamer. Researchers reported finding a genetic basis for thrill-seeking in 1996. The notion that homelessness is caused by genetic disease was the informed speculation of Daniel Koshland, a molecular biologist and former editor of the prestigious journal *Science*. ("Sequences and Consequences of the Human Genome," *Science*, October 13, 1989). For critiques of genetic reductionism and determinism, see Ruth Hubbard and Elijah Wald, *Exploding the Gene Myth*, Boston, MA: Beacon Press, 1993; and Richard Lewontin, *Biology as Ideology: The Doctrine of DNA*, New York: HarperPerennial, 1991. Also see Richard Strohman, "The Coming Kuhnian Revolution in Biology," *Nature Biotechnology* 15, March 1997, pp. 194–200.

6. Rick Weiss and Deborah Nelson, "Victim's Dad Faults Gene Therapy Team," *Washington Post*, February 3, 2000.

7. W. French Anderson at the "Engineering the Human Germline" symposium, http://www.ess.ucla.edu:80/huge. In April 2000 the international press reported that French researchers had successfully treated three babies affected by SCID, a rare gene-related immune deficiency disorder. While their work raises the prospect that "gene therapy" might help in some situations, caution and scepticism are still warranted. Even French Anderson commented that "because you correct SCID doesn't mean you can correct any other disease," adding, "If you can't correct SCID, you can't correct anything else." Gina Kolata, "Scientists Report the First Success of Gene Therapy," *New York Times*, April 28, 2000, p. 1.

8. In India and China, prenatal screening to ensure boy babies has become common. Demographers estimate 100 million "missing" – that is, aborted or killed – girls. "Sex-Selective Abortion," *Bulletin Briefing from the Women's Health Education Program*, vol. 1, no. 5, November 1997. A new sex-selection technology involving sperm-sorting is now used in the US, in spite of unknown risks to the resulting children. Lisa Belkin, "Getting the Girl," *New York Times Magazine*, July 25, 1999, pp. 26–31, 38, 54–5. For more on disability rights and human genetic technologies, see Gregor Wolbring, "Science and the Disadvantaged," Edmonds Institute Occasional Paper, Edmonds, WA, 2000.

9. Arthur Caplan, "If Gene Therapy is the Cure, What is the Disease?" www.med.upenn.edu/%7Ebioethic/genetics/articles/1.caplan.gene.therapy.html#pleaseleave, 1992.

10. Jeremy Rifkin, *The Biotech Century: Harnessing the Gene and Remaking the World*, New York: Jeremy P. Tarcher/Putnam, 1999, p. 128.

11. Richard Hayes, "The Threat of the New Human Techno-Eugenics," May 1999. Available from the Techno-Eugenics E-mail List newsletter, teel@adax.com.

12. Silver is not the first scientist to raise this prospect. In a 1966 article titled "Experimental Genetics and Human Evolution" (*American Naturalist*, September–October 1966, pp. 519–31), Nobel laureate Joshua Lederberg "saw the need to look at ... the status of clones as chimeras and 'subhumans.'" Cited in James M. Humber and Robert F. Almeder, eds, *Human Cloning*, Totowa, NJ: Humana Press, 1998. On the engineering of "inferior" humans, see Rachel Fishman, "Patenting Human Beings: Do Sub-Human Creatures Deserve Constitutional Protection?" *American Journal of Law and Medicine*, vol.

XV, no. 4, 1993, pp. 461–82.

13. The following quotes are from Silver's popular account of a designer-baby future, *Remaking Eden: Cloning and Beyond in a Brave New World*, New York: Avon Books, 1997, pp. 4–11.

14. For the summary report and for the following quotations, see http://www.ess.ucla.edu:80/huge.

15. Rick Weiss, "Scientists Seek Panel's Advice on In-Womb Genetic Tests," *Washington Post*, September 25, 1998, p. A2.

16. Examples include: *ABC Nightline*, August 19, 1999; David Jefferis, *Cloning: Frontiers of Genetic Engineering*, New York: Crabtree Publishing Company, 1999; Glenn Zorpette and Carol Ezzell, "Your Bionic Future," *Scientific American Presents*, vol. 10, no. 3, Fall 1999, p. 5; *Time* magazine cover, "The I.Q. Gene?" September 13, 1999.

17. Francis Fukuyama, "Second Thoughts: The Last Man in a Bottle," *The National Interest*, Summer 1999, pp. 16–33; quotations on pp. 28 and 33.

18. Barbara Katz Rothman, *Genetic Maps and Human Imaginations: The Limits of Science in Understanding Who We Are*, New York: W.W. Norton, 1998, pp. 35–9.

19. Consider, for example, the case of the pig, engineered to express human growth hormone, which turned out to be "[e]xcessively hairy, lethargic, riddled with arthritis, apparently impotent, ... slightly cross-eyed [and] could hardly stand up." Andrew Kimbrell, *The Human Body Shop*, New York: HarperCollins, 1993, pp. 175–6.

20. Mario Capecchi at the "Engineering the Human Germline" symposium, http://www.ess.ucla.edu:80/huge.

21. Huntington F. Willard, "Human Artificial Chromosomes Coming Into Focus," *Nature Biotechnology* 16, May 1998, pp. 415–16.

22. "A Safer Way of Altering Genes Will Make Engineering Humans More Tempting than Ever," *New Scientist*, October 23, 1999.

23. Gregory Stock, "The Prospects for Human Germline Engineering," January 29, 1999, http://www.heise.de/tp/english/inhalt/co/2621/1.htm.

24. Ibid.

25. R.C. Lewontin, "Confusion Over Cloning," *New York Review of Books*, October 23, 1997.

26. "Dolly Goes to Market: World Patents on Sheep Clones Include Humans," RAFI press release, May 8, 1997, at http://www.rafi.org.

27. See, for example, "Voices from Roslin: The Creators of Dolly Discuss Science, Ethics, and Social Responsibility with Arlene Judith Klotzko," *Cambridge Quarterly of Healthcare Ethics*, vol. 7, no. 2, Spring 1998, pp. 121–40.

28. Rick Weiss, "Clone Defects Point to Need for Two Genetic Parents," *Washington Post*, May 10, 1999, available at http://www.rage.org.nz/cloning-abnormalities.html; Philip Cohen, "Bigger, not Better," *New Scientist*, January 23, 1999, http://www.newscientist.com/nsplus/insight/clone/bigger.html; Eddie Lau, "Dark Side of Cloning Perplexes Scientists," *Sacramento Bee*, August 18, 1999.

29. "Theological Letter Concerning the Moral Arguments," June 8, 1983, presented to the US Congress.

30. Rifkin has continued to write and speak about human genetic technologies, presenting his opposition most fully in his 1998 book *The Biotech Century*. Also see David Suzuki and Peter Knudston, *Genethics: The Clash Between the New Genetics and Human Values*, Cambridge, MA: Harvard University Press, 1989; Gina Maranto, *Quest for Perfection: The Drive to Breed Better Human Beings*, New York: Scribner, 1996; Kimbrell, *The Human Body Shop*; and Rothman, *Genetic Maps and Human Imaginations*.

31. Contact CRG at 5 Upland Road, Suite 3, Cambridge, MA 02140 or http://www.gene-watch.org.

32. The Campaign's website is http://www.users.globalnet.co.uk/~cahge.

33. To subscribe to the free on-line newsletter of the Exploratory Initiative on the New Human Genetic Technologies, email teel@adax.com.

10

Cancer is (not) a Genetic Disease

BARBARA KATZ ROTHMAN

People are dying of tuberculosis, of gunshots, in car crashes and the occasional plane crash, of infections they picked up in hospitals, of pneumonia, of diabetes, of heart disease, of strokes, of weird viruses, of AIDS of course, of cancer certainly. We are all born in so remarkably similar a condition, and then we die in so many different ways, of so many different things.

Take heart disease: that kills a lot of people. Heart disease has a genetic component, undoubtedly, as probably everything does. There are forms of early-onset heart disease that run in families, which parallel the early-onset familial cancers. Diabetes too. But heart disease, stroke, diabetes – none of these things speaks to us the way that cancer does. If you want to understand illness as we enter a new millennium, if you want to understand the *body* as we now experience it, look to cancer. That is the disease we are using to write the new body. It is cancer that is telling us who we are, what we are made of, and what will become of us.

It is not, oddly enough, AIDS. There is an irony to AIDS. It is of course the contemporary plague, the disease you have to respond to when you talk about illness at the turn of the millennium. It attaches itself to two of the great concerns of America these days: sexuality and drugs. It is a disease, too, of identity politics, attaching itself to "communities" within. It is a disease one wants to compare to tuberculosis, a disease that takes the place of tuberculosis in our imagery. TB was a disease of fashion, its contrasting pallor and flush setting the style, and yet a disease of poverty – and so too with AIDS. It is what the artists die of, and the poor folks. This is a very rich and beautiful story, and if only it had a happy ending...

The "genetics" of HIV that people are learning is the takeover story. Like a computer virus, the infiltrator "takes over" the cells, rewrites the commands, and shuts down the system. The "genetics" here is not part of the larger morality play. It is just the mechanism the infiltrator uses. Polio, when you think about it, didn't eat away muscles. It, too, was a virus that rewrote cell instructions. But that is not the way people thought about it at the time.[1] The story, the plot line, is that something comes in and destroys. A very old story.

Still, it is cancer, and especially breast cancer, that is writing a new story, that moves us forward into the new thinking, the new geneticism. Breast cancer is the competing disease, the *other* plague of our time. The competition with AIDS comes right to the fore in the ribbon display: in response to the angry, flaming red ribbon of AIDS, breast cancer offers us the pink ribbon. If AIDS is the disease of dirty boys and their innocent victims, breast cancer is the disease of innocence, of mothers, grandmothers, aunts and sisters. It is a disease that grows at home, not out there in dirty places. It is the disease that both reflects and in turn reinforces the newer, genetic model of disease.

It wasn't always thus: breast cancer was, and not all that long ago, unmentionable. Cancer itself was hushed: the big C, the anonymous "long illness" mentioned in obituaries. Cancer was dirty and it hit dirty places – the breast, cervix, uterus, bladder, kidney, colon, prostate; privates of one sort or another – and places one didn't want to think about overmuch about: liver, pancreas, lung – the less pleasant organs, entrails and guts. But that isn't what cancer is now. Cancer is no longer a disease of organs and guts, a disease of the flesh of the body. Cancer is a disease of the cell, of the program of the body, a genetic disease.[2]

The idea, the philosophy, the ideology of genes is much older than what we are now calling genetics. People have long held the idea that we inherit from family. Our genesis, our beginning, is in our genes: our essence comes from those before us and is passed down to us. That idea is very specifically rooted in patriarchy: that we are of our fathers, that our fathers make us from their seed, and that we unfold from our fathers' loins while curled in the safety of our nurturing mothers. The patriarchal assumption places our essence in a seed, in the piece of our fathers that turns into ourselves. Nurturance, environment, the world itself can help or hinder, but the essence, the basic limitation on what we can and can never become, is written in the seed, engraved in the genesis.

When people think of things as genetic, as in our genes, that is the idea to which they hark back: that we carry within us part of those who preceded us, that aspects of our ancestors have made us, that pieces of them unfold in us. We are no longer a classic patriarchy; we recognize too the seeds of women, the genetics of the egg as well as the sperm. So when we have a grandmother's nose, a grandfather's talent for music, a mother's argumentative style, a father's sense of humour, we see these pieces of others in us, we see history repeating itself, and we say "It's in our genes."

When geneticists speak about genes, they are speaking of stretches or segments of DNA.[3] Those segments, when in the sperm and the egg, are indeed our "inherited genes." Those segments in a breast cell or a prostate cell or a lung cell are still genes to a geneticist, but not "genes" in that sense of history moved along from parent to child. When geneticists study cancer, they are observing, studying, trying to understand how the DNA segments operate in the cancerous cells. Very occasionally, that has to do with the DNA segments as they were "passed on," in the sense of inheriting a "gene for" cancer. But mostly what they are looking at are the mutations, the changes, that turn cells of individual people from normal cells of the body, ordinary breast, prostate or lung cells, into cancerous cells spreading within the organ and metastasizing throughout the body.

From the perspective of the geneticist, it makes perfect sense to think of cancer as a genetic disease: it is a disease that occurs within the DNA of the cells, and an understanding of the changes within the DNA segments will perhaps provide an understanding of the cancer. But that is *not* the same as thinking of cancer as a "genetic disease" in the old sense that preceded modern genetics – as an inherited, passed-along-the-family disease. Very few cancers show any sign of being "inherited" or "genetic" in that sense.

It is not that we are all so stupid we can't understand that. We can certainly understand the differences between "genetics" as the study of the workings of the DNA within cells and "genetics" as the study of inheritance. The confusion is within the language, but the confusion is also within the presentation.

A number of people are now studying and have written about how "genetics" gets represented in the news. The pattern is pretty much what you would expect: a series of headlines about the discovery of "the gene for" this or that, and then a lot of qualifications, fudging, hemming and hawing about what exactly it all means, and oftentimes a (back-page) retraction at a later date.[4] Those of us who have made this "our issue" read those articles carefully, pounce upon the qualifications, the uncertainties, that follow the headline. We read all of it, and read between the lines as well, just as those who have made Bosnia, or the space shuttle, or welfare reform or school districting their issue can read, and read between the lines of, articles on those topics, understanding richly and fully what is and what is not being said. But for everybody else for whom that is not the issue they focus on, we see the headline, glance at the article, get a sense of the issue, and fold the paper as the train reaches our stop – or turn off the radio, zap the television, close the magazine and move on.

In the genetics articles, the sense has been given, over and over again, that there are "genes for" lots of things, and that cancer is a "genetic" disease. The general sense is that people have, or do not have, the genes for various cancers, and they are *essentially* (in their essence, in their genes) doomed or spared from the start.

A couple of years ago a news article on the radio declared that the gene for colon cancer had been found. I was in the car with a friend for whom this is not a big issue. I heard the news that a gene that accounts for a significant proportion of the 10 per cent of colon cancers that are believed to be genetic had been isolated. In other words, they seem to have found a cause of about 10 per cent of 10 per cent of colon cancers. The remaining 99 per cent were not explained by this gene. It was a public radio station, and a thoughtful report, and all of this was said; I heard it. My friend, a bright enough, thoughtful enough person, heard that they had found the gene for colon cancer. I pushed the conversation with her for a minute, and yes, she realized that they did say they found the gene for the inherited colon cancer. Whatever. Basically, the sense she got was that colon cancer was a genetic disease, and that they were closing in on the genes for it. Maybe it wasn't as close as she first thought, but that was the direction we were heading in.

It is not a stupid mistake: it is catching the general thrust of all of these stories. The frame for genetics stories is the finding of "the gene for": that is, if you will, the plot line. Then there are some holes to fill in, some finer stuff about what percentage or what variation of the disease, how common the gene is, and how useful this will or will not be for prevention or for cure. But the basic story line is there: they are finding the genes that cause disease, and cancer is a genetic disease.

BREAST CANCER

Breast cancer is not a "genetic disease." It is not there, born in, waiting to come forth. You wouldn't know that if you read the newspapers. The widely publicized "breast cancer genes," BRCA1 on the chromosome 17 and BRCA2 on the chromosome 13 together account for maybe 5 to 7 per cent of breast cancers, and 5 to 10 per cent of ovarian cancers. Those two genes together seem to account for 90 per cent of all of the inherited cases of breast cancer, all the breast cancers that "run in families." Other genes, not yet found, may account for the other 10 per cent of these inherited breast cancers, but all the inherited breast cancers together still leave 93 to 95 per cent of breast cancers unexplained.

When the genes were first found, it was claimed that women who had one of those genes had an 85 per cent chance of developing breast cancer. The genes were found by studying families that had a high rate of breast cancer. Once the research went beyond those few families, the number dropped rapidly. It is altogether possible that those original family members shared other risks – other genes, environmental factors. Now it seems that only about half of the women with BRCA 1 or 2 develop breast cancer in the course of their lives.[5] It turns out, then, that a very small number of women carry a gene that dramatically increases their chances of developing breast cancer.

Much has been made of this gene in the media: I have read, heard, been earnestly told that they have found the gene for breast cancer. Women who have had one grandmother develop breast cancer in her old age tell me they are "at risk" for breast cancer, it "runs in the family." A 70-year-old woman, aunt of one of my graduate students, was told by her family doctor to have both of her healthy breasts removed because her sister developed breast cancer. If it weren't so tragic, if people weren't so terrified, you could say this was getting silly. People are coming to believe that breast cancer is a genetic disease, a family disease, that some of us are doomed and some of us are safe. It is not so. Between 93 and 95 per cent of all breast cancers are not familial, are not "genetic" in that sense. And almost half the women who have one of these identified "breast cancer genes" (and maybe even more than that once the numbers are all in) won't get breast cancer.

What does happen to women who learn they have BRCA 1 or 2? What exactly are they supposed to do about it? Finding these genes has opened up a troubling can of worms for those affected women. While breast cancers usually develop later in life, the breast cancers associated with BRCA 1 and 2 often occur in women in their thirties and forties. Mammography is not very effective in finding cancers in younger women: their denser breasts make the growths harder to see. Besides, more frequent mammography starting at a younger age would mean more radiation exposure – not necessarily a smart thing to do to people who are at increased risk of cancer. So early detection, the first thing that pops into mind for a high-risk group, may not be a good thing and may not work all that well.

What is rather glibly called "prophylactic mastectomy," removing the breasts while they still seem healthy, is sometimes suggested. One doctor suggested it to a young woman I know, calling it "preventative medicine." "Preventative medicine," she replied, "is wearing a hat in the rain, not cutting off your breasts." There are no data to show that such extreme measures will work: no one knows how early in breast development the breast cancer cells start, and they may very well have seeded throughout the body before surgery. And no one knows how slight an amount of breast tissue left behind might be enough to develop a cancer. So even women ready to have both of their breasts removed in their youth cannot guarantee escaping breast cancer. American women face an added burden, in that they also risk losing their health insurance coverage, or being made unemployable because uninsurable, if word of their BRCA 1 or 2 status leaks out.

Some families want prenatal diagnosis, in order to abort affected daughters. A woman who has lost a mother, an aunt and a sister to breast cancer while they were still young might seriously consider selective abortion to avoid passing on what feels like a family curse to her daughter. Does that make any sense? This raises basic questions about how long a life has to be to be worth living, what the meaning of a life is, and who is to judge. My father died absurdly young of an unusual cancer. Should he have been aborted had his cancer death been predictable and not preventable? Not for my sake, certainly, but consider my grandmother – she was a bitter, sad old woman who lost both of her children to cancer, my father in his twenties, my aunt to lung cancer in her fifties. If those cancers had been "familial" cancers, what should my grandmother have done? And who could know? Who could judge?

Families react to this diagnosis of BRCA 1 or 2 in their families in about the same ways that families react to all of the hard stuff, some better than others, some with support, some with estrangement between the "affected" and "non affected" members. Guilt, grief, anger, sorrow – about what you'd expect. But all of this has nothing to do with the vast majority of women who have breast cancer. This has nothing to do with the "breast cancer epidemic," an epidemic of *awareness* even more than of disease.

CONTRADICTIONS

Cancer, at the level of the cell, is a process. You can't "explain it" by the presence of a carcinogenic stimulus – not everyone so exposed develops the cancer. And you can't "explain it" by the presence of predisposing mutations in the cell that started the cancer growing – not everyone with such "cancer genes" develops the cancer. Cancer occurs at the level of a cell. And cancer also occurs at the level of an atomic blast. It is caused by changes in the DNA. It is caused by industrialization. It is caused by oncogenes. It is caused by cigarettes.

Impinging on the cell are the stimulants, the carcinogens, the cancer-causers, and the cancer-protectors, that come from outside of that cell. In the broadest possible sense, they are environmental. Some are the environment of the body itself – oestrogens produced by the body that stimulate a given cell, for example. Some are environmental in the sense of coming from outside of the cell, outside of the body, like the viruses that stimulate some cancers. Some are from the environment of that particular body – smoke inhaled, coal tars, workplace chemicals, oestrogens that come from particular meats that are eaten by that particular person. And some are "environmental" in the sense that we more commonly use them, like radiation levels, pollutants that come from water, air and food that enter our bodies, penetrate our cells, stimulate our cancers.

Cancer represents the tension between thinking socially and thinking in-dividually, between an "environmental" ideology and a "genetic" ideology. Both ways of thinking are with us, and both frame cancer for us. On the one hand, there has been an increased awareness of cancer as being environmentally caused: questions about fat in the diet, about smoking and other lifestyle issues, as well as larger questions about industrial pollutants, power lines and the ozone layer. On the other hand, there is this newer image, this genetic thinking which asks us to think of the cancers as lurking in the cells of the individual. Both ways of thinking draw maps for us, shape our imagination as we think about cancer. Neither map precludes the other: environmental stimulants can goose predisposed cells into action, or can bring about the predisposition.

Even for the diseases of concern during earlier periods, susceptibility varied, of course. Not every child exposed to polio got polio; not every person exposed to TB got TB. Some susceptibility follows the social map: children are differently placed; and malnutrition, harsh living conditions, different patterns of exposure to different diseases all shape the child's response. And some response was more individual: I am now coming across data that suggest a susceptibility to polio might lie on chromosome 19. The explanation for the diseases lies in the external "cause," the bacillus for TB and the virus for polio, but there is also an explanation to be had in individual response. In the best of all possible worlds, all lines of research would be followed.

But we never have lived and do not now live in the best of all possible worlds. TB itself is rising again as the environmental stresses that made for vulnerability return, and as the bacillus becomes immune to the treatments. It begins to look like poverty *does* cause TB. As to cancer, just a few years ago the bulk of cancer money was going into viral research: the answer to everything seemed to keep coming. Now it goes into genetic research. The social epidemiology of cancer, the role of industrial capitalism, gets glossed over as cause moves deeper and deeper inside the individual.

The TB bacillus and the polio virus did not have the financial backing of big investors, of government support, the advantages of public relations firms behind them. The viruses that appear to stimulate some cancers are similarly politically unprotected. Other kinds of environmental stimulants do have backing. Tobacco is probably the easiest to see, as the tide begins to turn, and because tobacco appears to be such an individual choice. Tobacco is still a legitimate business in the United States; its advertising and sales are still legitimate business expenses. Joe Camel was at least a recognizable if formidable enemy.

Whatever, whoever it is that destroys the ozone layer, or lets chemicals slip into my water, air and food is more diffuse. I'm oftentimes not sure just who I should be angry at. The sun – the very sunlight, source of enormous joy to me – is an increasing threat. The public health approach has been largely aimed at *me*, at individuals to "Just say no" to the sun too, along with all the other things that are bad for us. There are of course activist, political approaches to cancer that look up, blame the destroyers of the ozone layer, the polluters of the air and the water. But mostly we seem to be in an era of looking down – telling individuals to eat more wisely, slather on the skin protectors, wear hats, stop smoking, and go live somewhere else if you don't like the air in your neighbourhood.

A friend in a public health Master's programme was writing a paper for a course. She had to develop a programme to encourage teenagers to cover up from the sun. This is what they *teach* in a public health program? What if the public health movement had, back in its earlier incarnation, directed its efforts not at improving the water supply, but at getting individuals to boil their water, to "Just say no" to unboiled water? Perrier and Evian are not public health measures.

But it is worse than that, this new thinking. It's not only about encouraging individual change (and boiling water was an immediate and necessary solution). What if the earlier generation of public health workers had spent their research money on figuring out which arm of which chromosome holds the "gene for" susceptibility to cholera? Do we want them to have figured out which people were most susceptible and what was wrong with them that made them so vulnerable? I have no idea what individual, cellular, internal factors account for variations in susceptibility to cholera. And I don't care. I opt for a safe water supply.

And so, I believe, it can be with cancer. Of course individual susceptibility varies. And for a few relatively rare cases – the small percentage of breast or colon cancers that "run in families" – that susceptibility is so great that it does need to be considered. But for most of us some susceptibility is there, a given. Expose us to carcinogens and we will, as a population, as a people, as a community, develop more cases of cancer. Remove those dangers, and we will, as a population, as a people, as a community, be healthier.

To switch our attention from power lines to cell lines in the search for the causes of cancer, to keep looking down instead of up, is dangerous. To individualize the disease obscures the role of the social world in causing cancer. Cancer is the product of mutations. But it is also the product of society. We need more than one map if we are to imagine solutions.

NOTES

1. For a study of ideas about polio in contrast to AIDS, see Emily Martin, *Flexible Bodies: The Role of Immunity in American Culture from the Days of Polio to the Age of AIDS*, Boston, MA: Beacon Press, 1994.

2. Any discussion of the "meaning" of cancer owes a great deal to Susan Sontag, *Illness as Metaphor*, New York: Vintage, 1979.

3. For an explanation of genetic sciences for the non-scientist, see Ruth Hubbard and Elijah Wald, *Exploding the Gene Myth*, Boston, MA: Beacon Press, 1993.

4. See, for example, Peter Conrad, "Public Eyes and Private Genes: Historical Frames, New Constructions and Social Problems," *Social Problems*, vol. 44, no. 2, May 1997, pp. 19–154.

5. The numbers on breast cancer are still undergoing refinement. A very small percentage of all breast cancers appear to be "explained" by BRCA 1 or 2. The vast majority of breast cancers, including the far more common post-menopausal breast cancers, are not "familial." See Kenneth Van Golen, Kara Milliron, Seena Davies and Sofia D. Merajver, "BRCA-Associated Cancer Risk: Molecular Biology and Clinical Practice," *Journal of Laboratory and Clinical Medicine*, vol. 134, no. 1, July 1999, pp. 11–18.

11

If Cloning is the Answer,
What was the Question? Genetics
and the Politics of Human Health

SARAH SEXTON

Dolly the cloned sheep was an unexpected scientific triumph. In replicating an adult mammal for the first time in 1996, Dolly's creators at the Roslin Institute in Scotland overturned long-established assumptions about cell biology and cell differentiation.[1] But Dolly was a public relations disaster too. The public worldwide was shocked. The idea of evil megalomaniacs creating row upon row of identical copies of themselves seemed no longer the stuff of futuristic fiction but an imminent possibility. The public demanded to be reassured that the Dolly techniques would never be applied to humans. Now many scientists and biotech companies became alarmed. Even if they had no corporate interest in duplicating human beings, they feared that public concern about cloning could undermine confidence in gene testing, gene therapies and a range of medicines and vaccines which use genetic technologies and knowledge – products on which many biotech and pharmaceutical companies were, and still are, pinning their financial futures. It was time for some serious PR damage limitation. Scientists, medics and industry representatives went before the public to emphasize the potential benefits of human cloning techniques in particular and medical biotech research in general. Assiduous attempts were launched to "educate" the public into a view of health as something which demands the therapies that the biotech industry is seeking to provide.

A first step in allaying public fears has been to deny categorically that the techniques which yielded Dolly might be used to produce humans. Replicating human beings, insist governments, biotech companies, scientists and medical bodies, would be abhorrent. But they do not deny that the techniques might be applied to humans for other purposes. Far from it. Human cloning techniques, they explain, could be used for other "morally unobjectionable"[2] or "beneficial" purposes: to produce organs or tissues to replace or rejuvenate failing and diseased parts of the body; to assist in cancer research; to develop products to slow or reverse ageing; to test new pharmaceuticals; to test embryos before implanting them in women's wombs during in vitro fertilization (IVF). According to a British government-sponsored consultation, this "therapeutic cloning"[3] should never be confused with "reproductive cloning" – even if both involve research into how to replicate human embryos and how to engineer human beings genetically.[4]

This powerful and neat distinction between end uses has allowed legislators and scientific bodies to outlaw (in theory) the replication of human beings while at the same time countenancing the research that would make it possible. Once this research has accustomed the public to the idea of cloning for spare organs and the like – the *Financial Times* estimates a period of five years may be adequate[5] – moratoria on human replication could be reviewed. By that time, it is assumed, the promise of, for example, curing "hundreds of thousands of sufferers of Parkinson's disease" with an "injection of nerve cells grown in a laboratory dish"[6] would make the unbreakable links between medical or therapeutic cloning and the replication of humans seem unworthy of concern. The medical benefits being claimed for cloning technology cannot be achieved, in other words, without also laying the basis for replication of human beings and for human genetic engineering.

CONFUSING MECHANISMS WITH CAUSES

The possibility that human cloning techniques (and indeed genetic research more generally) could help to cure, treat or slow the progression of many diseases for which at present nothing much can be done is perhaps the most widely advertised benefit of these techniques. Cancer, Parkinson's and Alzheimer's are often mentioned. For example, it is argued that cloning human embryos could extend knowledge into how cancer cells begin to multiply uncontrollably with a view to finding treatments to stop them. Similarly, Parkinson's disease, a progressive brain disorder caused by the death of a certain class of brain cells, might be blocked, it is thought, by transplanting certain embryo cells into the cranium. Transplanting embryo cells into the pancreas of a diabetic, by the same token, could encourage production of insulin.

What is invariably left out of such descriptions is what causes cells to behave abnormally or to cease functioning in the first place. While molecular genetics has certainly provided much information about how cell behaviour and DNA alterations are implicated in cancer, "to say that 'DNA alteration is at the heart of cancer induction' ... confuses mechanism with cause," says molecular biologist Bonnie Spanier. "It does not necessarily follow," she concludes, "that genetic research is the best approach for understanding what causes cancer or how to prevent it": "A large majority of human cancers are influenced or promoted by environmental carcinogens in our workplaces, in air, water, and food, in such cultural habits as sunbathing and tobacco use, and in our social conditions such as poverty and stress."[7]

Similarly, the underlying cause of the worldwide increase in the incidence of diabetes – the World Health Organization (WHO) projects a more than twofold increase in incidence of the disease by 2025, with up to 300 million people being affected – cannot be that people are suddenly sprouting "diabetes genes." If certain individuals are indeed genetically predisposed to the disease, something must be triggering its growing incidence.

It may well be, as Harvard biologist Ruth Hubbard suggests, that it "is far easier and more convenient for scientists to pretend they will conquer cancer by studying the molecular transformations of genes and cells" than to press for a lowering of exposures to carcinogens and other pollutants.[8] Nonetheless, it would surely be more rational and efficacious to improve individual and societal health and to alleviate suffering by pursuing this latter course than to clone human embryos for cancer molecular research or to undertake speculative programmes of genetic alteration. It is largely through obscuring the wider causes of these diseases that human embryo cloning techniques come to seem beneficial, plausible and reasonable.

ENVIRONMENTS ARE EVERYWHERE

A heightened focus on the cellular causes of disease ignores the economic, political and social forces that contribute to cell misbehaviour. So, too, privileging the role of genetic anomalies in causing disease involves downplaying the importance of the "environment" of the genes. Such anomalies may be inherited from one or other or both parents, present at birth, or acquired later in life. Again, this may obscure wider causes of ill health – in ways that favour cloning or genetic technologies and knowledge as "self-evident" solutions, but that in reality fail to address the underlying reasons for the condition.

For example, the inference that genetic testing of cloned embryos in IVF procedures – and indeed of prenatal testing in general or of embryonic/fetal genetic engineering in future – will lead to a healthy infant seems plausible only if the role of a number of environments is ignored, that of the gene, the egg and the mother, for instance, and if infant health is considered during pregnancy but not afterwards.

Genes are not simply "objects" – distinct and independent units that "determine" biological outcomes. They are more part of complex dynamic interdependent processes between all the small and large molecules and ions in a cell, which are in turn affected by interactions with adjacent cells. Moreover, a gene may behave differently depending on its location on the chromosome and the presence or absence of other genes. There are few genes that result in a specific genetic condition irrespective of their environment.

The environment of the egg is similarly downplayed. The egg cytoplasm provides mitochondrial DNA – packages of DNA inside a cell that are entirely separate from the chromosomes in the nucleus.[9] This mitochondrial DNA is now believed to be implicated in some diseases and some body processes, such as ageing, one of the proposed targets of research into human embryo cloning. Other components of the egg's cytoplasm also become part of the resulting embryo and play a major role in directing its development.

Even less attention is typically paid to ensuring that mothers have adequate nutrition, housing and income and a domestic life free from stress, violence and abuse before, during and after pregnancy.[10] Nor is much attention usually paid to the variety of toxins which may exist in their workplaces, homes or neighbourhoods – aside from those in cigarette smoke or alcoholic drinks, to which responsibility for exposure can be assigned to pregnant women. Yet these environments can be at least as critical as an embryo's nuclear DNA endowment to an infant's immediate and lifetime health. For instance, some reports suggest that babies of women living near toxic waste dumps have a one-third higher risk of birth defects.[11]

Finally, an emphasis on genes and good maternal behaviour obscures the fact that infant health is only partly a matter of what happens before birth. No matter how many prenatal genetic tests are undertaken, a healthy baby is not guaranteed because the tests are not foolproof and because other events may happen. Most disabled people become disabled because of what happens to them after birth, not because of genetic conditions. Conversely, finding a genetic predisposition in an embryo to a disease or condition does not mean the child will develop the disease in later life. Explanations of ill health that ignore these wider "environments" almost invariably lead to reductionist – and misleading – accounts of disease causation.

Social influences, often the most powerful determinants of health, tend to be obscured by a genetic focus. Inferences that the applications of human cloning and genetic engineering are critical to improving our health ignore these findings. Poorer people in developed countries, for instance, have annual death rates anywhere between twice and four times as high as richer people in the same society.[12] A health study in New York's Harlem found that, at most ages, death rates were higher than in rural Bangladesh. In Brazil, infant mortality rates varied between different areas of the same city from 12 per 1,000 live births to 90 per 1,000 live births.[13]

Such health inequalities, according to British sociologist Richard Wilkinson, cannot be attributed solely to differences in medical care or different genetic susceptibilities between social classes, and are only partly explained by individual health-related behaviour (smoking tobacco, drinking alcohol, taking narcotic drugs, lack of exercise, poor diet). They are due, rather, to the "effects of the different social and economic circumstances in which people live"[14] – including unemployment, poverty, bad housing and environmental pollution. "Much more important than the small differences medicine can make in survival from cancers and heart disease are differences in the incidence of these diseases."[15] All the broad categories of causes of death in developed countries – heart disease, respiratory illness and cancer (some of the main targets of biotech research) – are related to income distribution, argues Wilkinson. Wilkinson found that the healthiest societies were not the richest, but those that had the smallest income differences between rich and poor. Inequality and relative poverty have absolute effects: they increase death rates.

Privileging the role of genes not only plays down the contribution of social factors to ill health, disability and disease. It also adds weight to the idea that ill health is an individual misfortune – to be tackled through making individuals aware of their genetic predispositions and then recommending to them individual programmes of risk-minimizing behaviour.

Geneticization militates against making efforts for change which would be good for everyone's health, irrespective of their genetic predispositions. For instance, it creates an atmosphere in which a "safe workplace" is to be achieved not by cleaning up toxic production systems but by "weeding out the so-called susceptibles"[16] or putting the onus on them to prevent their "predispositions" from becoming reality. A worker at a major US car manufacturer comments:

> For years, companies have been saying that workers' diseases are not caused by what we work with in the plants, but by smoking, diet, lack of exercise, and other problems with our life-style. Now they're saying it's the workers' genetic heritage.[17]

Assigning genes to behaviours such as alcoholism and violence serves a similar social function. Ruth Hubbard concludes:

> By erasing the social context, genetic predictions and labels individualise our problems, blame the victim ("If you get sick, it's because you have bad genes") and are authoritarian ("You should have had your genes tested and done what the doctor said").[18]

Under the genetic model, the "right" to be born healthy becomes not a reason to clean up the environment but an argument for not implanting or carrying to term embryos and fetuses which do not pass their gene tests.

CLONES FOR SALE

Suppose, however, that human embryo cloning did yield some of its speculative benefits such as replacement organs, new cancer drugs, medicines to slow the onset of old age, or embryonic tests and treatments for some genetic diseases. Would they not make it worth putting aside any qualms that they would be paving the way for the replication of humans and of human genetic engineering? As David Tracy of the University of Chicago's Divinity School says, "opponents of human cloning (as I am) cannot afford to ignore the benefits that such cloning might provide for all humankind."[19]

Yet the economic imperatives driving cloning and genetic research and underpinning health care more generally suggest that "all humankind" will not benefit. Because of such imperatives, it is not surprising that the human body has become a "resource to be 'mined', 'harvested', patented and traded commercially for profit as well as scientific and therapeutic advances."[20] Even in the early days of mammal cloning research in the 1970s and 1980s at the

University of Wisconsin it was the "economic promise of cloning" to multiply embryos from prized cattle costing $500 to $1,500 apiece that provided the impetus.[21]

The main aim of the Roslin Institute, which produced Dolly, and PPL Therapeutics, the company formed to raise funds and commercialize research at Roslin, is to produce pharmaceutical drugs in the milk of animals more cheaply than drugs that can be produced by existing methods. The vision is to create "flocks and herds of living medicine factories"[22] or "bioreactors," as Roslin calls them. Ronald James, a director of PPL and a former venture capital portfolio manager, had the idea in the early 1990s that there were "riches to be made by any company that could figure out cheap, reliable ways to make valuable protein drugs ... that cost hundreds of pounds per dose,"[23] not least because "genetically-engineered animals can be used to make products on a scale no chemical factory could achieve."[24]

The Roslin group has also engineered cows with human genes to produce what the group claims is human-like milk. But while the milk from Roslin's genetically engineered cows may contain the major whey protein found in human milk, it is impossible to engineer genetically many of the anti-viral, anti-parasitic and anti-infective properties of human breast milk – which protect an infant from many diseases until its own immune system is developed – or all of its nutritional components. A mother's milk is tailor-made for her baby – in contrast to any genetically engineered version – and "delivered" in a uniquely safe way.

The tie-up between the US Geron Corporation and Roslin Bio-Med (under which Roslin Bio-Med becomes a wholly owned UK subsidiary of Geron) (see Chapter 9 of this volume) hopes to create human cells and tissues that can be used to repair organs damaged by degenerative diseases, laboratory cultures of heart, skin or blood cells (derived from embryonic stem cells) on which to test new pharmaceuticals, and genetically engineered cloned animals to provide human blood products and organs for transplantation.[25] Meanwhile, scientists at the University of Hawaii, by producing tens of cloned adult mice using a "relatively efficient" version of the Dolly technique,[26] have opened up "the possibility of creating made-to-measure mice on a commercial scale"[27] – ideal for testing pharmaceutical drugs. *Financial Times* journalist David Pilling comments: "It is a little like Henry Ford and the car; he did not invent the car; he worked out how to mass produce it – and this was what made all the difference."[28]

An economics-driven impatience with nature reveals itself in dismissive descriptions of "natural" human reproduction, which is labelled "remarkably inefficient"[29] because of the millions of eggs and sperms that "go to waste." "Most embryos die before a woman is even aware she is pregnant," goes another lament.[30] A female human fetus is described as having a "stockpile" of some 7 million eggs in its ovaries, although the average woman releases only 400 eggs in her lifetime, most of which go "unused." It is only a short step from thinking in such language to being able to say that aborted fetuses will only "go to waste" if their brain cells are not transplanted into people with Parkinson's disease or their ovaries mined for immature eggs. A *Financial Times* editorial notes breezily that "no real ethical dilemma exists" over the use of human embryonic stem cells as "they come from embryos that would otherwise be thrown away" or are simply a "by-product" of in vitro fertilization or come from "foetuses that are already aborted."[31]

FREE MARKET "CHOICE"

The speculative applications of human embryo cloning are more easily made to seem "beneficial" when the environmental, social, economic and political causes of ill-health and disease are obscured, and when the "benefits" are presented in an abstract way which hides issues of access and commercialization. Yet these purported benefits also receive support from more general contemporary attitudes towards health, death, life and children.

While many opponents of genetically engineered foods focus on the matter of consumer choice, it is sometimes argued that these concerns do not apply to genetic medicine. Health care "consumers," it is said, do have a choice about whether to avail themselves of germline therapy or organs grown in genetically engineered pigs.[32] Yet closer examination of current medical realities reveals that it may be more difficult to avoid such "choices" and their consequences than might first appear.

Take, for example, prenatal screening, which is often presented in impeccable feminist language as something which "enhances women's choice." After all, no one forces pregnant women to screen their fetuses, nor, if the test indicates the presence of a certain gene or chromosome abnormality, to undergo an abortion. Yet the "context in which testing and termination decisions are taken" is one full of social pressure and lacking in "balanced information for pregnant women." As British sociologist Tom Shakespeare notes: "Thirty per cent of obstetricians would not give a woman a test for Down's if she did not agree to have a termination after a positive diagnosis. Only 32 per cent of obstetricians reported counselling pregnant women non-directively."[33]

A woman's agreeing to the genetic testing of her unborn baby – or agreeing to abort it as a result – may thus be less an expression of choice than an instance of conformity, a response to coercion, or even a co-opting of her needs to fit established biomedical goals. If a woman chooses to continue her pregnancy after her fetus has been diagnosed as having certain "unwanted" genes or anomalies, moreover, it becomes easier to maintain that it is her responsibility to raise and look after the child without expecting any support from the state or society. Several decades ago some commentators were already arguing that carrying to term a fetus believed to be "genetically defective" could be considered fetal abuse. In July 1999, Bob Edwards, the British embryologist who helped pioneer in vitro fertilization, informed his colleagues that it would soon be a "sin" for parents to give birth to disabled children. "We are entering a world where we have to consider the quality of our children."[34]

As more and more genes are identified, women will have more and more types of "disabilities" to divide into acceptable and unacceptable, normal and abnormal.[35] They will come under increasing social pressure to bring their "choices" about whether to terminate their pregnancies or not into line with what their dominant society currently regards as normal or defective.

It is small wonder that some disabled people see prenatal screening as "yet another form of social abuse" which "reinforces the general public's stereotyped attitudes about people with disabilities" and is bound to result in increased "job discrimination, barriers to obtaining health insurance coverage, cut-backs on public support programs, and other similar negative actions."[36] "We know the real territory which genetics assumes as its own," says Bill Albert of the British Council of Disabled People – "the quality of our lives":[37]

> I would say to people who say that genetics is about removing illness and suffering from the world that I am somebody who they might think of as ill. The only way they could remove my illness from the world is by removing me. And at the moment, that's the only way they can remove most things, most so-called disabling conditions. I don't want to be removed from the world, I don't want my fellow disabled people to be removed from the world, and that's the basic argument. Because there is no therapy except screening, and screening is about eliminating people."[38]

It may be argued that such worrisome social consequences need not follow on from human cloning technologies which produce genetically engineered embryos, replacement organs or new drugs. Yet these technologies also presuppose and reinforce far-reaching political and cultural changes. Each implanted egg, for instance, presupposes a complicated and largely hidden social infrastructure for extracting industrial quantities of "surplus" ova from women undergoing fertility treatments or, potentially, from aborted female fetuses. This system may not only put pressure on women to have later abortions so as to obtain an intact fetus from which ovaries could be extracted, or to donate eggs in return for help in having their own baby through IVF; it is also intimately linked to a system of experimentation on lab-produced embryos propagated via cloning embryos which are genetic extensions of living individuals. It is small wonder that participants in a Wellcome Trust research project into public perspectives on human cloning regarded the use of cloning technology in medical research as "good" only until they found out what was actually involved in practice: "as the participants' awareness increased, so did their concern and apprehension."[39]

As a "solution" to infertility, human cloning technologies accentuate, as IVF itself does, a general feeling that some children are more worth raising than others. As social critic Barbara Ehrenreich points out:

> Millions of low-income babies die every year from preventable ills like dysentery, while heroic efforts go into maintaining yuppie zygotes in test tubes at the unicellular stage. This is the dread "nightmare" of eugenics in familiar, marketplace form – which involves breeding the best-paid instead of the "best."[40]

While resources are poured into esoteric techniques for providing genetically related children to women and men who might not otherwise be able to have the number of children they want, they are withheld from investigations into what might be causing infertility in the first place. This is happening at a time when chemicals which mimic the action of oestrogen are increasingly believed to contribute to the decrease in sperm count and quality, and when the incidence of the sexually transmitted disease, chlamydia, which scars and blocks women's Fallopian tubes, is on the rise in Britain, particularly among young women.

PUBLIC DEBATE – BUT OF WHAT?

In the wake of the storm of controversy over genetically engineered crops, it has become clear to proponents of human embryo cloning techniques and genetic engineering that they need at least tacit public acceptance of their projects in order to proceed. As one University College London report concluded: "The rate at which the general public can be reassured about the underlying technology is likely to be the single most important factor influencing the rate of uptake of genetic technology for health care."[41]

If previous experience with "sensitive" new technologies (such as IVF was once) is any guide, the crucial benefits that are likely to be stressed are those which would potentially accrue to individuals who can elicit public sympathy and demonstrate the existence of "demand": the distressed woman who cannot have children; the young, promising sufferer of a rare and fatal disease; the accident victim dying for want of an organ transplant; perhaps even the young girl whose social life is crippled by shyness. If enough real-life stories of individual tragedies which could supposedly be averted through scientific progress can be played out one after the other on news programmes or documentaries, it will seem churlish to ask questions about public health systems, inequity, distribution, exploitation, racism, eugenics and corporate control, all of which will recede safely into the background. Journalist Anne McElvoy points out, "scientists have been round this course many times before, and will respond by presenting the most persuasive examples of the benefits of their work."[42]

Nor will it be easy to bring up the awkward fact that many of the new treatments do not achieve their goal, or that they result in new problems. For example, genetically engineered human insulin, which has been available since the early 1980s, is often touted as evidence that gene research yields benefits. The two companies which manufacture the product, Novo Nordisk and Eli Lilly, have denied that it might have negative effects. Yet many patients claim their lives have deteriorated after switching to the genetically engineered version. Up to 20 per cent of diabetics taking it can no longer control their symptoms and can go into comas without warning. Doctors and specialists have, by and large, ignored patients' distress and dangerous symptoms. Only as a result of campaigning by insulin-dependent diabetics has animal-derived insulin once again been made available – and even then, only in some countries.

Similarly, even after two decades of use, it is not often reported that IVF still boasts at most a 20 per cent success rate, bringing "intense disappointment" for women who "walk away from the IVF clinic childless";[43] nor that the number of British children born through intra-cytoplasmic sperm injection (whereby a single sperm is injected into an egg) with birth defects appears to be twice that of children conceived naturally; nor that five-year survival rates for heart and liver transplant recipients are still only 64 and 55 per cent, respectively.[44] Where "failures" are mentioned, it is likely to be only in those contexts in which they can be used to justify yet more research. Thus IVF practitioners are likely to refer to the low success rate of IVF only when they are arguing that more embryo research is needed in order to increase it.

One side effect of the PR-like focus on benefits to certain individuals is likely to be increased squabbling among different sectors of society over which diseases genetic research efforts should focus on – the "my-disease-is-more-important-than-your-disease" syndrome. Questions about how everyone's health might benefit from basic and affordable public health, disease prevention and pollution control measures may well be obscured in the smoke raised by such disputes. As Ruth Hubbard stresses, although high-tech treatments can turn out to be a "real boon" to a limited number of individuals, they unfortunately "drain resources away from the kinds of public health and medical measures that could improve the health of a much larger number of people."[45]

A WIDER CONVERSATION

Critically, public debates on where to "set limits" or how to "ensure access to benefits" divert attention from broader questions of what kind of wider health care system people want, and from questions about the nature of health and disease.

If cloning is the answer, what was the question? If the question was how to improve health and quality of life for all, cloning and human genetic engineering are probably not the answers. If the question is how to keep an existing health care system and industry going, then it scores nine out of ten. Most debates and consultations fall into the trap of assuming that the status quo – whether embryo research or the sale of human body parts – is acceptable and that the only questions needing examination are the supposedly new ones.

For centuries, societies have contended with issues about who gets born and raised and who does not; who gives birth and who does not; who raises children and who does not; who lives and who dies. Eugenicists have always argued that parents try to give their children the best start in life by providing them with education, food and so on: why not also try to give them the "best" genes possible? Human embryo cloning and related techniques add no new concepts to this argument but, if realized, could provide new resources of power and control – ones which would be made available not just to prospective parents, but to state and medical institutions as well.

This is not to suggest that debates about benefits are not an important and necessary part of the "ethical" discussion. But by themselves, they do not encourage the essential "larger conversation about the fabric of social relations that sale of biotechnologies feeds on or promotes; nor about the dimensions of the common good that biomedical research can or should serve; nor about the sort of communal relations that exchanges of certain goods, labour, expertise, and services might reflect or produce; nor about determining criteria for deciding which possible objects of 'equitable access' are deserving of communal resources."[46]

The potential of cloning technologies may not be realized – indeed it is not even close to being realized. Even if it were, however, and even if the technologies were legalized, it is unlikely that their use in producing babies and spare parts, or for other ends, would become widespread.

Attempts to promote cloning technologies, however, will affect us all. No aspect of human existence will remain unaffected by discoveries in human genetics – irrespective of the new science's predictive accuracy or therapeutic efficacy. In their increasing claims on our attention and our resources, the new technologies will shape the way nearly everyone thinks. In that sense, it is not the spectre of cloned humans that should give pause as much as what the readiness to clone humans says about the way society is being organized.

NOTES

1. Dolly was born on 5 July 1996, but was not publicly announced until 23 February 1997 in the *Observer* newspaper, and a few days later in the scientific journal *Nature*, because the Roslin Institute wanted to obtain a patent on the nuclear cloning technique. The patent, eventually granted in January 2000, covers all mammals, which would include humans. See G. Kolata, *Clone: The Road to Dolly and the Path Ahead*, Harmondsworth: Penguin 1997; R. McKie, "Scientists Clone Adult Sheep," *Observer*, February 23, 1997, p. 1; I. Wilmut et al., "Viable Offspring Derived from Fetal and Adult Mammalian Cells," *Nature* 385, February 27, 1997, pp. 810–13, reprinted in M.C. Nussbaum and C.R. Sunstein, *Clones and Clones: Facts and Fantasies about Human Cloning*, New York and London: W.W. Norton, 1998, pp. 21–28.

2. W.B. Schwartz, *Life Without Disease: The Pursuit of Medical Utopia*, Berkeley and Los Angeles: University of California Press, 1998, p. 80.

3. "Cloning Issues in Reproduction, Science and Medicine: A Consultation Document," Human Genetics Advisory Commission and Human Fertilisation & Embryology Authority, London, January 1998 (report of consultation published in December 1998).

4. Research into human cloning techniques facilitates the genetic engineering of humans – adding or replacing genes in a cell's nucleus before, via the cloning techniques, it becomes part of an embryo. The main aim of the Dolly research was not to replicate sheep but to engineer mammals with human genes. As *New Scientist* has said, "It's not cloning we should be worrying about, but its sister technology, genetic engineering". See "Into the Clone Zone," *New Scientist*, May 9, 1998, p. 25; D. King, "Why Human Cloning Research Should Not Be Funded," *GenEthics News* 21, December/January 1998, p. 8.

5. "Time to Act on Cloning," editorial, *Financial Times*, January 12, 1998.

6. "Brave New World ... Needs Brave New Ethics," *Guardian*, editorial, December 9, 1998, p. 23.

7. B.B. Spanier, *Im/partial Science: Gender Ideology in Molecular Biology*, Bloomington and Indianapolis: Indiana University Press, 1995, p. 109.

8. R. Hubbard and E. Wald, *Exploding the Gene Myth: How Genetic Information is Produced and Manipulated by Scientists, Physicians, Employers, Insurance Companies, Educators and Law Enforcers*, Boston, MA: Beacon Press, 1993, p. 92.

9. Thus clones produced by nuclear transfer, as Dolly was, are not clones in the sense of having DNA identical to that of another organism. As mitochondrial DNA can come only from the egg, it would be possible for a woman to replicate herself genetically through the process which produced Dolly only if she used the nucleus from a cell of her body tissue and her own egg. A human being cloned from a man's adult tissue would not, in fact, be his genetically identical clone because the mitochondrial DNA of the offspring would come from the donated egg.

10. The use of ultrasound techniques has contributed to a perception of women as simply "fetal environments". The conceptual separation of the fetus from the woman has been facilitated further by various new reproductive technologies, including the physical separation of the egg and embryo from the woman and its isolation in the laboratory. The language and concepts of fetal rights, constructed in opposition to maternal rights, have added to this separation. See E. Martin, *The Woman in the Body: A Cultural Analysis of Reproduction*, Boston, MA: Beacon Press, 1987; B. Duden, *Disembodying Women: Perspectives on Pregnancy and the Unborn*, Cambridge, MA: Harvard University Press, 1993.

11. F. Abrams, "Toxic Dumps: Official Inquiry Ordered," *Independent*, March 8, 1999, p. 9.

12. R. Wilkinson, *Unhealthy Societies: The Afflictions of Inequality*, London and New York: Routledge, 1996, p. 3.

13. Ibid., p. 53.

14. Ibid., p. 7.

15. Ibid., p. 67.

16. E. Draper, *Risky Business: Genetic Testing and Exclusionary Practices in the Hazardous Workplace*, Cambridge: Cambridge University Press, 1991, quoted in M. Lock, "Breast Cancer: Reading the Omens," *Anthropology Today*, vol. 14, no. 4, August 1998, p. 11.

17. Ibid.

18. Hubbard and Wald, *Exploding the Gene Myth*, p. 74.

19. D. Tracy, "Human Cloning and the Public Realm: A Defense of Intuitions of the Good," in Nussbaum and Sunstein, *Clones and Clones*, p. 193.

20. D. Nelkin and L. Andrews, "Homo Economicus: Commercialization of Body Tissue in the Age of Biotechnology," *Hastings Center Report*, vol. 28, no. 5, September–October 1998, p. 30.

21. Kolata, *Clone*, pp. 133, 136.

22. D. Green, "From Cloned Sheep to Cash-cow," *Financial Times*, July 5–6, 1997, p. 9.

23. Kolata, *Clone*, p. 182.

24. T. Radford, "Cloned Goats Join Fight against Illness," *Guardian*, April 28, 1999.

25. T. Radford, "Cloning Boom with US Tie," *Guardian*, May 5, 1999.

26. In this cloned mice technique, the nucleus of the adult cell was injected directly into the egg, whereas with Dolly the nucleus of the adult cell was kept in a laboratory culture and "starved" back to its undifferentiated state. The Hawaiian mice technique reinforces the importance of recognizing the role of the egg cytoplasm in encouraging embryonic development. See T. Wakayama et al., "Full-Term development of mice from enucleated oocytes injected with cumulus cell nuclei," *Nature* 394, July 1998, p. 369.

27. D. Pilling, "Dolly Finds New Playmates," *Financial Times*, July 23, 1998, p. 20.

28. D. Pilling, "Send in the Clones," *Financial Times*, July 25–6, 1998.

29. P. Cohen, "Dolly Helps the Infertile: Cloning Technology Yields an Important Spin-off for IVF," *New Scientist*, May 9, 1998, p. 6.

30. Ibid.

31. "Cell Celebration," *Financial Times*, editorial, November 9, 1998.

32. Use of the term "choice" strengthens the perception of food, health and medical products and technologies (including children) as consumer goods. Sometimes, however, not even the appearance of choice is present. Many of those who have to take life-saving medicines every day do not have a choice over the type of medicine they take. In several countries, for example, insulin is supplied by just one or two companies which no longer supply animal-derived insulin, but only the genetically engineered variety. An estimated 20 per cent of diabetics who take genetically engineered insulin have experienced life-threatening episodes as a result. See website of Insulin Forum Switzerland: homepages.iprolink.ch.tenscher/englvers/english.htm.

33. T. Shakespeare, "Down's but Not Out," *Guardian*, November 5, 1998, p. 8.

34. L. Rogers, "Test-tube Revolution," *Sunday Times*, July 4, 1999.

35. G. Landsman, "Reconstructing Motherhood in the Age of 'Perfect' Babies: Mothers of Infants and Toddlers with Disabilities," *Signs*, vol. 24, no. 1, Autumn 1998, p. 71.

36. D. Kaplan and M. Saxton, "Disability, Community and Identity: Perception of Prenatal Screening. Most people will experience some form of disability, either permanent or temporary, over the course of their lives," *GeneWatch* (US), vol. 12, no. 2, April 1999, p. 6. These authors stress that equating disability with suffering and gene testing with avoidance of harm enables prenatal testing to be directed towards eliminating disabled people before they are born rather than addressing fundamental social causes of disability discrimination and the resulting lowered socio-economic status of people with disabilities.

37. B. Albert, "If you tolerate this, your children will be next," *Health Matters* 36, Spring 1999, p. 12.

38. Bill Albert, British Council of Disabled People, quoted in T. Shakespeare, "Eugenics? Slipping Down the Slope," *Splice*, vol. 5, no. 2, December 1998/January 1999, pp. 4–5. For a copy of British Council of Disabled People's disability movement statement on genetics, see http://www.bcodp.org.uk/genetics.html.

39. Wellcome Trust, *Public Perspectives on Human Cloning*, London: The Wellcome Trust, 1998, p. 25. Also available at http://www.wellcome.ac.uk.

40. B. Ehrenreich, "The Economics of Cloning," in B. Ehrenreich, *The Snarling Citizen: Essays*, New York: Farrar, Straus & Giroux, 1995, p. 39.

41. M.H. Richmond, "The Implications of Genetics and Genomics for Healthcare and the Pharmaceutical Industry," School of Public Policy, University College London, London, January 1999, p. 78.

42. A. McElvoy, "Worcester Woman and Woking Man Favour Cloning. So that's OK then," *Independent*, February 3, 1999.

43. "No Issue," BBC Radio 4, January 5, 1999.

44. Figures from UK Transplant Support Service, cited in R. Dobson, "Body Swap," *Guardian*, July 14, 1998, Section 2, p. 16.

45. Hubbard and Wald, *Exploding the Gene Myth*, p. 112.

46. L.S. Cahill, "The New Biotech World Order," *Hastings Center Report*, vol. 29, no. 2, March–April 1999, p. 47.

12

Eugenic Tendencies in Modern Genetics

DAVID KING

There is much confusion about eugenics in today's discussions about the ethics of human genetic research. The association of the subject with full-scale genocide seems to produce an inability to think clearly on both sides of the debate. It is true that the word is sometimes used as a blunt instrument to silence those who argue for the benefits of current research. On the other hand, there is a converse tendency to avoid any discussion of the subject for fear that it will provoke "hysteria."

The dominant tendency is to view eugenics as a purely historical phenomenon, and to minimize its relevance to current debates. Scientists view eugenics as causing public fear of genetics, and see this fear as an expression of ignorance or misunderstanding. The conventional view is that the eugenics movement of the first four decades of the twentieth century was based on "bad science," and was the result of a unique set of social and political circumstances. The implication is that now that we know so much more about genes, and have witnessed the horrific consequences of eugenics, we will not make that mistake again.

There is some truth in this. Early eugenic schemes were undoubtedly based on a crudely deterministic view of the role of genes, a view which is now scientifically discredited, though resurgent in popular culture. While eugenics has acquired a very bad name in liberal political culture, this has not prevented its re-emergence in its traditional form in China, for example. Events of the last decade remind us of the ease with which "ethnic cleansing" can happen. In the advanced post-industrial democracies, however, the resurgence of state-sponsored eugenics programmes is probably not a major threat.

It would be a serious mistake, though, to assume that eugenics itself is no longer a threat. Eugenics must be understood as more than just a right-wing attempt to get rid of "undesirable" people using spurious scientific justifications. It is necessary to examine tendencies both within genetics itself and in liberal, market-oriented societies that are likely to lead to a resurgence of a form of eugenics which, although less brutal than the earlier version, may be no different in some of its consequences.

GENETICS AND EUGENICS

As part of its legitimating ideology, science has promoted the idea that all knowledge is inherently good, and that the scientific urge is simply curiosity – the discovery of knowledge for its own intrinsic interest. This is the story told to children, who are indeed very curious about the world, in order to encourage them to embark upon scientific careers. The reality, as Francis Bacon, the key philosopher of the Scientific Revolution of the sixteenth and seventeenth centuries, emphasized, is that the purpose of acquiring knowledge is to control the world.

The idea that continually expanding knowledge, and ever greater possibilities of control, constitute progress, and are necessarily good, has become the key defining feature of modern Western societies since the Renaissance. As sociologists such as Weber and Foucault have observed, parallel to the creation of new knowledge has been a gradual process of rationalization and increasing control over nature and over society in the form of scientific management or bureaucracy. An example from the early twentieth century is Taylorism, the attempt to apply scientific management to industrial processes through "time and motion studies" of production-line workers. Taylor captured something crucial to scientific modernism when he argued, "In the past, the man has been first. In the future, the System must be first."

In the case of genetics, the managerial tendency is expressed through eugenics, which, at its root, is the urge to tidy up the accidents and messes that arise from sexual reproduction. Eugenicists argue for "improvement" of the overall human gene pool, but what really appalls them is that the whole business of human reproduction is out of rational control, and is left to chance. The eugenicists of the early twentieth century often pointed out the care we take over the genetics of our crops and domestic animals: how can we do this, and yet be so careless about human reproduction, they asked. This desire to bring human reproduction under scientific control was the common factor between the right wing and the socialist eugenicists, who saw eugenics as a progressive and humane aspect of modernization. In essence, eugenics is a form of technocracy, an attempt at social management based on the knowledge of a scientifically qualified elite. It is no surprise that Henry Ford was a key devotee of eugenics. Harry Laughlin, the linchpin of the US eugenics movement, summarized eugenics neatly in the *Birmingham Mail* in 1913 as "simply the application of big business methods in human reproduction."[1]

Eugenics is not an "abuse" of genetics: genetics and eugenics are inseparably linked. Some form of eugenics is an inevitable consequence of the advance of the science of genetics, although the political popularity of overt eugenics programmes will vary according to time and place. From this perspective, the popular eugenics movement of the early twentieth century was a highly damaging false start for eugenics. An auspicious set of political circumstances propelled it prematurely into the light, with disastrous consequences for its reputation. But the underlying social and psychological conditions for eugenics did not go away: the eugenics movement drew on a huge base of popular support, because the urge to eliminate reproductive uncertainty is deep-rooted. In a society whose central vision of itself is steady progress towards a future of prosperity, based on greater control over nature, it is close to unthinkable that technologies which allow us not to leave things to chance should lie unused.

LAISSEZ-FAIRE EUGENICS

After the Second World War, state-sponsored eugenics programmes became unpopular, although eugenics laws still persist on the statute books of some American states. The post-war liberal resolution of the eugenics problem was to declare that the problem with eugenics was that outsiders (i.e., the state) were interfering with what should be a free choice. Instead of coercion, parents are given genetic counselling, which, at least in English-speaking countries and Northern Europe, is supposed to be non-directive – that is, counsellors are not supposed to offer advice or tell their clients what to do. This practice has been buttressed by the swing in medical ethics since the 1960s away from paternalism and towards patient autonomy in the doctor–patient relationship. It is even suggested that the sole purpose of providing prenatal testing and counselling is to provide parents with greater reproductive choice.

In reality, the existing regime is eugenic both in purpose and outcome; the aim is clearly to reduce the number of births of children with congenital and genetic disorders. In some English-speaking and Northern European countries, official statements often assert that the aim of the system is purely to allow parents more informed reproductive choice.[2] On the other hand, equally authoritative sources, including the British Royal Society of Physicians, clearly indicate that the purpose of offering screening is to reduce the number of births of congenitally disabled children.[3] Although many genetic counsellors express discomfort with the practice, arguments for the introduction of genetic screening programmes are often couched in terms of financial benefits to the state.[4] In Britain, many obstetricians refuse amniocentesis unless the woman agrees to a termination in advance, showing a lack of willingness to support parental choice.

Research shows that many women are simply unaware that prenatal tests have been performed, indicating that the level of informed consent is low.[5] Many women feel that once they have agreed to testing they should opt for termination should abnormality be found, since otherwise there was little point in undergoing testing.[6] Thus, routine testing becomes a kind of conveyor belt leading to termination, which requires considerable conviction to resist.

The prevailing emphasis on free parental choice, rather than guaranteeing that eugenics will not return, itself opens the door to a different form of eugenics. The liberal consensus, in assuming that parents' decisions about reproduction are private, personal matters – mainly about whether they feel they could cope with a disabled child – is divorced from the social realm. This view misses the degree to which the personal, as the feminist slogan has it, is political. In fact, such decisions are strongly affected by many social factors which militate against the birth of genetically disabled children. The combination of such factors amounts to a strongly eugenic pressure on both parents and doctors. Among these factors are:

Disability oppression   There are several aspects to the way this affects parental reproductive decisions.[7] First, able-bodied people receive negative images of people with disabilities and general misinformation about what their lives are like. Doctors are particularly likely to be misinformed since they often see only the most severe cases of a particular disability in clinical situations. Parents receiving prenatal genetic counselling are rarely put in touch with people who actually live with the genetic disorder in question. Further, parents will be aware of the material aspects of disability oppression: insufficient welfare provision, inadequate access and discrimination. Lack of adequate welfare services, in particular, will not only affect the child but may create financial problems for the family, as well as increased stress. Parents will also be concerned about how society will view their decision either to give birth to a disabled child or to refuse a test.

Sexism   Having a child is a serious burden in a society which radically under-resources parenthood; this is the traditional reason for most abortions. Women, who still bear the vast majority of responsibility for childcare, are sharply aware that the extra burden of caring for a disabled child will fall on them.

Medical pressure   There is considerable evidence that genetic counselling is not non-directive in practice. Wertz and Fletcher conducted a survey of the attitudes and practices of nearly three thousand geneticists and genetic counsellors in thirty-seven countries.[8] Their results show clearly that only geneticists in English-speaking countries and Northern Europe (ENE) can make any claim at all to non-directiveness and questioning of eugenic thinking. In Eastern and Southern Europe, the Middle East, Asia and Latin America, geneticists not only hold eugenic ideas, but see no problem in directing their clients in accordance with those ideas. For example, 13 per cent of UK geneticists agreed with the clearly eugenic suggestion that "An important goal of genetic counselling is to reduce the number of deleterious genes in the population." In Eastern and Southern Europe this rises to an average of 50 per cent, and in China and India to nearly 100 per cent. An average of 20 per cent of ENE geneticists feel that, given the availability of prenatal testing, it is not fair to society knowingly to have a child with a serious genetic disorder. In the rest of the world, majorities of geneticists supported this view, rising to nearly 100 per cent in some countries.

Wertz and Fletcher's research details what geneticists say they think and do in response to a questionnaire. Figures derived from such answers almost certainly underestimate the degree to which counsellors contravene their professional norms in practice, particularly in counselling those facing more serious conditions. Their conclusions are supported by studies in which genetic counselling sessions were videotaped. These revealed a high level of directives by genetic counsellors. Most disturbingly, the level was highest when clients were from lower socio-economic groups.[9]

Eugenic assumptions   As I have argued, there is a general eugenic assumption in modern Western societies, in favour of preventing reproductive uncertainty, wherever possible. In arguing that the purpose of the existing regime is eugenic, I am not suggesting the existence of some conspiracy, on the part of doctors, scientists or the state. However, structural factors and social pressures guarantee that allowing parents a "free choice" results in a systematic bias against the birth of genetically disabled children that can only be called eugenic. Kitcher has dubbed the current situation "laissez-faire eugenics."[10] By influencing the genes of the next generation according to a particular set of dominant social values, it is no different in essence from the earlier eugenics, even though it is less violent and direct in its execution. The unchallenged existence of such a form of eugenics is, in itself, a major aspect of disability oppression.

In noting this reality, I am not attempting to deny parents' personal experience, free will, or the existence of countervailing social forces and resources (which are often religious). The present system is not anywhere near as harmful as the earlier state-sponsored eugenics. Many people who wish to avoid the birth of disabled children will experience it as highly beneficial. It is, nonetheless, eugenic, and would no doubt have pleased the earlier eugenicists.

THE CURRENT SITUATION

Today a new set of circumstances, including scientific, technological and social changes, point to a radical and disturbing expansion of laissez-faire eugenics over the next few years. The major development is, of course, the stunning success of molecular biology since the advent of recombinant DNA technology. The successes of molecular biology have given it enormous prestige, even an aura of invincibility. As Nelkin and Lindee have documented, there is a gradually developing "common sense" genetic determinism, which itself has extremely disturbing potential consequences.[11]

The technical development that is most likely to make widespread human genetic selection a realistic possibility is pre-implantation genetic diagnosis (PID), in which a single cell from an embryo is removed for genetic testing. PID offers far greater possibilities for selection than prenatal testing and abortion, because it does not involve abortion of an established pregnancy. PID is performed in conjunction with *in vitro* fertilization (IVF), which produces, on average, ten embryos, of which only two or three are implanted. Such embryos have a far lower moral and emotional weight than a fetus. Because abortion is not involved, if PID were to become widely available it would deepen the social pressure on parents to avoid genetic disorders and undergo testing. While there is still widespread sympathy for a woman who refuses prenatal testing or abortion, it is unlikely there would be such sympathy for parents who rejected the opportunity of PID, simply on the grounds of wishing to leave things to chance.

The very fact that the embryo is outside the woman's body, and before the onset of pregnancy, means that the argument for the prospective mother's right to choose is weakened, and the decisions over the embryo's fate become much more open to intervention from male partners, doctors and society at large. In particular, the site of PID, in the IVF clinic, means that doctors will have a much greater say than in prenatal testing. In the IVF clinic, decisions over which embryos to implant are controlled by medical expertise, in contrast to the genetic counselling situation where the couple decides – at least in theory. PID thus represents a new intensification of the medical surveillance of human reproduction, which would institute a far more preventive regime than currently operates for prenatal testing.

The number of embryos is a key factor here. Once presented with the genetic data for each of ten embryos, it will be very difficult for a parent to ignore the data and refuse to select. PID allows not merely the elimination of clearly harmful alleles, but the selection of embryos carrying the "best" combination of alleles. This is equivalent to a shift from negative to positive eugenics. There is a danger that if PID were to become widespread, it would encourage a culture of choosiness regarding embryos.

Scientists often suggest that an expansion of genetic selection is unlikely, because genetic tests will be never be sufficiently accurate to predict predisposition to disease. There are two issues here. The first is the adequacy of the testing technology, which will have to assess variations in many different genes. In fact, there are already gene chips that can look at major mutations in a hundred different genes at once.[12] Genetic testing technology is continually improving and becoming cheaper, and it would be a mistake to expect this not to continue. The second point is that it may indeed be difficult to predict health status accurately, much less behavioural and personality characteristics, from genetic tests. I believe that the jury is still out on this issue, but it seems quite likely that several genes that substantially affect predisposition to important diseases will be found. Precise predictions will probably never be possible, but for people trying to select embryos precision may not be necessary. Neither will they be concerned about the uncertainties due to environmental factors. Since in PID they are obliged to select two or three embryos from a pool of ten, they will accept a certain amount of inaccuracy, and simply try to select the "best" embryos that they can find, according to the information presented to them on the embryo's genetic predispositions.

It is also suggested that PID is far too invasive to become commonplace, involving as it does the risks, rigours and low efficiency of IVF. This may be a short-sighted view, however. The technology of *in vitro* oocyte maturation is developing rapidly, and it appears likely that in a few years it will be relatively simple to perform an ovary biopsy, containing hundreds of eggs, which can be frozen and matured at will. This will replace the current dangerous and unpredictable procedure of hormonal stimulation of ovaries. If oocyte maturation succeeds, IVF may become a much more accessible choice for normally fertile couples. Given the possibilities that it holds for selection, PID may become the technology of choice for the conscientious couple who want to make sure they give their baby the best start in life.

At the same time as these scientific and technological developments are occurring, the political climate, particularly in the United States, is changing in a number of ways which will encourage eugenic outcomes. First, there is an increasing pressure in all countries to reduce the cost of health care. Politicians are looking for ways to cut health care budgets, and it is clear that preventing the birth of disabled children, with lifelong health care costs, is very cost effective. The crucial decisions, about the introduction of screening programmes, will be taken by health care bureaucrats, under the influence of accountants. Financial pressure on families will be exacerbated by the fear that they or their children may become uninsurable unless they take genetic tests.

In largely private health care systems, such as in the United States, doctors come under the direct influence of Health Maintenance Organizations and associated biotechnology companies anxious to make a rapid return on their investment. In the United States, where genetic tests are unregulated, companies are already selling tests which have not received ethical approval from patients' or medical organizations.[13] Some ethicists, including some in the pay of biotechnology companies, argue that individuals have a right to any personal information that can be obtained by genetic testing. To refuse them this information is unacceptable paternalism, it is argued.

CONCLUSION

The combination of rapidly developing genetic and reproductive technology, free market capitalism, a liberal medical ethics and underlying eugenic patterns of thought are creating the conditions for a dramatic expansion of laissez-faire eugenics. Until now this been largely a matter of preventing the birth of babies with Down's syndrome, spina bifida and a few rare single-gene disorders. In the future it seems likely that controlling the genes of our offspring may become an integral, even central part of most people's reproductive experience. It is possible to envisage a situation, in perhaps ten years' time, in which middle-class professional couples will routinely undergo prenuptial or preconceptual, and pre-implantation or prenatal, genetic testing for genes predisposing to major diseases, as well as some more common single-gene disorders. It will soon become common sense that sex is for fun, but having a baby is a serious matter, not to be left to chance. Although there will be a certain amount of state involvement, through welfare cuts and the piecemeal and surreptitious introduction of genetic screening programmes, change will mainly be enacted through the reproductive decisions of individuals. Everything will be done in the name of better health for our children, but the result, the shaping of the gene pool by social pressures and prejudices, will be no different in essence from the former eugenics.

What can be done about this? Do we even want to do anything? For many people, the prospect of bringing reproduction under scientific control and improving the health of our children is wholly positive. If nobody is forcibly sterilized, what is the problem? It is necessary to state clearly why an expansion of laissez-faire eugenics is to be feared. The first reason is its effect on the liberation of disabled people. Philosophers such as John Harris have been at pains to emphasize that there is no inconsistency between compassion and respect for disabled people and attempting to prevent their birth. Although this may be true in the abstract, what it misses is that in the real world, and in the minds of prospective parents, the two things are intimately related. It is difficult to believe that in a society which had overcome its fears of disability, and truly accepted disabled people as equal members of the community, there would be such an interest in prenatal screening.

At present most able-bodied people suffer from massive fear about disability, coupled with ignorance, misinformation and negative images. If people were aware of the reality of the lives of people with Down's syndrome, for example, they would be much less likely to abort such fetuses, disability rights activists argue. To come in touch with such realities, and overcome fears of disability, will take a prolonged process of listening to disabled people and learning to accept their judgements of which lives are worth living. If, instead, we choose to ignore this obligation and proceed blindly with an expanding programme of genetic testing, ignoring the other eugenic social pressures which militate against the birth of genetically disabled people, we are essentially abandoning their struggle for liberation.

Second, selecting the "best" among multiple embryos sets up a new relationship between parents and offspring. Such children are likely to feel that the essence of themselves in an important sense no longer belongs solely to them, since it has been overseen by their parents, using the all-seeing eye of genetic technology. They are no longer a gift from God, or the random forces of nature, but selected products, expressing, in part, their parents' aspirations, desires and whims.

Third, in the long term, there are good biological reasons for not allowing market forces to shape the human gene pool. Many people, for example, would prefer their children not to have tendencies towards being fat; being overweight is a risk factor for many diseases including heart disease and diabetes. Should we allow free access to genetic testing for obesity? We do not have sufficient understanding of human biology to allow us to judge the evolutionary value of genes, which in existing societies produce disease: in short, we lack the wisdom to play God.

Fourth, as Silver has recently expounded,[14] a logical consequence of a system of free-market eugenics in societies where large disparities of wealth and social class continue to exist is a gradual polarization of society into a genetically privileged ruling elite and an underclass (see Chapter 9 in this volume).

Fifth, under a laissez-faire regime it will be impossible to maintain a strict dividing line between serious medical conditions, more trivial conditions and non-medical characteristics such as appearance and aptitudes (assuming that genes influencing these characteristics can be found). Even now, it is accepted that women using private sperm banks wish to know the educational achievements, ethnic origins and athletic abilities of their sperm donor. The existence of a burgeoning market for cosmetic surgery, and the prescription of growth hormone to normal short children with no hormone deficiency are further pointers towards future trends.

Finally, the practice of laissez-faire eugenics would make it easy for governments subtly to influence the process, not to eliminate particular social groups in the old-fashioned way, but to further goals of national policy, such as increased competitiveness and a lower health care budget. Laissez-faire eugenics could easily collapse into state-managed eugenics.

So, to return to the question, what is to be done? The first step must surely be to recognize the reality of laissez-faire eugenics and the seriousness of its impending expansion. It is time for a critical look at the way that the concept of reproductive freedom is based upon free-market economics and a return to nineteenth-century market-centred liberalism. The refusal to recognize the existence and harmful consequences of laissez-faire eugenics parallels free-market economists' refusal to concede that the unrestrained working of market forces can ever have harmful consequences for society as whole. It is time for a more thoughtful debate about these issues.

At the very least, the right to freedom from interference in any decisions made after a test has been taken should be maintained. However, there is no basis for an automatic right of access to all tests that science has developed. It is perfectly possible to maintain essential rights to freedom from outside coercion without allowing a testing free-for-all. Society must assert its right to exert some control over the development of genetic testing and screening.

Further, there must be a broad public debate on the question of which conditions justify termination of pregnancy. In current debates, the views of people with disabilities are marginalized and rarely heard: instead, the debate is heavily dominated by the views of scientists, doctors and medical ethicists. Such debates should be led by people with disabilities. My intention is not to propose legal bans or to censure morally those who opt for termination of disabled pregnancies. Ideally, such a debate would result in guidelines, with a certain amount of room left open for flexibility in particular cases. In the interim, we should adopt a policy of erring, if it is an error, on the side of caution: prenatal genetic tests should be offered only for disorders which are fatal in childhood, or in which there is a very high degree of pain and suffering.

As Kitcher has argued, a second precondition to making the world safe for genetic testing is a commitment to combat those forces which make freedom of reproductive choice little more than a fiction. Some of those forces derive obviously from free-market liberalism, such as the penchant for cutting welfare and health care budgets. We must also renew our commitment to combating disability oppression and the sexism that dictates that the burden of care falls predominantly on women. The current situation is positive in that it offers us the opportunity to reopen fundamental debates. Do we really believe that the diversity of human life is an important value worth preserving? Is freedom of reproductive choice an untouchable absolute? Is the avoidance of suffering our overriding moral value? And if we have the tools, can we resist the opportunity to take control of something that looks messy and very uncertain? Tackling the latter question involves engaging with a fundamental critique of science, modernity and our ideas of progress.

NOTES

1. Quoted in Garland Allen, "Eugenics and its Social and Economic Context," paper delivered at the symposium "Eugenic Thought and Practice: A Reappraisal toward the End of the Twentieth Century," Jerusalem and Tel Aviv, 1997.

2. For example, Nuffield Council on Bioethics, *Genetic Screening: Ethical Issues*, London, 1993.

3. *Pre-natal Diagnosis and Genetic Screening: Community and Service Implications*, London: Royal College of Physicians, 1989.

4. N.J. Wald, A. Kennard, J.W. Densem, H.S. Cuckle, T. Chard and L. Butler, "Antenatal maternal serum screening for Down's Syndrome: results of a demonstration project," *British Medical Journal*, 305, 1992, pp. 391–4; J. Murray, C. Cuckle, G. Taylor and J. Hewison, "Screening for Fragile X syndrome," *Health Technology Assessment*, vol. 1, no. 4, 1997; "Consensus Development Statement: Genetic Testing for Cystic Fibrosis," Bethesda, MD: National Institutes of Health, 1997.

5. T.M. Marteau, M. Johnston, M. Plenicar, R.W. Shaw and J. Slack, "Development of a self-administered questionnaire to measure women's knowledge of pre-natal screening and diagnostic tests," *Journal of Psychosomatic Research* 32, 1988, pp. 403–8.

6. J. Green and H. Statham, "Psychosocial aspects of prenatal screening and diagnosis," In T. Marteau and M. Richards, eds, *The Troubled Helix: Social and Psychological Implications of the New Human Genetics*, Cambridge: Cambridge University Press, 1996, pp. 140–63.

7. T. Shakespeare, "Back to the Future? New Genetics and Disabled People," *Critical Social Policy* 44, 1995, pp. 22–35; T. Shakespeare, "Choices and Rights: Eugenics, Genetics and Disability," *Disability and Society* vol. 13, no. 5, 1998, pp. 665–81; R. Hubbard, "Abortion and Disability: Who Should and Who Should Not Inhabit the World?" in L. Davis, ed., *The Disability Studies Reader*, New York: Routledge, 1997.

8. D.C. Wertz, "Eugenics is Alive and Well," paper delivered at the symposium "Eugenic Thought and Practice: A Reappraisal toward the End of the Twentieth Century," Jerusalem and Tel Aviv, 1997; D.C. Wertz, "Society and the Not-so-new Genetics: What Are We Afraid Of? Some Predictions from a Social Scientist," *Journal of Contemporary Health Law and Policy* 13, 1997, pp. 299–346.

9. S. Michie, F. Bron, M. Bobrow and T.M. Marteau, "Nondirectiveness in Genetic Counseling: An Empirical Study," *American Journal of Human Genetics* 60, 1997, pp. 40–47.

10. P. Kitcher, *The Lives to Come: The Genetic Revolution and Human Possibilities,* Harmondsworth: Penguin, 1996.

11. D. Nelkin and M.S. Lindee, *The DNA Mystique: The Gene as a Cultural Icon,* New York: W.H. Freeman, 1995.

12. M. Southern, "DNA Chips: Analyzing sequence by hybridization to oligonucleotides on a large scale," *Trends in Genetics* 12, 1996, pp. 110–14; Editorial, "To Affinity ... and Beyond," *Nature Genetics* 14, 1996, pp. 367–70.

13. *GenEthics News* 12, 1996, p. 3.

14. L. Silver, *Remaking Eden: Cloning and Beyond in a Brave New World,* London: Weidenfeld & Nicolson, 1998.

If Pigs Could Fly, They Would:
The Problems with Xenotransplantation

ALIX FANO

According to the United Network for Organ Sharing, a quasi-governmental organization that coordinates human organ and tissue donation in the US, some 4,000 Americans die each year waiting for transplantable organs.[1] Compared to the number that die from heart disease (726,974), cancer (539,577), pneumonia/influenza (86,449), AIDS (16,516), and by suicide (30,535),[2] this may not seem high. But with over 60,000 people on transplant waiting lists in the US, 180,000 worldwide,[3] and a perceived chronic shortage of human organs and tissues, researchers, drug companies and health officials in the US and elsewhere are proposing a radical "solution": using genetically altered animals, such as "humanized" pigs and nonhuman primates, as "donors."

The prospect of commercial cross-species transplantation (xenotransplantation) has created huge financial incentives for multinational drug and biotechnology companies. Novartis, Baxter Health Care and their many subsidiaries that dominate the field have already invested $100 million in research,[4] and stand to make billions from sales of transgenic pig parts and expensive anti-rejection drugs. A 1996 Salomon Brothers UK report, *The Unrecognized Potential of Xenotransplantation* (commissioned by Sandoz, a Novartis subsidiary and maker of anti-rejection drugs), estimated that annual profits could soar to $6 billion by the year 2010. In their defence, these multinationals claim that xenotransplantation will create an unlimited supply of organs and cells on demand for needy patients. Robert Michler, chief of cardiothoracic surgery at Ohio State University Medical Center, claims that xenotransplantation is a "revolutionary technology that will one day benefit many millions of patients." Even though it is admittedly "extraordinarily unimaginably expensive," he believes it will eventually become cost-effective.[5] But is this true? And is the technology, which is being developed with substantial public and private financing, worth the risks and costs?

In 1984, at Loma Linda University Medical Center in California, transplant surgeon Leonard Bailey performed an experimental baboon heart transplant on Baby Fae, an infant born with a congenital heart defect. No attempt was made to find a human heart for her, though the director of California's Organ Procurement Agency said one would have been available. In less than two weeks, Baby Fae rejected the baboon blood, her veins and arteries became blocked, and she died.[6] The controversy set the xeno field back for many years, and Bailey put his four other cross-species trials on hold.

Today, most researchers agree that ethical issues and disease risks have virtually precluded the use of chimpanzees and other apes as organ donors. In April 1999 the US Food and Drug Administration (FDA) issued non-binding guidelines instituting a de facto – but not permanent – ban on the use of nonhuman primate xenografts because of "significant infectious disease risk."[7] The issuance of the guidelines was the likely result of pressure from noted virologists like Jonathan Allan of the Southwest Foundation for Biomedical Research in San Antonio, Texas. Allan and others have elucidated the public health risks of using nonhuman primates as source animals in xenotransplants.[8] First, nonhuman primates carry viruses that are deadly to humans, including Herpes B, Marburg and Ebola. Second, patients risk infection with new viruses that previously only infected nonhuman animals. Some scientists now believe that HIV-1 originated in chimpanzees;[9] and in 1992, a 35-year-old HIV patient received a baboon liver amid great publicity. An analysis of his tissues later revealed that he had contracted baboon cytomegalovirus, a virus thought to be species-specific.[10]

PROGRESS OR HUBRIS?

Since 1906, 82 humans have received whole organs from chimpanzees, baboons, pigs, goats and other animals, and a majority have died from infections and complications related to hyperacute rejection within hours or days of their surgeries.[11] Today, however, biotech companies claim that they have bred "germ-free" pigs with human genes whose organs are less likely to be rejected by the human body. Pigs, they say, are the source animals of choice, because they breed quickly, have been extensively farmed, and have organs that are allegedly "similar" in size to ours.[12]

But are those *scientific* justifications for their use? A conventional pig heart put into a human will turn black and stop beating in about fifteen minutes. There is no reason to believe that organs from "humanized" pigs will be any less likely to be rejected by the human body, or that acute cellular and vascular rejection will ever be overcome. There exist basic interspecies differences in life span, heart rate, metabolism, blood type, blood pressure, circulation and coagulation, hormone systems, and response to disease, among other things.[13]

The author of an article in *Nursing Times* asks, "Can a pig's heart – normally on the same level as its head – pump enough blood to a human brain 15–18 inches above? Will a pig kidney filter human blood effectively, or will the pig's different uric acid metabolism lead to biochemical aberrations? And will the human recipient's immune system work in a transplanted pig organ?"[14] Past experience would suggest not. Yet Goran Klintmalm, a transplant surgeon at Baylor University Medical Center in Dallas, Texas, says that clinical trials with whole pig organs will involve hearts and kidneys.[15] It is also likely that the massive doses of anti-rejection drugs needed to suppress the human immune system in such operations would cause severe toxicity, increase the patient's chances of developing cancer, and facilitate the transmission of animal and human viruses to the patient.

Thousands of gruesome experiments between numerous species of animals, particularly pigs and baboons, using various combinations of toxic anti-rejection drugs and steroids, have killed lots of animals, but failed to tell us whether xenotransplantation is either safe or effective for humans. Scientists themselves admit that current animal models (rats, nonhuman primates), are very limited in their ability to predict the human scenario, largely due to differences in basic physiology and biology.[16] Animal experiments failed to predict the cascade of side effects – jaundice, blood clotting, kidney failure, lung failure – that killed an 18-year-old human gene therapy patient in September 1999.[17] Nor have limited human trials provided statistically significant or reliable information about safety or efficacy. Nevertheless, researchers around the world continue their efforts undaunted.

Most transplant surgeons, eager to move into clinical trials and disenchanted by the poor track record of whole organ xenotransplants, have thus far focused their efforts on other forms of xenotransplantation. Over the last twelve to fifteen years, several hundred patients from different nations have been injected with fetal pig cells allegedly to treat the symptoms of epilepsy, diabetes, Parkinson's[18] and Huntington's disease, and to alleviate chronic pain associated with cancer. Burn patients and kidney and liver failure patients have had their blood "filtered" and recirculated (or perfused) through pig spleens, kidneys and livers.[19] And pig organs have been used as temporary "bridges," for brief periods of time, to keep acutely ill patients alive until human organs become available.[20]

PIGS, PRIMATES AND PLAGUES

The increasing economic investment in xenotransplantation is unsettling given the acknowledgement by the US Department of Health and Human Services (HHS),[21] the World Health Organization (WHO)[22] and eminent scientists that the technology could transmit potentially deadly animal viruses, both known and unknown, to patients and the general public. A report commissioned by the Organization for Economic Cooperation and Development (OECD) has stated, "it is unlikely that risk [of transmitting infectious agents] can be reduced to zero."[23] Placing animal organs directly into humans circumvents all the natural barriers designed to prevent infection, and represents a unique route for the transfer of potentially infectious animal viruses to humans. Many doctors fear that xenotransplantation could unleash an AIDS-like virus as easily transmissible as the common cold. French virologist Claude Chastel has said that xenotransplantation will create a "new infectious Chernobyl."[24]

Indeed, pigs can carry bacterial, viral, fungal, protozoal and helminth (parasitic worm) pathogens, as well as prion proteins, implicated in "mad cow disease."[25] Numerous retroviruses and fungal infections, in humans and animals, are latent, or can incubate for long periods of time in healthy individuals, but may become activated after host defences are weakened by immunosuppression. A broad spectrum of viruses in pigs (rhabdoviruses, influenza viruses, picornaviruses, paramyxoviruses) cause a wide range of symptoms in humans, including fever and general malaise, sores on the face or feet, neurological disorders, meningitis and death. Pigs also act as efficient "mixing vessels" for viruses from other mammals and birds.[26]

The swine flu epidemic of 1918, which likely jumped from birds to pigs and then to humans, killed between 20 and 40 million people worldwide.[27] In 1998–99, the novel Malaysian "Nipah" encephalitis virus, which may have originated in fruit bats, jumped from pigs to humans, infected 269 people, killed 117, left dozens brain-damaged, and led to the mass slaughter of a million pigs, as well as several hundred cats and dogs.[28]

Known pig viruses include the porcine endogenous retroviruses (PERVs) that have infected human cells. A retrospective study of 160 patients exposed to pig cells produced troubling results: 30 patients whose blood was perfused through pig spleens tested positive for PERV DNA; 23 had pig cells circulating in their bodies eight and a half years after treatment; and 4 patients injected with pig cells produced antibodies against pig PERV – suggesting a potential active infection by pig viruses. Although the study's authors claim that there is no conclusive evidence of human infection by PERV, they admit that "PERV infection [cannot] be excluded."[29] But PERVs are not the only concern. There may be dozens of unknown viruses, waiting to be discovered. Scientists admit that "our understanding of the retrovirology of xenotransplant source animals is incomplete," and "little or nothing is known about the pathogenic potential of endogenous retroviruses introduced directly into other species."[30] A pig virus, contracted via xenotransplantation, could spread to other humans undetected, causing a global health pandemic.

Officials from the US Centers for Disease Control (CDC) have estimated that, before Legionnaire's disease was identified in 1976, between two and six thousand deaths per year were incorrectly attributed to pneumonia. Similarly, although the HIV virus was identified in 1983, researchers now say they have traced its origins back to the 1930s.[31]

Should it occur, the widespread adoption of xenotransplantation will inevitably place an enormous strain on our global public health and disease surveillance systems. Despite globally coordinated efforts, we are still having trouble controlling infectious diseases like AIDS, hepatitis and malaria. How would our health care system cope with the consequences of infection by an unfamiliar virus? How would health officials contain it if they didn't know what to look for? How would carriers of a new xeno-related virus in the general population be identified? What if there were thousands, or tens of thousands, of carriers? Would special facilities have to be built to accommodate them? These questions have yet to be answered by US public health authorities, who continue to approve clinical xenotransplant trials. Can they be trusted to assess the safety of xenotransplants and protect the public from harm that may result?

Can scientists involved in clinical xenotransplant trials – many in partnerships with industry, some perhaps seeking fame and glory – be trusted? Recently, the *Washington Post* reported that scientists and drug companies failed to notify the US National Institutes of Health about six deaths that occurred in gene therapy experiments between 1998 and 1999. Despite, or perhaps in light of these recent events, companies are increasing efforts to weaken federal reporting requirements.[32]

THE XENO BEAST:
OUR COLLECTIVE BURDEN

What about the costs of xenotransplantation? The OECD has reported that "[t]he economic impacts of the development of xenotransplantation have not been adequately addressed."[33] A 1996 US Institute of Medicine report estimated that xenotransplantation would boost annual US transplant expenditures from almost $3 billion to $20.3 billion.[34] But a more recent cost–benefit analysis, using industry figures for projected xenograft use worldwide, revealed that first-year implementation costs for xenotransplants in the US could reach $35 billion by 2010, and $90 billion worldwide. First-year follow-up costs would be almost $4 billion in the US and almost $10 billion worldwide.[35] There are also the hidden costs of housing, breeding, feeding, medicating, testing and disposing of the remains of herds of transgenic animals; of hiring skilled hospital personnel, surgical staff, infectious disease experts and veterinarians capable of properly monitoring xenograft patients, their close contacts and source animals; of government-mandated patient registries, blood and tissue archives, programmes and technologies to screen for new viruses; and unpredictable medical and legal costs associated with disease outbreaks.

In May 1998, in the *British Medical Journal*, a group of international physicians pointed out that poverty, and resultant lack of access to basic health care and adequate sanitation, are the world's number one health problem. In the US – which prides itself as having some of the most advanced medical technologies in the world – almost 50 million Americans (two-thirds of them minorities, children and the working poor) lack basic health care.[36] According to Richard Nicholson, editor of the *Bulletin of Medical Ethics*, even if xenotransplantation works perfectly, and setting aside public health concerns, the technology would allegedly increase overall life expectancy by only 0.02 per cent.[37] The OECD agreed that the development of xenotransplantation "may conflict with efforts to keep medical costs down ... [and] may not be consistent with striving for humane and fair medicine."[38] Is it equitable, therefore, to allocate scarce resources to a relatively small number of acutely ill patients at the expense of a majority, particularly if the technology in question may not only be ineffective, but dangerous?

"TWO LEGS BAD"

Xenotransplantation raises serious animal welfare concerns for the animals that are genetically modified and bred as organ factories in unnatural, sterile environments, those used in virus research, and those killed in gruesome pre-clinical experiments. There are few, if any, guidelines governing the breeding of transgenic animals used in research, and countries differ with respect to standards for the treatment, transportation and housing of animals, as well as personnel qualifications and record-keeping.

What is known is that the techniques used to breed transgenic pigs and harvest their parts are invasive, cruel and deadly. *New York Times* science reporter Sheryl Gay Stolberg, who visited a Nextran[39] pig-breeding facility, described the sight of a pregnant female pig, on her back on an operating table, her head suspended over a bucket to catch any vomit, "her insides splayed out on a blue-paper surgical drape as a scientist rearranges the DNA of her unborn young." The mother pig's eggs are flushed out of her ovaries, human genes are inserted into each egg, and the manipulated embryos are implanted back into her. Perhaps one out of twenty piglets will be born with human genes. The rest will be destroyed.[40] Because genetic engineering techniques are imprecise, it is unknown how many thousands of animals may be bred and killed as byproducts because they do not carry the genetic trait(s) desired by a company.[41]

British reporter Anthony Browne writes that, in order to obtain pig neuronal cells to treat Parkinson's disease patients, "pregnant genetically modi-fied sows are slaughtered, their foetuses are chopped out, their heads cut off, and their brain cells sucked out and injected into the heads of humans."[42] For a single transplantation of pig pancreatic islet cells into a diabetic patient, 39–100 fetuses (4–8 litters) were used.[43] Since hundreds of thousands, millions, and even billions of cells are required for some of these treatments,[44] such a wholesale slaughter of sows and young piglets would become the norm if xenotransplantation were to become widely adopted.

Pigs raised for xenotransplants spend their entire lives in sterile, confined areas with no exposure to grass or sunlight. Overcrowding leads to stress, vicious fighting and cannibalism.[45] Some corporations want to clone pigs for xenotransplantation, but data derived from cloned and transgenic animals have revealed that such animals often have weaker immune systems, and may be born with physical disabilities which cause them great pain. These include chronic arthritis, swelling of joints, enlarged organs, blindness and respiratory ailments.[46]

REGULATING XENOTRANSPLANTATION

There are virtually no regulations governing the field of xenotransplantation.[47] There has been a push, however, by the US, WHO, OECD and industry towards harmonization of xenotransplant guidelines across countries – a way to legitimize the field in the eyes of the public, to "ensure that xeno-transplantation is developed in conformity with accepted ethical and legal standards,"[48] to establish multilateral disease surveillance procedures, and to provide guidance for less developed countries that may adopt the technology without establishing adequate safeguards.

On January 29, 1999, in the name of the precautionary principle, the Parliamentary Assembly of the Council of Europe unanimously adopted a recommendation on xenotransplantation which, among other things, called for a legally binding moratorium on clinical (human) trials.[49] Instead of taking a position on the proposed moratorium, however, the Committee of Ministers of the Council of Europe decided to set up a Working Party on Xeno-transplantation, which was mandated to draft guidelines on xenotransplantation within three years, under the joint responsibility of the Steering Committee on Bioethics and the European Health Committee. The Working Party was to address public safety concerns, and the preconditions – legal or otherwise – which must be met before clinical trials can proceed.[50]

Most governments actively engaged in xenotransplant research, including the USA, Canada, Britain, Spain, Sweden, Switzerland and the Netherlands,[51] have drafted their own recommendations and/or guidelines – some of them draconian and mind-bogglingly complex – outlining requirements for clinical trials, disease-monitoring procedures and animal health standards. These are continually evolving. Some countries, like Sweden, the Netherlands and France, have expressed more caution about moving to clinical trials than others. But none of the existing guidelines is legally binding; all would be vulnerable to human error, particularly in the tracking of experimental subjects; their implementation would be cost-prohibitive; and all would be open to legal challenge for being unduly invasive of privacy and restrictive of liberty.

Most guidelines call for rigorous lifelong monitoring of xenotransplant patients, family members, friends, social contacts and health care workers. US guidelines have proposed indefinite source animal and patient tissue archiving, though the American Society of Transplant Surgeons has objected to this, due to the magnitude of the task.[52] Patients must also agree to be quarantined indefinitely if they are found to be harbouring an active infection.

The UK has issued more than sixty detailed recommendations regarding the development and implementation of xenotransplantation. Draft guidelines, issued at the end of October 1999, proposed that xenograft patients be barred from having children and donating blood, that they never again have sex without the use of barrier contraception, and that their future sexual partners be registered with, and monitored by, medical authorities.[53] It may well be legally impossible to force or coerce unwilling individuals to comply with these procedures and there is no legal framework authorizing health agencies to enforce them.[54] And yet, if these people fail to cooperate, the entire surveillance and public health system would be undermined.

There is another point to consider. All current guidelines are being proposed in hindsight, because humans have been receiving cells, tissues and organs from animals for decades. Yet, shockingly, there are no national or international registries of all patients, dead or alive, who have heretofore received xenotransplants, and it is possible that these people may have already engaged in risky behaviours and/or donated blood. The ramifications of blood supplies contaminated with HIV, hepatitis and Creutzfeldt-Jakob ("mad cow") disease have been deadly. Could an as yet unknown porcine virus already be lurking in the blood supply, undetected by current commercial testing methods?

The author of a law article suggested that the US draft xenotransplantation guidelines of 1996 "failed to provide sufficient instruction for minimizing the risks of unknown infections." The problem is: "there is no way to account for unknown pathogens."[55]

PUBLIC ATTITUDES UNCLEAR

While terminally ill patients may be willing to undergo any procedure to stay alive, imposing risks of infection on healthy members of the general public without their informed consent raises serious legal questions because it could constitute a form of involuntary human experimentation. And yet it may be impossible to obtain the public's consent, for both practical and social reasons.

Although industry has spent large sums of money on public-relations campaigns to promote the alleged necessity and desirability of xenotransplantation, the public remains wary. A British poll revealed that 69 per cent of Britons favour a ban on xenotransplants to allow for a full public debate on the technology's use.[56] Some suggest that nurses may have a particular aversion to xenotransplants for fear of becoming infected with a zoonotic virus.[57] A Swedish study revealed that 77 per cent of respondents preferred organs from living human donors, and only 40 per cent were prepared to accept an animal organ.[58] A 1997 European survey on biotechnology indicated that xenotransplantation was perceived to involve unacceptable risks and that the use of transgenic animals as a source was seen as "morally dubious."[59]

Americans have, thus far, largely been exposed to favourable news stories about xenotransplantation. In a study done by the National Kidney Foundation in 1998, and funded by Novartis, it was not surprising to find that 62 per cent of Americans would allegedly favour xenotransplants; however, "opposition to [the technology] was strongest among the best informed."[60] Other articles implied that Americans, in fact, had "visceral reactions against xenotransplants."[61]

Yet public scepticism and aversion to xenotransplantation has not prevented researchers from going forward. Limited clinical xenotransplant trials have been performed in Canada, France, Germany, Israel, New Zealand,

Russia and Sweden,[62] and basic research is ongoing around the world. But the US is perhaps the technology's most enthusiastic proponent. The US Department of Health and Human Services (HHS), comprising the FDA, the CDC and the National Institutes of Health (NIH), is devoting considerable resources to funding basic research, approving and overseeing clinical xenotransplant trials, studying zoonotic viruses and developing guidelines.

Within the last ten years, the FDA has approved numerous clinical xenotransplant trials on hundreds of patients, including baboon livers into patients with HIV infection and hepatitis;[63] baboon bone marrow into an AIDS patient;[64] fetal pig neuronal (brain) cells into epileptics[65] and patients with neurological diseases; pig pancreas cells into diabetic patients; cow adrenal cells into the spinal cords of cancer patients;[66] and the use of whole pig livers,[67] and pig liver cells, to perfuse the blood of patients with acute liver failure.[68]

The CDC is actively collaborating with industry and research centres to monitor xenotransplant patients. It is developing diagnostic assays to detect pig and baboon cells and viruses in those patients, and to detect viruses in the animals themselves.[69]

Over the last several decades, the NIH has dispensed tens of millions of dollars to researchers, in the US and abroad, developing xenotransplantation. One researcher at Massachusetts General Hospital in Boston received over $15 million between 1992 and 2000 to study the immune response involved in xenograft rejection between pigs and baboons.[70] A team of researchers at Duke University in North Carolina received almost $2.5 million in 1997–98 to transplant pig hearts into monkeys and other animals in an effort to elucidate the "immunological barrier to cardiac xenotransplantation." In 1997 the NIH published an announcement on the Internet soliciting "applications [from domestic and foreign researchers] to enhance [the] ability to transplant organs and tissues across species barriers (xenotransplantation)."[71] The NIH even promotes xenotransplantation to children on its website under the guise of "science education."[72]

The US Department of Commerce's Advanced Technology Program is also dispensing millions of dollars in grants to corporations developing xenotransplantation.[73] It gave PPL Therapeutics – the multinational corporation that helped to clone Dolly the sheep – $2 million in 1999 for a research program to clone pigs for xenotransplants, and to reduce the risk of "rapid rejection of pig organs when transplanted into humans."[74]

NO XENOTRANSPLANTATION WITHOUT REPRESENTATION

Xenotransplantation has been developed with substantial public financing, but little if any regulation or public debate. The technology's critics have been virtually shut out of decision-making processes, and xenotransplant researchers have been allowed to chair government committees and dictate public policy.

The Campaign for Responsible Transplantation (CRT) – an international coalition of physicians, scientists and ninety public interest groups – is seeking a ban on xenotransplantation, pointing to the irrationality of public health authorities that promote a technology which, they acknowledge, could unleash viral epidemics.[75] CRT is considering legal action against US agencies for what it views as their violation of mandates to protect the public health and prevent disease. Moreover, CRT believes that they are ignoring existing alternatives. Aggressive investment in population-based prevention programmes – encouraging people to exercise and cut down on high-fat foods, cigarettes and alcohol – could prevent many of the diseases that result in the need for transplantation, and thereby shrink the number of people on transplant waiting lists.[76] Reports have shown that many governments, including those of the USA[77] and Canada,[78] are doing a poor job of retrieving human organs, and that a large number are being buried or burned.[79] In contrast, several European countries have increased their organ donation rates after passing "presumed consent" laws, which assume that every citizen is an organ donor unless they "opt out."[80] The feasibility of enacting this legislation in the US and Canada has not been evaluated. Moreover, cultivating human tissue and cells for transplantation would provide a much safer source than animals.

If pigs could fly out of xenotransplant laboratories, they surely would. Alas, they cannot. The power to change things is in our collective hands. Concerned citizens, many already opposed to genetically engineered foods, must now rise up to protest against the development of biotechnologies, like xenotransplantation, which harm humans and animals alike, and threaten life on earth as we know it.

NOTES

1. See http://www.unos.org. That number is expected to rise steadily, at least in developed countries, where diets high in fat, excessive smoking and alcohol consumption, and sedentary lifestyles contribute to rising disease rates.

2. These statistics are from 1997: http://www.cdc.gov/nchs/fastats/deaths.htm.

3. Geoffrey Cowley, "A Pig May Someday Save Your Life," *Newsweek*, January 1, 2000, pp. 87–9.

4. Lauran Neergaard, "Animal–Human Transplant Ban Urged," Associated Press, January 21, 1998.

5. From proceedings, Food and Drug Administration, Center for Biologics Evaluation and Research, Biological Response Modifiers Advisory Committee, Xenotransplant Subcommittee meeting, June 4, 1999. Source: http://www.fda.gov/ohrms/dockets/ac/99/transcpt/3517t2.rtf.

6. Tony Stark, *Knife to the Heart*, London: Macmillan, 1996, pp. 164–9.

7. Fritz H. Bach, "Putting the Public at Risk," *Bulletin of the World Health Organization*, vol. 77, no. 1, 1999, p. 66.

8. Jonathan S. Allan, "The Risk of Using Baboons as Transplant Donors: Exogenous and Endogenous Viruses," *Annals of the New York Academy of Sciences* 862, 1998, pp. 87–99; Jonathan S. Allan, "Xenotransplantation at a Crossroads: Prevention versus Progress," *Nature Medicine*, vol. 2, no. 1, January 1996, pp. 18–21; Frederick A. Murphy, "The Public Health Risk of Animal Organ and Tissue Transplantation into Humans," *Science* 273, August 9, 1996, pp. 746–7.

9. Daniel Q. Haney, "Experts Conclude AIDS Virus Originated from Chimps," Associated Press, January 31, 1999.

10. Susan Gordon, "Baboon Virus Transmitted along with Liver Transplant," Reuters Health, October 7, 1999.

11. A.S. Daar, "Animal-to-Human Organ Transplants – A Solution or a New Problem?" *Bulletin of the World Health Organization*, vol. 77, no. 1, 1999, p. 55.

12. While heart valves from slaughterhouse pigs have been used for several decades, these valves are treated with a chemical called glutaraldehyde which renders them biologically inert. Moreover, synthetic valves are now replacing pig valves.

13. M.E. Breimer, "Physiologic Incompatibilities in Discordant Xenotransplantation," *Transplantation Proceedings* 31, 1999, pp. 905–8; Simon J. Crick et al., "Anatomy of the Pig Heart: Comparisons with Normal Human Cardiac Structure," *Journal of Anatomy* 193, 1998, pp. 105–9.

14. Richard Nicholson, "If Pigs Could Fly," *Nursing Times*, vol. 93, no. 6, February 5–11, 1997, p. 20.

15. Kathy Fieweger, "Firm Sees Trials of Pig Organ Transplants," *Yahoo News Online*, February 15, 2000, at http://dailynews.yahoo.com/h/nm/20000215/sc/health_transplant _1.html.

16. Ibid.

17. Sheryl Gay Stolberg, "New Information on Gene Patient's Death Fails to Resolve Mystery," *New York Times*, December 2, 1999, p. A32.

18. Researchers claim that Parkinson's patients injected with porcine neuronal cells have shown clinical improvement. T. Deacon et al., "Histological Evidence of Fetal Pig Neural Cell Survival after Transplantation into a Patient with Parkinson's Disease," *Nature Medicine*, vol. 3, no. 3, March 1, 1997, p. 350; J. Schumacher et al., "Neuronal Xenotransplantation in Parkinson's Disease," *Nature Medicine*, vol. 3, no. 5, May 1, 1997, p. 474. However, some of these studies have only involved single patients and are not statistically significant.

19. K. Paradis et al., "Search for Cross-Species Transmission of Porcine Endogenous Retrovirus in Patients Treated with Living Pig Tissue," *Science* 285, August 20, 1999, pp. 1236–41.

20. Sheryl Gay Stolberg, "Could this Pig Save Your Life?" *New York Times Magazine*, October 3, 1999, pp. 46–51.

21. *US Federal Register*, vol. 61, no. 185 September 23, 1996, pp. 49920–32.

22. World Health Organization, *Global Aspects of Emerging and Potential Zoonoses: A WHO Perspective, Emerging Infectious Diseases*, vol. 3, April–June 1997.

23. Elettra Ronchi, Biotechnology Unit, *Advances in Transplantation Biotechnology and Animal to Human Organ Transplants (Xenotransplantation)*, Paris: OECD, 1996, p. 74.

24. "Animal-Organ Transplants Could Lead to New AIDS, Group Warns," Reuters News Service March 19, 1998; Claude E. Chastel, "The Dilemma of Xenotransplantation," *Emerging Infectious Disease*, vol. 2, no. 2, April–June 1996, p. 155.

25. Dominic Borie et al., "Microbiological Hazards Related to Xenotransplantation of Porcine Organs into Man," *Infection Control and Hospital Epidemiology*, vol. 19, no. 5, May 1998, pp. 355–65; Fred Pearce, "BSE May Lurk in Pigs and Chickens," *New Scientist*, April 6, 1996, http://www.last-word.com/nsplus/insight/bse/pigs101.html.

26. Borie et al., "Microbiological Hazards," pp. 356–62.

27. Maggie Fox, "Deadly 1918 Virus Resembled Common Swine Flu," Reuters, February 15, 1999.

28. Craig Skehan, "Pig-Borne Epidemic Kills 117," *Sydney Morning Herald*, online, April 10, 1999; Alvin Ung, "Tropical Killer Virus is First of its Kind; Experts Stumped," Associated Press, April 8, 1999.

29. Paradis et al., "Search for Cross-Species Transmission," pp. 1236–41.

30. Jonathan P. Stoye et al., "Endogenous Retroviruses: A Potential Problem for Xenotransplantation," *Annals of the New York Academy of Sciences* 862, 1998, p. 68.

31. Lawrence K. Altman, "AIDS Virus Originated around 1930, Study Says," *New York Times*, February 2, 2000, p. A15.

32. Deborah Nelson and Rick Weiss, "NIH Not Told of Deaths in Gene Studies; Researchers, Companies Kept Agency in the Dark," *Washington Post*, November 3, 1999, p. A1.

33. Ronchi, *Advances in Transplantation Biotechnology*, p. 78.

34. US Institute of Medicine, *Xenotransplantation: Science, Ethics and Public Policy*, Washington, DC: National Academy Press, 1996.

35. Alan Berger, "Animal-to-Human Transplants: A Frightening Cost–Benefit Analysis," Sacramento, CA: Animal Protection Institute, October 1999.

36. "Number of Americans Without Health Insurance Coverage Increases in 1997, Census Bureau Reports," US Census Bureau, September 28, 1998.

37. Mae-Wan Ho, *Genetic Engineering: Dream or Nightmare*, Dublin: Gateway Books, 1998, p. 180.

38. Ronchi, *Advances in Transplantation Biotechnology*, p. 79.

39. Nextran is a subsidiary of Baxter Health Care.

40. Sheryl Gay Stolberg, "Could this Pig Save Your Life?" pp. 47, 49.

41. Andy Coghlan, "Hidden Sacrifice," *New Scientist*, May 8, 1999.

42. Anthony Browne, "When Science Makes a Pig's Ear of it," *New Statesman*, November 15, 1999.

43. C. G. Groth et al., "Transplantation of Porcine Fetal Pancreas to Diabetic Patients," *Lancet* 344, November 19, 1994, p. 1402.

44. Some 400,000 fetal pig cells were used to treat a man with severe epilepsy at Beth Israel Deaconess Medical Center in Boston, Massachusetts. Richard A. Knox, "Boston Doctors Transplant Pig Cells into Epilepsy Patient," *Boston Globe*, January 21, 1998, p. A13. Between 330,000 and 1,020,000 pig fetal islet-cell clusters were used to treat diabetic patients, and the number was gradually increased during the study to "increase the chances of engraftment." A. Tibell et al., "Pig-to-human islet transplantation in eight patients," *Transplantation Proceedings*, vol. 26, no. 2, April 1994, pp. 762–3. A device used to support patients with acute liver failure uses 5 billion pig liver cells in each treatment. Zorina Pitkin of Circe Biomedical quoted in FDA Xenotransplant Subcommittee minutes, June 3, 1999, Washington DC, at http://www.fda.gov/ohrms/dockets/ac/99/transcpt/3517t1.rtf.

45. An undercover video of transgenic pigs raised for xenotransplantation research obtained by the British Union for the Abolition of Vivisection (BUAV), London, in July 1999 clearly demonstrated this behaviour. See also Palmer Holden and John McGlone, "Animal Welfare Issues: Swine," *AWIC Bulletin*, vol. 9, no. 3–4, Spring 1999, pp. 9–11.

46. Michael W. Fox, *Beyond Evolution*, New York: Lyons Press, 1999, pp. 108–15.

47. S.J. Chae et al., "Legal Implications of Xenotransplantation," *Xenotransplantation* 4, 1997, pp. 132–9.

48. World Health Organization, *Report of WHO Consultation on Xenotransplantation*, October 28–30, 1997, Geneva, p. 15.

49. "Assembly Calls for More Information on Benefits and Risks of Animal Transplants," Council of Europe Press Office, Strasbourg, January 29, 1999.

50. Bart Wijnberg and Didier Houssin, "Xenotransplantation and the Council of Europe," Council of Europe, Strasbourg, September 1999.

51. See http://www.oecd.org/dsti/sti/s_t/biotech/xenosite/country.htm for summaries of each country's legislation.

52. Jodi K. Fredrickson, "He's All Heart ... and a Little Pig Too: A Look at the FDA Draft Xenotransplant Guideline," *Food and Drug Law Journal*, vol. 52, no. 4, 1997, p. 446.

53. "Animal Transplant Patients Face Child Ban," BBC News Online, October 25, 1999. Source: http://news.bbc.co.uk/hi/english/health/newsid_484000/484665.stm.

54. Chae et al., "Legal Implications of Xenotransplantation," p. 136.

55. Fredrickson, "He's All Heart," pp. 429–51.

56. Marie Woolf, "Two Thirds Fear Animal Transplants to Humans," *Telegraph*,

December 2, 1999.

57. Suzanne D. Fullbrook and M.B. Wilkinson, "Animal to Human Transplants: The Ethics of Xenotransplantation (2)," *British Journal of Theatre Nursing*, vol. 6, no. 3, June 1996, pp. 13–18.

58. Margareta A. Sanner, "Giving and Taking – to Whom and from Whom? People's Attitudes toward Transplantation of Organs and Tissue from Different Sources," *Clinical Transplantation* 12, 1998, pp. 530–37.

59. "Europe Ambivalent on Biotechnology," *Nature* 387, June 26, 1997, pp. 845–7.

60. Declan Butler, "Poll Reveals Backing for Xenotransplants," *Nature* 391, 22 January 1998, p. 315.

61. Paul Recer, "Patient: Animal Organs Not Accepted," Associated Press, February 14, 1998.

62. Paradis et al., "Search for Cross-Species Transmission," pp. 1236–41.

63. T.E. Starzl et al., "Baboon-to-Human Liver Transplantation," *Lancet* 341, January 9, 1993, pp. 65–71.

64. Maria Goodavage, "Implanted Baboon Cells Fail to Help AIDS Patient," *USA Today*, February 8, 1996, p. A1.

65. "Fetal Pig Cells May Help Epileptics," Associated Press, December 9, 1998.

66. "Current Transplant Experiments," Associated Press, January 21, 1998.

67. Sheryl Stolberg, "Could this Pig Save Your Life?" p. 50; Deborah Spak, "Northwestern Joins Baxter in Transgenic Pig Liver Trial," Press Release, Northwestern University, October 6, 1997.

68. Angeles Baquerizo et al., "Characterization of Human Xenoreactive Antibodies in Liver Failure Patients Exposed to Pig Hepatocytes after Bioartificial Liver Treatment," *Transplantation* 67, January 15, 1999, pp. 5–18; Rush–Presbyterian–St. Luke's Medical Center, Illinois, "First Use of Bioartificial Liver in Chicago Region…" PR Newswire, January 26, 1999.

69. "Working Groups Develop Policies on Xenotransplantation," *CDC/NCID Focus*, vol. 7, no. 1, January–February 1998, at http://www.cdc.gov/ncidod/focus/vol7no1/xenotran.htm.

70. A.B. Cosimi, "Tolerance – Approach to Cardiac Allotransplants and Xenotransplants," Federal Grant #3P01HL8646–22S1. From Computer Retrieval of Information on Scientific Projects (CRISP) database at http://crisp.cit.nih.gov.

71. The announcement was posted as part of HHS's "Healthy People 2000" initiative. See "Mechanisms of the Immune Response to Xenotransplant Antigens," at http://grants.nih.gov/grants/guide/pa-files/PA-97–089.html.

72. See http://science-education.nih.gov/newsnapshots/xeno/discovery.html.

73. See for example, http://www.atp.nist.gov/www/comps/briefs/99011113.htm.

74. PPL Therapeutics press release, "PPL Therapeutics Wins $2,000,000 ATP Award for its Xenotransplantation Program," October 7, 1999, http://www.ppl-therapeutics.com/press/99–ppl-29.html.

75. See http://www.crt-online.org.

76. See, for example, R.J. Deckelbaum et al., *Circulation: The Journal of the American Heart Association* 100, July 27, 1999, pp. 450–6.

77. R. W. Evans, "The Potential Supply of Organ Donors: An Assessment of the Efficacy of Organ Procurement in the United States," *JAMA* 267, 1992, pp. 239–46.

78. Tim Harper, "Critics Slam Organ Donor Proposal," *Toronto Star*, October 16, 1999.

79. Patricia M. La Hay, "Organ-Donation Bill Would Pay for Life Insurance," *The Philadelphia Inquirer*, November 22, 1999, p. B1.

80. I. Kennedy et al., "The Case for 'Presumed Consent' in Organ Donation," *Lancet* 351, May 30, 1998, pp. 1650–52; M.F.X. Gnant, "The Impact of the Presumed Consent Law, Quadruplication in the Number of Organ Donors," *Transplantation Proceedings*, vol. 23, no. 5, October 1991, pp. 2685–6.

14

Reproductive Technology:
Welcome to the Brave New World

ZOË C. MELEO-ERWIN

Women Helping Women. Egg Donors Needed for Infertile Women. If you are under 34 and healthy, you could have the satisfaction of helping someone in a very special way.

Center for Reproductive Medicine, Colorado[1]

The whole business seems to treat human reproduction as something analogous to a car. Do we want it to be bright or blue? How do we make it sleeker? It turns reproduction into a technological procedure, more and more divorced from human intimacy and relationships.

Jean Bethke Elshtain[2]

When Aldous Huxley's novel *Brave New World* hit bookshops in 1932, the reading public was introduced to a future in which children were no longer the result of sexual intercourse. Through laboratory procedures, embryos were produced from *in vitro* fertilization, grown in vats and finally "decanted." Their creation and growth took place entirely outside women's bodies and was controlled largely by men through science and technology. At each stage of the embryos' development, various substances were added to the artificial wombs to help them grow; each "class" of embryos received substances designed to bring about the characteristics particular to their societal rank. Laboratory technicians would then clone a large number of children out of the cells of one fetus. Therefore, from one designed embryo, ten, twenty, even one hundred identical people all equally suited to the labour of their rank could be grown and decanted.

While many considered *Brave New World* to be a fantastical and frightening science-fiction story, others perceived Huxley's warning about the direction in which society was heading. Twenty years later when Huxley rewrote the introduction to the book he stated, "Then, I projected it six hundred years into the future. Today it seems quite possible that the horror may be upon us within a single century."[3] Huxley was correct in asserting that developments in modern science and technology were leading us to the "Brave New World" far faster than he originally anticipated. However, our arrival at the border of this dystopian future occurred only sixty years after Huxley wrote his original forewarning and forty years after he rewrote the book's introduction. Through the science and techniques of what is called "reproductive technology" the Brave New World is upon us.

HELPING THE INFERTILE

In 1978 two British scientists, Robert Edwards and Patrick Steptoe, helped bring about the birth of Louise Brown, the world's first successful test-tube baby. The Browns were unable to conceive a child in the regular manner, and thus were "assisted" in their reproduction by Edwards and Steptoe, who took sperm and egg from them, fertilized the egg in the laboratory through what is called *in vitro* (literally, "in glass") fertilization (IVF), and implanted the embryo in Lesley Brown's womb. Today, reproductive technologies and procedures, including pre-implantation genetic testing, egg donation and embryo adoption, are used, supposedly to help the infertile have children, prevent and cure disease, and even "improve upon" the human genome.

Increasing numbers of infertile couples are drawn to fertility clinics with the hope of being able to have a child of their own. When a couple has tried and "failed" to conceive a child, they often undergo fertility testing. For a man, this procedure is quite simple, involving blood and semen tests as well as a complete physical examination. Testing to determine whether or not a woman is infertile is much more invasive and painful. There is the Sims–Huhner test to determine if sperm can survive in a woman's vagina after intercourse, the Rubens or tubal insufflation test to determine if her Fallopian tubes are blocked, a hysterosalpinography x-ray or laparoscopy procedure to determine if one or both Fallopian tubes are blocked, and a hysteroscopy to locate possible polyps, fibroids, adhesions, cysts and other deformities.[4]

These fertility-test procedures range from embarrassing and demeaning to painful and invasive, and are very stress-provoking for both partners involved.[5] However, the aforementioned procedures are only the beginning for a woman who is undergoing fertility testing. If it is found that she has blocked Fallopian tubes, cysts, polyps or other deformities, there is still the need for surgery to correct the matter. Then, she can begin treatment with reproductive technologies to promote pregnancy. While some reproductive technologies artificially introduce sperm into a woman's uterus to fertilize the egg, most involve *in vitro* fertilization. Sperm and ova are mixed in a Petri dish, fertilized *in vitro*, and the resulting embryo(s) is (are) placed into the woman's uterus. Unlike sperm, however, which is easily obtained, the eggs must be surgically removed from the woman's body. She is placed on drugs to stimulate her hormones for increased fertility.[6] The fertility drugs induce what is called "superovulation," which means that during the woman's menstrual cycle her ovary produces an average of thirteen but as many as forty ova, rather than the normal one ovum.[7] As ovulation approaches, she undergoes a number of tests, including an ultrasound test, and once ovulation occurs she is admitted to the hospital for egg removal.[8]

After the eggs are removed by tearing them from the ovarian walls in procedures like TUDOR (Trans Vaginal Ultrasound Directed Oocyte Recovery), they are finally mixed with sperm. After further hormone therapy, the resulting embryos, sometimes three or four at a time, are placed in the woman's uterus through a tube which is inserted into her vagina. Other embryos may be frozen for use in a later procedure if the first are unsuccessful. More often than not, none of the embryos implants, and the woman must undergo this three-week process several times over.[9]

Each year, increasing numbers of couples around the world enter fertility clinics. No matter which partner of a heterosexual couple is infertile, it is the woman who must undergo the invasive reproductive technology treatments. In many cases, the woman is perfectly fertile, but submits to various procedures because her male partner is unable to impregnate her through natural sexual intercourse. Couples enter fertility clinics because of the strong desire for a child; however, the vast majority leave fertility clinics childless. Andrew Kimbrell, a critic of reproductive technology, states that only 10 to 14 per cent of couples in the United States that enrol in fertility clinics actually succeed in a live birth.[10]

Many couples are lured to fertility clinics by misrepresented success rates. Clinics boast the success of their pregnancy rates, not the number of live births. Often, advertised pregnancy rates are altered to include those who undergo hormonal changes but do not necessarily become pregnant or give birth; these women are considered to have experienced an early pregnancy that resulted in miscarriage.[11] The media perpetuate the exaggerated statistics by focusing only on the successes of reproductive technology, describing them as "miraculous" and portraying doctors and scientists in a godlike manner. Couples who might have otherwise adopted, or lived childless and fulfilling lives, become seduced by the promise of bringing home a brand new baby of their own. The vast majority do not.

DONORS AND SURROGATES

Many couples undergoing fertility testing decide that they prefer not to use their own egg and/or sperm to create an embryo. This occurs for many reasons. One partner may have a genetic disease that the couple do not want passed on to their offspring. In other cases, women who are considered "too old" to conceive naturally will seek out the eggs of younger women. Still others look to donors for entirely superficial reasons. Many fertility clinics, such as the Bionetics Foundation, Inc. and the Repository for Germinal Choice, offer "elite" ova and sperm which have supposedly come from the best, brightest and most beautiful in society. These fertility clinics not only sell biological materials and reproductive technology, but market the notion that characteristics such as beauty, intelligence, athletic ability and economic success – all culturally defined categories, to be sure – are genetic in origin and able to be passed on to future generations. In each case, a couple "shops" for their egg and sperm, carefully reviewing the photographs and profiles of donors and selecting the most appealing choice.

Procedures which involve donor materials for *in vitro* fertilization include the use of sperm donation, egg donation and embryo adoption, in which an embryo is created through IVF from both paid donor egg and sperm and is implanted into the womb of the female client. The procedures by which eggs from donors are acquired are the same as those described in the last section. The difference, however, is that women who donate their eggs and undergo humiliating, uncomfortable, often painful and potentially dangerous procedures are not doing so out of the hope that they might become pregnant and bear a child. Women who donate their eggs almost always undergo the procedure to make money.[12]

Such women have little idea what they will experience for the money they will earn.[13] Women who "donate" their eggs must mix and administer hormone treatment by injection into their buttocks daily, are required to abstain from sexual intercourse for long periods of time, and risk bleeding, infections, swollen ovaries, cysts and possible ovarian and breast cancer from the super-ovulation hormonal drugs. These risks increase with each egg donation procedure a woman undergoes.[14]

Some reproductive technology advocates have claimed that egg donation actually benefits women and the feminist cause by virtue of the fact that women are sole owners of a "precious resource."[15] However, a critical view of this practice shows that for the physical risks involved, the possibility of prolonged injury (for which egg donors are legally not eligible to be compensated), and the potential for psychological damage, the financial compensation to women is clearly not adequate. In addition, their financial gain is not comparable to that of the reproductive technology industry, which has in fact appropriated control of this "precious resource." Although women who donate their eggs are sometimes described as altruistic "gift givers," their situation can be seen as a continuation of the patriarchal tradition of using women's bodies as a medium of exchange.[16] This, along with the serious health complications involved, has led many nations, including Germany and Japan, to forbid egg donation.[17]

The rise of the reproductive technology industry has brought a revived interest in the practice of surrogacy for infertile couples. Traditionally a surrogate mother was inseminated with the male client's sperm so that the child would share some genetic relationship with the couple. Now it is possible to use both donated egg and sperm (screened for particular characteristics) or an embryo from the couple seeking a child. The ova and sperm, whether from donors or the clients, are retrieved through the procedures described previously, and the resulting embryos are implanted in a surrogate mother. While this surrogate mother will certainly have a biological and emotional relationship with the developing fetus, she will not share any genetic relationship with it.

The surrogate mother is screened to ensure that she is responsible and will provide a good "environment" for the child which she has been paid to carry and deliver. To ensure the "quality" of the child, she must undergo prenatal testing, including ultrasound and amniocentesis. The contractual agreements mandate that the surrogate provide a favourable environment for the developing fetus, and any action considered dangerous to the developing fetus is grounds for breaking the contract. This means that the surrogate will not be paid, may be subject to legal actions from the couple which hired her (and/or the fertility clinic), and finally will be faced with – even required to have – an abortion. If it is too late in the pregnancy, she may be faced with birthing an unwanted child that is not hers.

While a substantial majority of those who hire surrogate mothers have high incomes and advanced education, the majority of surrogate mothers earn just above the poverty line.[18] Although the surrogate is paid, she is only paid if she "delivers" an "acceptable product." The little she is paid does not compensate her for twenty-four hours a day for nine months and exposure to numerous drugs, tests and treatments. Furthermore, surrogacy reduces women to their biology and defines their bodies as vessels to be used for economic purposes. The use of women in general, and particularly poor and working-class women, to carry and bear the children of wealthy couples has been termed "bioslavery."[19]

Surrogacy brokers are able to offer their affluent clients the option of paying less for a surrogate if they choose to hire a woman from a developing country. John Stehura, president of the Bionetics Foundation (the same Bionetics Foundation that offers "elite" donor sperm and ova) has remarked that poor women from developing countries will jump at the chance to earn half of what their counterparts in industrialized nations make for the same service.[20] The racism and classism inherent in these arrangements is obvious.

Because of the dangerous and demeaning nature of surrogacy, particularly when it takes advantage of women in developing countries, most Western nations have banned the practice.[21] In the United States this practice is entirely legal. As Kimbrell explains:

> The United States is a country deeply committed to the ideology of the market. Many US legal scholars and economists defend surrogacy on the grounds that if a market system is to survive, contracts must be sacrosanct, more so, apparently than motherhood. They also argue that an open market in babies would enhance public good by more equally distributing babies from those who have them (often the poor) to those who do not and can afford to buy them.[22]

Not only is surrogacy entirely legal in the United States, but it is a completely unregulated business. The surrogacy businesses and the reproductive technology industry in general have turned reproduction into a profitable commercial endeavour. Such businesses take advantage of the social and

economic system and promote the use of a "breeder" class of women who
rent their wombs and sell their eggs in order to survive.

HUMAN GUINEA PIGS, LIVING LABORATORIES
AND BIOLOGICAL FACTORIES

The patriarchal nature of this science and technology was inherent in the
reproductive technology industry right from its inception. Lesley Brown, the
mother of the first test-tube baby, had no idea that she was the first woman
to bear a child through *in vitro* fertilization. She had in fact been led to
believe that there were many other women who had undergone the same
tests and procedures.[23] Lesley Brown would not have died from being infertile; however, she might have died from serving as an unknowing human
guinea pig.

Abuses of this sort are even more pronounced in Southern and developing
countries. By the time reproductive technologies become available in the
industrialized nations, they have already been tested on women from developing countries. Drugs, contraception, injections and reproductive technology procedures are often tested on such women with little regard for the
unintended effects. The experimental nature of these products and procedures
has no doubt caused numerous women pain, suffering and even death.

The same "pharmacrats," as Gena Corea refers to them,[24] are testing new,
experimental forms of contraception on women of colour from Southern and
developing countries – and directly or indirectly promoting their sterilization
– even while they are actively engaged in promoting IVF and other fertility
programmes for affluent women in the industrialized world. Thus, while
affluent, usually white, women are being offered new ways to have children,
those in the developing countries are being subjected to new ways to ensure
that they don't have them; these technologies are clearly being used to promote
the births of some babies and not others. Many see this phenomenon as a
modern-day form of racist population control. For example, in some areas of
the developing world, particularly in Africa, over 40 per cent of women have
never borne a child by the age of 45. Such women are often infertile due to
the high incidence of venereal disease in their regions.[25] Those involved in
promoting IVF and other reproductive technologies for women in industrialized nations are not addressing the need to offer infertile women in developing countries the same "choice." Nor do they offer treatment to cure
sexually transmitted diseases and make condoms available. Rather, they speak
of the "population problem."

CONTROLLING REPRODUCTION

Jocelynne A. Scutt states that "[d]evelopments in human medicine often follow
developments in the plant and animal world."[26] She feels that this phenomenon is nowhere more evident than in reproductive technology. Reproductive

technology techniques which are now used to control women's biology were first developed and tested on livestock. Experiments on animals were most definitely motivated by an economic incentive, as they allowed for a great increase in the reproduction of "high stock" animals. As (mostly male) scientists gain "competence" with such experiments done on animals, and can do so in a completely objectified manner, they gain the experience necessary, and set the ideological and scientific precedent to conduct these experiments on women.

No one would question that animal reproductive technology only benefits those who stand to profit from it economically. Human reproductive technologies, on the other hand, are rationalized as being for the benefit of women. They are described as offering women greater "choice." However, as we have seen, these technologies do not benefit women.[27] Rather, they are dangerous experiments which are performed on women either forcibly or by persuasion. Jocelynne Scutt states:

> To assume that freedom from nurturing during gestation and from giving birth, through the intervention of technology, should grant women freedom and equality also assumes that technology is not patriarchal – or at least that women will control (or be in equal control of) biological technology. Both assumptions are incorrect. Genetic and reproductive engineering exist in a world where women and men are not equal; genetic and reproductive engineering arise out of and are incorporated into that reality.[28]

It is men, not women, who largely develop, administer, control and profit from the businesses of genetic and reproductive technologies. While there certainly are women involved in these processes, their numbers and power are far from equal to that of men. Reproductive technologies have been researched, developed and applied (and in some cases enforced) in a world where the social relations between men and women are still often based on domination, oppression, repression and violence. Feminist philosopher Maria Mies put this succinctly when she stated, "It is a historical fact that technological innovations within exploitative and unequal relationships lead to an intensification, not attenuation, of inequality, and to further exploitation of the groups concerned."[29]

Beside promoting uncertain and often harmful reproductive technologies, the pharmaceutical industry consistently overlooks real explanations as to why infertility exists and may be increasing. Infertility is often connected with occupational and environmental hazards, stress, smoking, alcohol, drugs, sexually transmitted diseases, abortion, contraceptive methods, and abdominal surgery. If we are truly concerned about infertility, we should pursue investigations of how environmental and social conditions promote it, and take steps to stop that which harms us.

Similarly, reproductive and medical biotechnologies ignore the real causes of human suffering, namely poverty, lack of access to medical care, lack of nutritious food, stress, oppression, which have very simple remedies. Instead they promise to cure these phenomena through engineering the human genome. Genetic and reproductive technologies sidestep positive social change and perpetuate negative social change by offering us another "techno-fix." Techno-fixes sustain the growing vicious cycle of technological dependence: technology is created which has unanticipated negative effects, so more technology is created to fix the damage. The next experimental technology has negative effects as well, and the process repeats itself endlessly. With the advent of reproductive and genetic engineering, which are radically new forms of technology, the consequences of perpetuating the cycle of technological dependence will likely be even more serious.

Reproductive and genetic technologies do, however, offer the promise of great money-making potential, power and control for their corporate owners. In understanding the gene revolution, Maria Mies states, we must understand that these new technologies "have been developed and produced on a mass scale, not to promote human happiness, but to overcome the difficulties faced by the present world system in continuing its model of sustained growth, of a lifestyle based on material goods and the accumulation of capital."[30] Corporations look to the generativity of nature and women as fodder for continued profit and power. We are being vivisected, recombined and subjected to total quality control, says Mies, and are being used to further the capitalist and patriarchal system by being treated as no more than part of the industrial process.[31]

Reproductive and genetic technologies are not designed to help the infertile, cure the sick, improve upon evolution or offer greater choice. This industry seeks power over nature, women, children and human evolution in the pursuit of increasing profits. It perpetuates violence against women by further taking away women's control over their bodies, their self-worth, and their freedom. It is important to see genetic and reproductive technologies as the culmination and ultimate expression of an ideology of conquest and control over women and nature. If we do not recognize this and take action against it, dystopian scenarios of "decanting" children will no longer be fictional. They will be reality.

NOTES

1. Center for Reproductive Medicine advertisement in *Westworld* (Denver, CO), February 29–March 6, 1996, p. 105.

2. Cited in Andrew Kimbrell, *The Human Body Shop: The Engineering and Marketing of Life*, San Francisco: HarperSanFrancisco, 1994, p. 118.

3. Aldous Huxley, *Brave New World*, New York: Perennial Library, 1978, p. xiv.

4. For all too graphic descriptions of these procedures, see Ramona Koval and Jocelynne A. Scutt, "Genetic and Reproductive Engineering: All for the Infertile," in Jocelynne A. Scutt, ed., *The Baby Machine: Reproductive Technology and the Commercialisation of Motherhood*, London: Green Print, 1990; or Gena Corea, *The Mother Machine: Reproductive Technologies from Artificial Insemination to Artificial Wombs*, London: The Women's Press, 1988.

5. In this respect, the tests may lead to greater infertility by being stress-provoking,

as stress has been linked to the inability to conceive.

6. At this time, nothing is known conclusively about the long-term effects of these drugs. There many be an increased risk of ovarian cysts, breast cancer, irregular menstrual cycles, or perhaps cancer from using them.

7. Jesse McKinley, "The Egg Woman," *New York Times*, Sunday, May 17, 1998, Section 14, p. 12.

8. Marion Brown, Kay Fielden and Jocelynne A. Scutt, "New Frontiers or Old Recycled: New Reproductive Techology as Primary Industry," in Scutt, ed., *The Baby Machine*, p. 91.

9. Ibid., p. 92.

10. Kimbrell, *The Human Body Shop*.

11. Jocelynne A. Scutt, "Introduction," in Scutt, ed., *The Baby Machine*, p. 3.

12. According to the guidelines of the American Fertility Society, women are not supposed to be paid for the actual eggs, but rather for the "expenses, time, risk, and inconvenience associated with the donation." However, as Andrew Kimbrell points out (*The Human Body Shop*, p. 84), there is really little difference between being paid for the eggs and the manner by which they are retrieved.

13. Egg donors typically receive between $3,000 and $5,000 of the $15,000 to $20,000 that egg "adoption" costs. See McKinley, "The Egg Woman," p. 12.

14. There are no federal or state laws regulating the number of times a woman may donate her eggs. The American Society of Reproductive Medicine has suggested no more than ten. Cited in McKinley, "The Egg Woman," p. 12.

15. See Gina Kolata, "In the Game of Cloning, Women Hold All the Cards," *New York Times*, February 22, 1998, p. 6.

16. For an excellent discussion on this subject, see Janice Raymond, *Women as Wombs: Reproductive Technology and the Battle over Women's Freedom*, San Francisco: Harper SanFrancisco, 1993.

17. McKinley, "The Egg Woman," p. 12.

18. The disparity in class and education level between surrogates and those who hire them can be interpreted a number of ways, but it is most likely that as a larger number of women enter the professional workforce they will chose to have children later in life. Because of this decision, many may run into fertility problems and seek surrogates as a solution. Other professional women may desire motherhood but not pregnancy, and therefore hire surrogates literally to labour for them. Statistics are cited in Kimbrell, *The Human Body Shop*, p. 108.

19. Ibid., p. 100.

20. Raymond, *Women as Wombs*, p. 144.

21. Kimbrell, *The Human Body Shop*, p. 112. It is important to note that because of this ban, the surrogacy industry in the United States and elsewhere has been able to tap into the resulting international demand. Raymond reports that surrogate brokers, involved in what she deems "global trafficking in women," advertise for buyers who cannot legally obtain a surrogate in their own countries. *Women as Wombs*, p. 143.

22. Kimbrell, *The Human Body Shop*, pp. 112–13.

23. Scutt, ed., *The Baby Machine*, p. 9.

24. Corea uses the word "pharmacrat" to denote those involved in, controlling and profiting from the professional reproductive and genetic engineering industries. She takes the word from Dr Thomas Szasz, Professor of Psychiatry at the State University of New York. Dr Szasz coined the word "pharmacracy," meaning the political rule by physicians which results when medicine becomes not a method of healing but a means of social control when allied with the state. See Gena Corea, "Women, Class and Genetic Engineering: The Effect of New Reproductive Technologies on All Women," in Scutt, ed., *The Baby Machine*, pp. 136, 148.

25. Ibid., p. 144.

26. Scutt, ed., *The Baby Machine*, p. 5.

27. In 1971, feminist author Shulamith Firestone stated that reproductive technology would liberate women from patriarchal oppression. She believed that only when women were liberated from pregnancy and childbirth would they be free. In response to her critique, many feminists have argued, as Scutt does, that such analyses assume that reproductive technologies are in and of themselves neutral technologies. Instead, such technologies can be seen as designed to gain greater control over reproduction, not to empower women.

28. Cited in Scutt, ed., *The Baby Machine*, p. 2.

29. Vandana Shiva and Maria Mies, *Ecofeminism*, London: Zed Books, 1993, p. 175.

30. Ibid., pp. 174–5.

31. Ibid., p. 186.

15

Is Violence in Your Genes?
The Violence Initiative Project:
Coming Soon to an Inner City Near You

MITCHEL COHEN

Unemployment runs in the genes just like bad teeth.
Richard Herrnstein, Princeton University, co-author of *The Bell Curve*[1]

In the 1980s, a small but influential group of prize-winning scientists were trying to prove that black children were, on average, less intelligent than white children. Intelligence, they said, ran in the genes of racial groups. Their evidence? They reviewed a series of infamous studies of twins who had been separated at birth and raised far from each other, studies designed to prove that behaviour is genetically determined. They also compiled results from IQ tests across the country and subjected them to statistical analysis, which, they claimed, showed a differential between racial groups. This alleged differential, they went on, was due to genetic factors.[2] These scientists – all of whom were white – concluded that black people were genetically inferior to whites when it came to intelligence.

And yet these scientists made a number of critical errors trying to squeeze their round "facts" into square holes. First, the variation of IQ scores *within* each racial group far outstripped the differential *between* racial groups. Second, what exactly did these tests measure? Were they culturally or racially biased? A large body of work during the 1970s and early 1980s had discredited the notion that standardized IQ tests measured anything "objective" at all.[3] Lastly, Sir Cyril Burt's "twin studies," on which many racially charged IQ assertions had been based, turned out to have been fabricated, and the benighted scientist's reputation now lies in tatters, his name disgraced.

Although such racially based intelligence theories never fully disappeared, they were roundly discredited scientifically.[4] In fact, as Professor Gerald Horne of the University of California at Santa Barbara writes, "[R]esearch is never 'neutral.' Who asks the questions, what questions are asked and what ones ignored, who pays for the research, who interprets the results are all subjective decisions outside the realm of 'pure science.' The bias is built in."[5] It was not until the publication of *The Bell Curve* in the early 1990s, by Herrnstein and Murray, that some of the same theorists re-emerged and, with new jargon, proceeded to offer the same white-supremacist views in the guise of new "scientific research."[6]

Today, some scientists are again proposing biologically determinist explanations for behaviour, but instead of focusing on a genetic basis for "intelligence" they now substitute the more *au courant* "violence" – a hereditary characteristic of black and Latino people, they say. The recent biology-and-crime movement was kicked off by the publication in 1985 of *Crime and Human Nature* by James Q. Wilson and Richard Herrnstein.[7] A major media campaign followed, leading in 1992 to a report by the National Academy of Sciences and the National Research Council. Titled *Understanding and Preventing Violence*, the report called for more attention to "biological and genetic factors" in violent crime, "new pharmaceuticals that reduce violent behavior," and studies of "whether male or black persons have a higher potential for violence."[8] Scientists now seek to control the alleged "genetic predisposition" of black children to commit criminal acts of violence, by medicating them before aggressive behaviour and violence ensue.

And so, under the aegis of the federally funded Violence Initiative Project, Gail Wasserman, a professor of child psychiatry at Columbia University, and Daniel Pine, a medical doctor associated with the same institution, have picked up where the discredited racially based intelligence theories of Jensen, Herrnstein, Eysenck, Shockley and Murray left off.[9] They led a team of researchers in performing numerous experiments, partly funded by federal tax dollars, on black and Latino children as young as 6 years of age.[10]

Dr Wasserman, in her funding proposal to establish a "behavioral disorders" centre at Columbia University's Department of Child Psychiatry, wrote that it is proper to focus on blacks and other minorities as they are overrepresented in the courts and not well studied.[11] In one such "study" at the New York State Psychiatric Institute, Wasserman and her cohorts took thirty-four healthy boys, aged 6 to 10, and administered the dangerous drug fenfluramine.[12] Fenfluramine is the primary ingredient in the diet drug "fen phen," which was banned by the US government.[13] The boys were all from impoverished families; 44 per cent were African-American and 56 per cent were Hispanic. They received intravenous doses of the drug fenfluramine hydroxide over a five-and-a-half-hour period, and blood was drawn hourly.

The clinicians hypothesized that they could counter the alleged racially inherited genetic predisposition to aggressive behaviour and violence by increasing levels of the neurotransmitter serotonin in the brain. Some studies have correlated low serotonin levels with aggressive behaviour. Despite the lack of clear evidence for a cause-and-effect relationship, the researchers hypothesized that genetically determined levels of serotonin are further reduced by socially adverse child-rearing practices in black and Latino families. Through medication, they say, they could increase serotonin levels, counter the negative effects of "adverse child rearing," and thus prevent the youths from committing acts of violence.[14] This despite the fact that most of the children had not committed any acts of violence at all.

Ninety per cent of adult subjects experience side effects from a single dose of fenfluramine, and here Wasserman and Pine were administering it to children. Effects of a single dose of fenfluramine on adults "frequently include anxiety, fatigue, headache, lightheadedness, difficulty concentrating, visual impairment, diarrhea, nausea, a feeling of being 'high' and irritability."[15] US Food and Drug Administration studies show that the drug causes severe heart valve damage in adults,[16] as well as a fatal heart condition known as pulmonary hypertension,[17] and, in animal studies, microscopic damage to brain cells lasting up to eighteen months. Yet the New York State Psychiatric Institute proceeded with administering fenfluramine to children, in doses eight times higher than that which caused damage in monkeys' brains, even after the drug had been banned in September 1997.

The children were selected because they (1) were black or Latino, and (2) each had older siblings who had been delinquents known to the Family Court. The children's names and addresses were – and continue to be – sorted and channelled by government officials on the public payroll at the Department of Probation and the New York City Board of Education, and passed along to the researchers. When local newspapers exposed this insidious involvement of public officials, a Board of Education spokesperson denied that students had been referred for the purpose of participating in research,[18] but the documents prove the Board was lying. In fact, reported *New York Newsday*, the Board's Committee on Special Education "worked closely with the researchers from the beginning."[19] Indeed, the original proposal submitted to the National Institute of Mental Health referred to the special education committee as "one particularly productive referral source," and noted that researchers had made "successful liaisons with a number of schools and agencies throughout the New York metropolitan area." A Department of Probation memo confirmed that agency's active participation in Dr Wasserman's "effort to identify early predictors of anti-social behavior."[20]

Similar experiments have been going on at Queens College and the Mount Sinai School of Medicine in New York, and at facilities throughout the United States, under the rubric of the National Violence Initiative Project, supervised and funded through the National Institute of Mental Health.[21] By claiming genetic predisposition, psychiatrists are able to tap into the hundreds of millions of dollars available for genetic research. That's the new claim for why people do anything: commit violent crimes, engage in homosexual or heterosexual behaviour, come down with certain cancers, even commit murder – it's in the genes. The money, in turn, has fuelled all sorts of similar projects, including new methods for genetically frisking prisoners for "bad DNA." Corrections officials are collecting samples of prisoners' DNA, computerizing the sequences and storing them for future use.

Far less money is spent on researching the effects of toxic byproducts of industrial production and chemical agriculture on the incidence of cancer and various immune-compromising diseases. Instead, funding – and thus the blame – is shifted to an individual's or group's biological makeup, their "genetic predisposition" towards cancer and other conditions.

Today, psychosurgery, lobotomies and electroshock "therapies" are making a comeback, as well as the medicalization of what are fundamentally socially or environmentally caused ailments and behaviours. New and dangerous genetically engineered drugs are being tested on prisoners, soldiers and mental patients institutionalized in asylums or warehoused in impoverished inner city slums, Indian reservations, and Third World countries.[22] To the giant pharmaceutical corporations, as well as to the government, these people are available as human guinea pigs. The Violence Initiative Project is the new face of what can best be described as Nazi "science" with an American accent.

CHALLENGING GENETIC REDUCTIONISM

Holistic thinking is, to say the least, not one of Western science's strong points. In the US there is a tendency to think in terms of cause and effect, every effect determined by its singular cause, every trait an expression of its singular gene. But scientists can rarely predict the full effects of altering a single gene on an individual organism, let alone the effects on larger ecosystems in which plants, animals and microorganisms evolve in symbiotic relation to each other. Experiments in gene therapy have caused the deaths of relatively healthy patients, and the Food and Drug Administration ordered a halt to human gene therapy experiments at the University of Pennsylvania – and a review of such experiments throughout the United States – after an inspection uncovered "numerous serious deficiencies" in patient safety during a clinical trial that cost an 18-year-old Arizona man his life.[23]

Critics of genetic reductionism, such as cell biologist Stuart Newman, have described multi-tiered and interactive mechanisms of development, cell morphogenesis and pattern formation, which rely on such non-reductionist factors as the position of a cell with respect to other cells.[24] The whole shapes the parts as much as the parts generate the whole. According to Newman, the interactions between living cells and their environment give rise to a vast number of possible developmental pathways. Genes can be seen as repositories of development that has already happened, rather than factors determining what is going to happen. Summing up one of his many investigations, Newman writes:

> Both cells and ecosystems can thus be analyzed as highly complex networks of large numbers of components undergoing mutually dependent changes in their relative abundances. But while this way of thinking is common among ecologists, it is not well suited to making precise predictions, and has failed to take hold to any significant extent in cell biology. Instead, the most common intellectual framework of cell and molecular biologists is a reductionist approach … [Understanding of the whole] is sacrificed in favor of exact knowledge of a more limited set of phenomena.[25]

The scientists working on the Violence Initiative Project are among those trapped in this reductionist framework. They start where the book *The Bell Curve* leaves off – the search for the gene or biological configuration that "causes" criminal behaviour and the assumption that low intelligence, poverty and criminal behaviour are the result of "deficient" genes.[26] The Violence Initiative is a well-funded attempt to assert a genetic predisposition to committing violent crime, and to paint this predisposition with a racial brush. Young white males are seen as "less violent," and thus not genetically predisposed to aggressive behaviour. Why? Because criminal records show that a much greater percentage of black people and Latinos are involved in the criminal justice system. The Violence Initiative Project seeks to provide scientific cover for such circular reasoning, and for the police and government repression that has led to the criminalization of more than one-third of all black males in the United States.

The race-based "biological theory of aggression" is neither new nor scientific. One champion of the Violence Initiative, Dr Frederick Goodwin, defended the "theory" before the National Health Advisory Council in February 1992:

> If you look, for example, at male monkeys, especially in the wild, roughly half of them survive to adulthood. The other half die by violence ... and, in fact, there are some interesting evolutionary implications of that because the same hyperaggressive monkeys who kill each other are also hypersexual, so they copulate more and therefore they reproduce more to offset the fact that half of them are dying.... [M]aybe it isn't just the careless use of the word when people call certain areas of certain cities jungles, that we may have gone back to what might be more natural, without all of the social controls that we have imposed upon ourselves as a civilization over thousands of years in our own evolution.[27]

Goodwin follows a long line of proponents of racial supremacy who have traded the Ku Klux Klan's white sheets for white lab coats. They argue that social problems are caused by biologically defective members of oppressed classes; society can be improved by identifying and eliminating the propagation of these "defectives." In the 1850s, Louisiana physician Samuel Cartwright described a mental disease of slaves called "drapetomania," which caused its victims to run away from their masters. A century later, American physicians Vernon Mark, Frank Ervin and William Sweet proposed that urban rebellions were caused by brain-damaged individuals who could be cured by psychosurgery (lobotomy). They received almost $1 million in federal funding.[28]

In the 1970s, O.J. Andy, director of Neurosurgery at the University of Mississippi, published reports on invasive surgeries he had performed on children who were said to be developmentally disabled; all were black. Dr Peter Breggin describes one of Andy's subjects, a 9-year-old boy said to be "hyperactive, aggressive, combative, explosive, destructive and sadistic."[29] Over a three-year period, Andy operated on the child on four different occasions and implanted electrodes in his brain. Andy concluded, in a 1970 article, that this "patient" was no longer combative or aggressive. In actuality, Andy had

mashed the child's brain, suppressing intellect and emotion, and disabled the
child by turning him into a vegetable. Dr Andy claimed, according to Breggin,
that "the kind of brain damage that could necessitate such radical surgery
might be manifested by participation in the Watts Uprising in 1965. Such
people, he diagnosed, 'could have abnormal pathological brains.'"[30]

FIGHTING BACK

The Coalition Against the Violence Initiative (CAVI) is leading the campaign
against such misuses of science. CAVI claims that aggressive, violent or criminal
behaviour is no more determined by genes than is the desire to study the
"inheritance" of violence or the "predisposition" to become a corporate lawyer
(which often runs in the family). One could argue that police, generals,
football players and many others have inherited a gene that predisposes them
to committing acts of violence, not to mention corporate executives and
politicians who murder with their pens. Yet, no one is studying their "genetic
predisposition" to aggressive behaviour. Capitalism itself is inherently violent:
the removal of the products of labour from the ownership and control of
those who produce it – which is the underlying basis for this system –
requires an enormous level of violence, and the system selects for those personalities capable of ministering to capital's needs. The society we live in validates
that violence, but it becomes part of the social substrate and most people
don't see it as abnormal or, for that matter, as violent. People shape social
conditions, and these in turn strongly influence their activities and behaviour.

The reductionist and biodeterminist approach exemplified by the Violence
Initiative Project is rampant in the scientific establishment's approach to social
ills. In November 1998, researchers distributed a memo to staff at George
Washington High School in Upper Manhattan announcing a survey to be
done on freshmen "at risk" for "negative behaviors." By now, we should all
have an idea of what "at risk" means. Youngsters so designated are to be sent
to the clinic run by Columbia Presbyterian and the Columbia School of
Public Health, for "assessment." CAVI sent a strongly worded letter to the
principal outlining its concerns and calling for the cancellation of the survey.
Members passed out leaflets to students and parents alerting them to the
dangers and advising them not to sign consent forms. Members of Lawyers
for the Public Interest also called the school, as did a number of individual
teachers whom the Coalition had contacted; eventually the principal cancelled
the survey.

Yet, despite occasional victories for those resisting genetic manipulation,
the Violence Initiative Project, along with the biotech industry, is charging
ahead full speed, dismissing all who dare to question both the Project's and
the industry's apparent willingness to sacrifice our lives and environment in
their rush for profits and social control of oppositional forces. In a brilliant
article, scathing about the project, Gerald Horne wrote:

Under the Initiative, researchers will use alleged genetic and biochemical markers to identify potentially violent minority children as young as five for biological and behavioral interventions – including drug therapy and possibly psychosurgery – purportedly aimed at preventing later adult violence.

The Initiative specifically rejects any examination of social, economic, or political questions, such as racism, poverty, or unemployment. Instead, this biomedical approach focuses heavily on the alleged role of the brain neurotransmitter, serotonin, in violence. Not coincidentally, this approach is favoured by many in the medical industry.[31]

Dr Peter Breggin, a leading analyst in the field, has observed that this approach "corresponds with the current financial interests of the pharmaceutical industry, since several drugs affecting serotonin neurotransmission have been submitted for approval to the Food and Drug Administration.... The controversial antidepressant, Prozac, is the first of these serotonergic drugs, and it has become the largest moneymaker in the pharmaceutical industry."[32]

Against this backdrop, NIH provided a hefty $100,000 grant for a conference entitled "Genetic Factors in Crime: Findings, Uses and Implications," sponsored by the Institute for Philosophy and Public Policy at the University of Maryland. The promotional brochure promised that "genetic research holds out the prospect of identifying individuals who may be predisposed to certain kinds of criminal conduct, of isolating environmental features which trigger those predispositions, and of treating some predispositions with drugs and unintrusive therapies."[33] Genetic research also gains impetus from the shortcomings of liberal "environmental" approaches to crime: deterrence, diversion and rehabilitation.[34] The "liberal paradigm" exchanges a sort of environmental or social/cultural determinism for the prevailing genetic determinist model. A more holistic, radical paradigm challenges the idea of "determinism" altogether, leading towards much more interactive, non-linear approaches, as developed by social movements in conjunction with socially engaged scientists. The failure of the liberal paradigm, and the dismissal of alternative models, have now allowed the focus to shift to exclusively genetic and medical "solutions."

Radicals, however, *have* succeeded in leaving a mark. In the 1970s we managed to beat back attempts by William Shockley and others to lay a pseudo-scientific basis for the racial inheritance of intelligence, and the disastrous policy implications being pursued at the time, by refuting each and every "scientific" assertion they made, as well as by exposing their funding sources. So, too, with the Violence Initiative Project, at least initially. Gerald Horne writes:

> The ensuing protest caused NIH to freeze conference funding – temporarily. The objections were led by enraged African Americans concerned that, in these dangerous times, such a project could easily be transformed into directed genocide. Their concern was not assuaged when it was revealed that Reagan appointee Marianne Mele Hall proclaimed that black and brown people are culturally or even genetically inferior. They have been conditioned, she said, "by 10,000 years of selective breeding for personal combat and the anti-work ethic of jungle freedoms," and were

therefore unfit for civic life. Great Society programs just "spoiled" them, she argued, "encouraging a sense of entitlements that led to laziness, drug use, and crime, particularly crime against whites."[35]

This brings us back to Goodwin. When we last left him, he was being chastised for making similar reference to jungles, comparing black people with monkeys. "By associating African Americans with monkeys and 'hypersexuality'," Horne writes, "Goodwin tapped into a wellspring of racist sentiment."[36] Dr Louis Sullivan, President George Bush's Health and Human Services Secretary, joined many others in criticizing Goodwin's remarks. But Goodwin's disfavour lasted barely a week. Sullivan soon rewarded Goodwin by appointing him head of the influential National Institute of Mental Health, a post not requiring Senate approval. His first project as head of this Institute was to approve the initial funding for the National Violence Initiative.[37]

THE RESULTS ARE IN

The long-awaited results of the "studies" on young children are now in, and they are exactly the reverse of what had been expected. The children, genetically and hormonally "predisposed" to aggression and violence due to low serotonin levels and bad parenting, turned out to have normal or elevated serotonin levels. Case closed? Guess again. Since Wasserman, Pine et al. had determined in advance what their conclusions were to be, and since their golden egg-laying goose needed to be coaxed yet again for further funding, they "explained" these results by inventing, out of thin air, the conclusion that *serotonin has the opposite effect in children as in adults*.[38] Perhaps, they continued, high serotonin in childhood leads to low serotonin in adults. Thus, they took a group of black and Latino youth with no history of trouble, in whom no expected abnormalities were found, whose serotonin levels were basically normal or slightly higher than expected, and waved their magic wand to draw more funds for their questionable research.

Similar projects are under way throughout the New York area and, indeed, throughout the US, as is resistance to them. The Coalition Against the Violence Initiative has targeted the New York Psychiatric Institute on a number of occasions. In 1999, a small group from the Coalition picketed a conference on "Mood and Anxiety Disorders in Children," where Daniel Pine was a featured speaker. Illustrating the incestuous arrangements that are increasingly common, the "scientific" programme was supported in part by a grant from Solvay Pharmaceuticals. Two activists were arrested and dragged out of the proceedings after attempting to hang a banner from the balcony. They were strip-searched by guards while awaiting arraignment, and faced criminal charges.

The Coalition fears that those most affected by Wasserman, Pine et al. – the parents and children – remain largely uninformed about the nature and outlook of the studies being conducted on them. Activists remain as sceptical as ever about the labelling of a large number of children, disproportionately minority and poor children, as having mental illness, and about the role of genetic explanations in legitimizing racial supremacist ideas and behaviour – in the name of "science." Those honestly concerned about children's mental health should take action to heal their environment, in the familial as well as the broader sense, rather than looking for genetic, hormonal and other causes of children's distress within the youths' biological makeup.

The interface between funding and research has always been an awkward match in the USA. The Violence Initiative Project is another way of deploying new levels of repression, using the rhetoric of "science" as a smokescreen for white supremacist ideology and control.

NOTES

1. Richard J. Herrnstein and Charles Murray, *The Bell Curve: The Reshaping of American Life by Differences in Intelligence*, New York: Free Press, 1994.

2. Russell Jacoby and Naomi Glauberman, *The Bell Curve Debate*, New York: Random House, 1995.

3. Stephen Jay Gould, *The Mismeasure of Man*, New York: W.W. Norton, 1981, pp. 234–320; also ibid.

4. Gould, *The Mismeasure of Man*.

5. Gerald Horne, "Race Backwards: Genes, Violence, Race and Genocide," *Covert Action Quarterly*, Winter 1992–93, p. 34.

6. Herrnstein and Murray, *The Bell Curve*. For a trenchant critique of Herrnstein and Murray, see Lori B. Andrews and Dorothy Nelkin, "The Ethical, Legal, and Social Implications of Human Genome Research," letter to *Science*, January 5, 1996. The authors conclude: "Since the lessons of genetics are not deterministic, they do not provide useful information on deciding whether or not to pursue various programs to enhance the capabilities of different members of society. Those decisions are moral, social, and political ones."

7. James Q. Wilson and Richard Herrnstein, *Crime and Human Nature*, New York: Simon & Schuster, 1986.

8. Albert Reiss, Jr. and Jeffrey Roth, eds, *Understanding and Preventing Violence*, Washington, DC: US National Research Council, 1993.

9. In 1999, the Novartis corporation – a key force behind the imposition of genetically engineered agriculture – featured the same Arthur Jensen as a key speaker providing "scientific" credibility for the company's genetic engineering efforts.

10. See, for example, Daniel S. Pine et al., "Neuroendocrine response to fenfluramine challenge in boys: associations with aggressive behavior and adverse rearing," in *Archive of General Psychiatry* 54, September 1997, pp. 839–46; Antonia New, "Serotonin Related Genes and Impulsive Aggression in PD," Grant #1R03MH57871–01, Mount Sinai School of Medicine, to National Institute of Mental Health, 1998; David Shaffer, "Center to Study Youth Depression, Anxiety and Suicide," Project #5 P30 MH43878–08, New York State Psychiatric Institute, 1997.

11. Gail Wasserman, Grant Application to the Lowenstein Foundation to Establish a Center for the Study and Prevention of Disruptive Behavior Disorders in Children at Columbia University Department of Child Psychiatry. The Lowenstein Foundation granted Wasserman $1.2 million to study the prediction and prevention of juvenile delinquency.

12. Pine et al., "Neuroendocrine response," pp. 839–40. Also, see Daniel S. Pine et al., "Platelet serotonin 2A (5 HT 2A) receptor characteristics and parenting factors for boys at risk for delinquency: a preliminary report," *American Journal of Psychiatry* 153,

1996, pp. 538–9, which describes a second experiment conducted on the same thirty-four boys.

13. "Questions and Answers Concerning the Department of Health and Fenfluramine," US Food and Drug Administration, November 13, 1997, at http://www.fda.gov/cder/news/fenqal11397.htm.

14. Wasserman, Grant Application.

15. Matthew F. Muldoon et al., "DL fenfluramine challenge test: experience in nonpatient sample," *Biological Psychiatry* 39, 1996, pp. 761, 765.

16. Gina Kolata, "Two Popular Diet Pills Are Withdrawn from Market," *New York Times*, September 16, 1997.

17. Cliff Zucker, Disability Advocates, Inc., and Ruth Lowenkron, Disability Law Center, NY Lawyers for the Public Interest, Inc., December 23, 1997 letter to Clifford C. Sharke, Chief, Assurance Branch, Division of Human Subject Protections, Office of Protection from Research Risks, Rockville, Maryland.

18. "Half Truths and Consequences: Did Doctors Mislead the Parents of Kids they Experimented On?" *Village Voice,* May 5, 1998; Douglas Montero, "Kid Drug Test Foes Picket New Hosp Site," *New York Post*, May 9, 1998; "Thugs in Bassinets: Teen Age Violence, Studies Suggest, Begins in the First Three Years of Life," *New York Times*, May 17, 1998; Douglas Montero, "Drug Test Kids May Have Been Forced: Kin Hoped Other Sibs Might Benefit," *New York Post,* June 12, 1998; Douglas Montero and Susan Edelman, "Ed. Board Referred Kids for Drug Study, *New York Post*, July 28, 1998; Kathleen Kerr, "Students Ended up in Study: Psych Referrals Became Part of Drug Research," *Newsday*, July 28, 1998, p. A3.

19. Kerr, "Students Ended up in Study."

20. Robert Stone, Branch Chief, Department of Probation, memo to Manhattan Family Intake and Investigation Probation Officer, August 30, 1991. The memo was leaked by Probation Officer Renee Jackson, who has been subjected to harassment, changing job assignments, and constant pressure ever since.

21. See New, "Serotonin Related Genes," among others.

22. Norplant, for example, was given to Haitians before it was approved for the US market; similarly, recombinant hepatitis vaccine was administered to the Lakota Sioux on the Pine Ridge reservation, Sabin oral polio vaccine in Africa, and genetically engineered anthrax vaccine to US soldiers beginning with the Gulf War in 1991.

23. Sheryl Gay Stolberg, "Gene Therapy Ordered Halted at University," *New York Times,* January 22, 2000. Stolberg writes: "The agency's action comes two days after its investigators completed a detailed inspection of patient records and laboratory data from the experiment that killed the Tucson man, Jesse Gelsinger … Mr. Gelsinger died of multiple organ failure caused by a severe immune reaction to an infusion of corrective genes and is the first person to have died as a direct result of gene therapy."

24. Stuart Newman, "Generic physical mechanisms of morphogenesis and pattern formation as determinants in the evolution of multicellular organization," *Journal of Bioscience*, vol. 17, no. 3, September 1992, pp. 193–215.

25. Stuart Newman, "Dynamic Balance in Living Systems," *GeneWatch* (Cambridge, MA), November/December 1985, pp. 12–13.

26. When they are pressed to concede that environmental factors do play some part, they argue that environment mostly serves to bring out inherited traits that are already present.

27. Warren Leary, "Struggle Continues over Remarks by Mental Health Official," *New York Times,* March 8, 1992, p. 34.

28. Mark, Ervin and Sweet described one young white male who had undergone brain surgery to cure his epilepsy and propensity for violent behaviour. They claimed that he was saved by psychosurgery (lobotomy), while his mother said that the doctors had turned him into a vegetable. See Barry Mehler, "In Genes We Trust: Where Science Bows to Racism," *Reform Judaism*, Winter, 1994.

29. Peter Breggin, "Campaign against Racist Federal Programs by the Center for the Study of Psychiatry and Psychology," *Journal of African American Men*, Winter 1995/96.

30. B.J. Mason, "Brain Surgery to Control Behavior: Controversial Options are Coming Back as Violence Curbs," *Ebony*, February 1973, p. 68.

31. Horne, "Race Backwards," p. 30.

32. Peter Breggin, "The Violence Initiative – a Racist Biomedical Program for Social Control," *The Rights Tenet*, Summer 1992 (Bethesda, MD: Center for the Study of Psychiatry).

33. Christopher Anderson, "NIH Under Fire," *Nature*, July 30, 1992, p. 357.

34. Vince Bielski, "Hunting the Crime Gene," *San Francisco Weekly*, June 15, 1992.

35. Micaela di Leonardo, "White Lies: Rape, Race and the Myth of the Black Underclass," *Village Voice*, September 22, 1992, cited in Horne, "Race Backwards."

36. Ibid.

37. Ibid.; also Mitchel Cohen, *The US Government's Secret Experimentation with Biological and Chemical Warfare*, New York: Red Balloon, 1995.

38. Pine et al., "Neuroendocrine response."

PART III

Patents, Corporate Power and the Theft of Knowledge and Resources

Now that we have addressed many of the specific problems with genetic engineering and other biotechnologies, both in agriculture and in medicine, our discussion has come full circle. The last several chapters have set the stage for where we are heading, by showing how the new genetic technologies heighten existing social inequities and injustices. Now we can begin to reach for the underlying roots of the problem and examine the ways in which biotechnology is itself an expression of fundamental inequities in society.

Technologies do not develop in a social vacuum. They are neither independent historical forces, nor mere value-free "tools" that can be enlisted in the service of any end we might desire. Technologies emerge from distinct social realities, from the institutions that create them, and the personal and professional outlooks of their developers. There is no "value-free" science or technology. In both the public and the private sectors, constraints of funding and a wide range of ideological, political and institutional factors shape what kinds of scientific and technical questions are considered intellectually and professionally legitimate to explore. They determine which of all the many potentially promising avenues are pursued and which are left to languish. Additionally, while social factors shape technological developments, choices along the road to technological innovation can also alter the course of our social evolution. It is a complex, dialectical relationship, which often defies the simple formulas of technological, economic and cultural determinists alike.[1]

Today's biotechnologies are products of a very particular social context: a market-dominated, hypercompetitive capitalist society in which virtually all the social decisions that most affect people's lives are driven by powerful, unaccountable political and economic institutions. In the private sphere these are huge, secretive corporations; in the public sphere they are faceless, unaccountable bureaucracies, most often acting in service to those same corporations. Other kinds of societies might choose to develop biological technologies, but they would likely be qualitatively different from the bio-technology that has emerged from today's context of corporate globalism, Western cultural hegemony and increasing worldwide inequality. Today's bio-technology is a triumph of a narrow, instrumental view of nature and em-bodies the hegemony of a thoroughly profit-centred social agenda. Just as nuclear technology a generation ago was a definitive symbol of the increasing militarization of society during the peak years of the Cold War, biotechnology is today the product of a post-Cold War society that is thoroughly enslaved to commercial interests. To oppose biotechnology effectively, it is necessary to expose the social, economic and ideological underpinnings of the global capitalist world-view that today's biotechnology is an expression of.

We have already discussed the ways in which the biotechnology industry seeks to commodify all of life, to absorb all that is alive into the sphere of products to be bought, sold and traded in the commercial marketplace. The institutions that are developing genetic engineering and other biotechnologies appear to be seeking nothing less than the private ownership of all life on earth. In today's world of global trade agreements and "intellectual property" regimes, this ownership is codified in the form of patents. Patents have been claimed by biotechnology companies and government agencies for everything from individual proteins, to microorganisms, seeds, laboratory animals and livestock, to many tens of thousands of human DNA sequences. The chapters in this section seek to explain the many dimensions of this profoundly dis-turbing situation: its institutional roots; the role of corporations and govern-ment agencies; the effects of this new "biopiracy"; and its implications for diverse communities of people worldwide. This section will also examine the conflicts of interest that have forestalled effective public regulation of genetic engineering in the United States, and demonstrate how the biotechnology industry has attempted to silence its critics and suppress public discussion of alternatives.

Many of these chapters focus specifically on the theft of biological re-sources and cultural knowledge from people around the world. Prospecting for useful plants from remote areas of the world has become a very lucrative business. Traditional healers are approached for information on how local plants are used for food and medicine, and indigenous peoples themselves are sought out for samples of blood and other tissues that might reveal genetic properties of interest to researchers. Since the most isolated populations are usually the most genetically homogeneous – and since conditions of poverty and undernourishment increase the likelihood that a genetic propensity to disease will be expressed within a person's lifetime – medical researchers often focus their "bioprospecting" efforts on remote indigenous populations.

Scientists have trekked to Saudi Arabia in search of genes "for" glaucoma, to Ghana and Nigeria to study diabetes, and to Mongolia for studies on congenital deafness. Rarely are communities informed about the full range of possible commercial uses for their genetic information, nor are they party to the multimillion-dollar agreements between research institutions and pharmaceutical companies that have become routine in the world of biotechnology. For example, one research group from the University of Toronto has been searching blood samples from a remote island in the South Atlantic for a gene that can be linked to asthma. These expeditions were funded by a California-based company called Axys Pharmaceuticals, which in turn sold its rights to the German pharmaceutical giant Boehringer Ingelheim for $70 million [2]

"Fraud, deception and bribery are being used to take samples from indigenous populations around the world," Debra Harry of the Nevada-based Indigenous People's Coalition on Biodiversity told panelists at a scientific symposium on the Human Genome Diversity Project (HGDP) in February of 1998.[3] As Victoria Tauli-Corpuz of the Philippines and Vandana Shiva of India illustrate in their chapters, indigenous representatives from Southeast Asia to India to Latin America have become some of the world's most articulate critics of the worldwide commercialization of genetic research.

Since the turn of the millennium, activists worldwide have also focused increasingly on the ways in which global institutions such as the World Trade Organization and the World Bank are actively promoting the development of biotechnology. The WTO, in particular, has become the primary vehicle through which the United States and other producers of genetically engineered crops are seeking to impose their policy agendas on other regions of the world. As described by several of the authors in this section, the WTO's TRIPs (Trade Related Intellectual Property Rights) provisions enable the US to pressure other countries to amend their laws to allow the patenting of DNA sequences and entire living organisms. The WTO has also made it possible for the US to challenge various countries' resistance to imports of genetically engineered commodity crops, claiming that they are an unfair restraint on trade.

The World Bank, for its part, has begun actively to insinuate biotechological "solutions" into its "development assistance" programmes, urging countries such as Mexico, Kenya and Indonesia to develop the scientific and regulatory infrastructure to facilitate genetically engineered imports.[4] The World Bank's Global Environment Facility has been supporting the bioprospecting of commercially useful plants, particularly in Latin America, since the early 1990s.[5] Meanwhile, grants of food aid from the United States to various international relief programmes are being channelled through major commercial distributors such as Cargill and Archer Daniels Midland, which own large stores of genetically engineered grain that have been rejected by many of their leading commercial customers.[6] It would appear that the consumers of last resort for GM crops may turn out to be those who are simply too desperate to refuse them.

This section begins with an analysis by Hope Shand of the Rural Advancement Foundation International (RAFI) of the institutional structures of the so-called "life science" industry. Shand has kept activists around the world abreast of rapidly changing developments in the corporate world of biotechnology for many years now, and has carefully documented every step toward the consolidation of corporate power over our food and our health care. She describes the corporate mergers that have created today's worldwide players in biotechnology, and the role of patents and intellectual property in reinforcing their social and economic power. Beth Burrows, who has participated in international negotiations on biosafety and life patenting for nearly a decade, then traces the origins of the patent system to the legacy of European colonialism. She explains how patents on living things have been promoted through various international agreements, describes the reactions of indigenous activists in North America when they first learned about the HGDP, and offers the disturbing tale of John Moore, probably the first medical patient to discover that his own cells had been patented.

Victoria Tauli-Corpuz then offers a comprehensive overview of the ways biotechnology has impacted on indigenous people worldwide. With examples from her own Igorot community in the Philippine Cordillera region, Corpuz reveals the implications of the HGDP and related commercial investigations in scientific, philosophical, economic and cultural terms. Michael Dorsey describes the origins of bioprospecting of useful plants, going back to the sixteenth century, and shows how developments in biotechnology have greatly accelerated this practice. He offers several contemporary examples, including from his own research on the US company Shaman Botanicals and their bioprospecting activities in the rainforests of Ecuador.

Vandana Shiva then describes the implications of biopiracy for traditional communities in India. She highlights the refusal of Northern researchers to acknowledge the contributions of traditional communities to the understanding of plants and how to use them, and demonstrates how the enforcement of intellectual property laws further marginalizes traditional peoples. Kimberly Wilson rounds out the discussion of patents by explaining the legal origins of the patenting of life in the United States; this is the very legal framework that is now being imposed on countries throughout the world, via the World Trade Organization's TRIPs mechanisms.

Jennifer Ferrara continues the discussion of the politics of genetic engineering by peering into the arcane world of US government regulation. She recounts the US government's consistent pattern of deregulation of genetically engineered products, and then describes the role of the public relations industry in pressuring regulators and suppressing information that is critical of genetic engineering. Finally, Mitchel Cohen explains the troubling effects of the biotechnology industry on free speech, civil liberties and dissent, in the United States and elsewhere. He reviews the history of legal repression against biotech critics, and shows how the ideological outlook of biotechnology itself fosters the suppression of critical voices. This overview of the politics and the international dimensions of biotechnology sets the stage for our final section, in which we explore the growing worldwide resistance to genetic engineering.

NOTES

1. General works on technology and society that are especially useful for understanding the social context of biotechnology include Murray Bookchin, *The Ecology of Freedom*, Montreal: Black Rose, 1991, esp. chs 9 and 10; David Noble, *Forces of Production*, New York: Knopf, 1984, and *The Religion of Technology*, New York: Knopf, 1997; and Bruno Latour, *Science in Action*, Cambridge, MA: Harvard University Press, 1987. See also Brian Tokar, *The Green Alternative: Creating an Ecological Future* (1987), rev. edn, Gabriola Island, BC: New Society Publishers, 1992, esp. chs 1 and 4.

2. Lisa Belkin, "The Clues are in the Blood," *New York Times Magazine*, April 26, 1998, p. 52; also see the special issue of *Cultural Survival Quarterly* devoted to the Human Genome Diversity Project, vol. 20, no. 2, Summer 1996.

3. Quoted in Brian Tokar, "The Human Genome Diversity Project: Indigenous Communities and the Commercialization of Science," an occasional paper of the Edmonds Institute, Edmonds, WA, June 1998. The symposium, sponsored by the American Association for the Advancement of Science, took place in Philadelphia in February 1998.

4. Luke Anderson, "The Engineering of Development – The World Bank and Biotech," unpublished ms, Washington, DC, April 2000; GRAIN (Genetic Resources Action International), "Investing in Destruction – The World Bank and Biodiversity," Barcelona, November 1996, at http://www.grain.org/publications/reports/bio8.htm.

5. GRAIN, "Investing in Destruction."

6. Declan Walsh, "America Finds Ready Market for GM Food – The Hungry," *Independent*, March 30, 2000.

16

Gene Giants:
Understanding the "Life Industry"

HOPE SHAND

In the 20th century, chemical companies made most of their products with non-living systems. In the next century, we will make many of them with living systems.

Jack Krol, DuPont board chairman[1]

Once, "power" was land – controlling what grew on it and what came from it. Then "power" became manufacturing – the smokestack industries.... Today "power" is life. And life is fast becoming the private preserve of transnationals and venture capitalists.

RAFI[2]

It is impossible to understand biotechnology without examining the power and global reach of the giant, transnational enterprises that are in the business of engineering, controlling, patenting and profiting from life. All parts of life – its products and processes, even its formulae – are being privatized. Market dominance combined with monopoly patents has given a steadily shrinking number of corporate titans unprecedented control over commercial food, farming and health.

In the mid-1990s, the US government estimated that transnational enterprises control one-quarter of global economic activity and that, in countries such as the United States, 40 per cent or more of all merchandise trade takes place between affiliated firms (i.e. between parent and/or subsidiary enterprises).[3] One measure of economic globalization is the worldwide level of foreign direct investment (FDI).[4] According to the United Nations, global cross-border mergers and acquisitions accounted for almost 60 per cent of all FDI in 1997.[5] Today, transnational enterprises account for at least two-thirds of world trade[6] The "buyer" and the "seller" are often divisions of the very same mega-corporation.

In this setting, competition becomes illusory as fewer, bigger enterprises and their global subsidiaries dominate world trade and finance. This is the reality of borderless economic globalization and the myth of "free" trade; it is also the driving force behind global trade institutions such as the WTO – and proposals such as the derailed Multilateral Agreement on Investment – that could ring the death knell for national sovereignty and electoral democracy. To conclude that transnational corporations rival the power of the nation-state is a gross understatement. *The Economist* reported that when corporate executives were negotiating the mega merger of financial giants Travelers and Citicorp in 1998, one of the negotiators mused: "Can anybody stop us?" The only response was "NATO."[7]

Corporate concentration is nothing new, but the pace of consolidation is accelerating rapidly. In 1990 the value of global mergers and acquisitions was $454 billion. In 1998, the total volume of mergers and acquisitions world-wide passed the two *trillion* dollar mark to a total of $2.4 trillion – a stagger-ing 50 per cent increase over the previous year. In 1999, the value of global merger activity soared to $3.1 trillion.[8]

How does biotechnology fit into this picture of corporate consolida-tion? Not only are buyers and sellers integrating, but vast industrial segments as different as agribusiness and health are using a host of new genetic tech-nologies to achieve a global technological integration that would have been inconceivable two decades ago. Today, traditional boundaries between pharma-ceutical, biotechnology, agribusiness, food, chemicals, cosmetics and energy sectors are blurring and eroding. Under the "life sciences" banner, trans-national firms are using complementary technologies to become dominant actors in all of these industrial sectors. The result – the "life industry" – poses a threat not only to national security but to the security of life.

What are the forces that are driving global technological integration and the formation of a handful of Gene Giants? The following discussion examines the brief history of commercial biotechnology and the major forces propelling the emergence of the global "life industry."

THE RISE OF COMMERCIAL BIOTECHNOLOGY IN THE USA

The commercial biotechnology industry is little more than a quarter of a century old. Genetic engineering as a commercial venture was launched in 1976 when Herbert Boyer, a University of California bacteriologist, joined forces with a young venture capitalist and formed the world's first company to commercialize genetic engineering – Genentech of South San Francisco, California. Boyer was one of many academic scientists who became a new breed of "bioentrepreneurs," wearing white lab coats instead of three-piece suits. When *Genetic Engineering News* published its first list of biotech's "molecular millionaires" in 1987, almost half of those listed were PhDs – and many made their fortunes working for biotech firms even while holding university positions.[9]

From 1979 to 1983, more than 250 small biotech firms were founded in the United States[10] – mushrooming to over 1,000 private and 345 public biotech firms in 1999.[11] Europe and Japan spawned world-class biotechnology research during the same period, but the dramatic growth of hundreds of biotech start-up firms was unique to the United States. The commercial biotech boom in the US was made possible because of two major factors. First, a long history of federal funding for basic science provided an infrastructure of trained scientists. Second, access to venture capital bankrolled embryonic firms, providing roughly $10 billion to US biotech companies since the mid-1970s.[12]

Nascent biotech firms provided early innovation and served as a litmus test for larger, corporate investors, who soon followed. Then and now, the survival of virtually all biotech firms depends on a wide range of partnering agreements – R&D contracts, joint ventures, equity investments, licensing and marketing agreements, amd so on – with larger pharmaceutical, food and chemical corporations. Most large corporations in the US and Europe did not begin major in-house biotechnology R&D until the early 1980s. DuPont, Hoffmann–La Roche and Ciba–Geigy (now Novartis) all announced R&D programmes for biotechnology and life sciences in 1981. Monsanto scientists genetically modified a plant cell for the first time in 1982.

SHAKE-OUT, AND A MILLENNIUM BOOM?

Consolidation in the biotechnology industry – either through bankruptcies or mergers and acquisitions – has been the norm for many years. The entrepreneurial start-up era for biotech may be drawing to a close as the giant corporations assume a more direct and dominant role. In 1997, a total of 345 public biotechnology companies in the US generated sales of $17.5 billion – but the vast majority of these companies operated in the red.[13] Almost half of the publicly traded biotechnology companies in the US are currently operating with two years' cash or less. In 1998, the top twenty publicly traded biotech companies accounted for approximately two-thirds of the total biotech industry market capitalization.[14] One of the mysteries, according to Sean Lance, the CEO of Chiron, "is how so many companies have survived for so long with so few financial resources."[15]

According to one industry analyst, "1998 had the most prominent number of bankruptcies that I've ever see in the business."[16] Equity financing in the biotech sector has "essentially shut down," according to Burrill and Company. In the last quarter of 1998, the nation's public biotech companies raised $242 million, compared with $2.2 billion during the same quarter in 1997. Analysts predict that the next few years will see a major shake-out in the industry; undercapitalized firms will either merge or fail.

In late 1999 and early 2000, however, biotech companies – primarily a select group of genomics firms – temporarily resurfaced as the darlings of Wall Street.[17] Genomics refers to the sequencing and patenting of the genetic code of plants, animals, microorganisms and humans. In early 2000 the stock prices of many genomics companies soared in a biotech buying frenzy dubbed "genomics giddiness."[18] The investor frenzy to acquire genomics stocks in late 1999 and early 2000 was likened to the rage over Internet stocks.

The explosive growth in biotech stocks was fuelled, in part, by the expectation that companies will exploit and profit from the vast amounts of biological data spewing from the Human Genome Project (HGP). The HGP is a fifteen-year, $3 billion project supported primarily by the US government and British partners to map the entire human genome, the 80,000–100,000 genes that exist within our DNA. The biotech companies and their pharmaceutical partners insist that genomics is on the verge of revolutionizing drug discovery.

But investors seem to ignore the fact that gene-based discoveries are still years away from commercial reality, most genomics companies have yet to make a profit, and patent portfolios covering thousands of bits and pieces of DNA have yet to be tested in the courts. According to industry analyst Jürgen Drews, the overall deficit of the biotech industry (both public and private companies) was $5.1 billion in 1998.[19] Between 1986 and 1999 only seventeen biotech companies reached a level of "sustainable profitability." Drews predicts that five companies will add their names to the list in 2000, and an additional fourteen will become profitable in 2001.[20]

LIFE PATENTS: REDEFINING HUMAN "INVENTION"

Several factors have propelled dramatic growth and consolidation in the biotechnology industry over the past twenty years, but chief among them has been the industry's successful bid to control access to the new biotechnologies. In the 1980s, the biotechnology industry and the US government successfully rewrote intellectual property laws to allow for exclusive monopoly control over all biological products and processes. Today, a handful of corporations that dominate commercial biotechnologies are staking far-reaching claims of ownership over a vast array of living organisms and life processes that are used to make commercial products.

Proponents of intellectual property (IP) argue that it is designed to promote innovation by rewarding inventors of new technologies, and that IP is essential because it enables companies to recoup their research investment. But there are clear winners and losers in the grab for life patenting. Biological resources have profound economic and social importance. Forty per cent of the world's market economy is based directly on biological products and processes.[21] The rural poor depend upon biological resources for an estimated 85–90 per cent of their survival requirements. As industrial IP systems extend worldwide, monopoly control over biological products and processes jeopardizes world food security, undermines conservation of biological diversity, and threatens further to marginalize the world's poor.

The patenting of life forms represents a radical departure from the scope of traditional intellectual property law. In addition to the basic criteria for patenting – novelty, usefulness and non-obviousness – there is a well-established doctrine in patent law that "products of nature" are not patentable. With the advent of genetic engineering, however, it did not take long to redefine what is considered human "invention" and legally patentable. Over the course of a single decade, the US government took giant steps to accommodate the corporate desire to patent life:

- In 1980 the US Supreme Court ruled in the landmark case of *Diamond* v. *Chakrabarty* that genetically engineered microorganisms are patentable.
- In 1985 the US Patent and Trademark Office ruled that plants (previously protected by plant breeders' rights) qualified under industrial patent laws.
- In 1987, the US Patent and Trademark Office ruled that animals are also patentable.

As a result of these decisions, virtually all living organisms in the US, including human genetic material, became patentable subject matter, just like any other industrial invention. As one industry analyst explains:

> Since 1980 it can no longer be said that something is not patentable just because it is living ... biotechnology has advanced so rapidly in recent years that there is now virtually no life form which does not have the potential as the subject of patent application.[22]

There is no doubt that patents are a powerful marketing tool, but they don't necessarily promote innovation. In fact, the monopolistic nature of the patent process can restrict innovation, limit competition and thwart new discoveries. Over time, intellectual property regimes have grown into mechanisms that allow corporations (not individual inventors) to protect *markets* rather than ideas. This trend complicates, makes more expensive, and slows the pace of scientific advancement in agriculture and health care. If one company "corners the market" with a strong monopoly patent, competitors may logically decide to orient their R&D sights elsewhere. In the words of Myriad Genetics, a company that has received a portfolio of patents related to breast cancer genes: "The identification and patenting of genes will present significant barriers to entry, and potentially permit substantial operating margins."[23] In January 2000 the United Kingdom's National Health Service discovered the high price of Myriad's exclusive monopoly patents – in this case, patents covering a diagnostic test to assess a patient's genetic predisposition to breast cancer. According to Tom Wilkie of the UK's Wellcome Trust, the royalties Myriad is demanding are set so high they will strain the budgets of some UK clinics, forcing them to curtail the use of this particular test.[24]

The life industry's control of patented genes and traits is creating legal barriers which make it difficult or impossible for smaller companies or public-sector researchers to compete, or to gain access to new technologies. One observer calls it "scientific apartheid." The power of exclusive monopoly patents is giving these companies the legal right to determine who gets access to proprietary science and at what price. Participation in the industry isn't possible unless a company holds patents or has the money to license them. In agriculture, for example, access to new technologies is legally restricted by a complex pedigree of patented gene traits. Pioneer Hi-Bred, the world's largest seed company (now a wholly owned subsidiary of DuPont), claims that one of its new, genetically engineered, insect-resistant corn hybrids requires access to thirty-eight different patents controlled by sixteen separate patent holders.[25] DuPont can afford to play that high-stakes game, but smaller enterprises will find it increasingly difficult to compete.

In today's knowledge-based economy, intellectual property assets have surpassed physical assets such as land, machinery or labour as the basis of corporate value.[26] The uncertainty and confusion over the application of patent law to living materials has resulted in immense legal battles between corporations that are competing for ownership of strategic genes, traits and biological processes. Life industry companies are securing and protecting information and technology via monopoly patents, and that quest is, in many cases, driving a restructuring of the industry. Demonstrating the value of intellectual property assets, the cover of Novartis's 1997 annual report announced that the company holds more than 40,000 patents.[27]

THE LIFE INDUSTRY EMERGES

The common denominator of our business is biology. The research and technology is applied to discover, develop and sell products that have an effect on biological systems, be they human beings, plants or animals.

Daniel Vasella, CEO of Novartis[28]

The agricultural and medical marketplaces are very different but at the research level there is growing commonality. Technologies such as gene sequencing, combinatorial chemistry and high-throughput screening are as relevant to the agricultural as to the human health section.

Sir David Barnes, chairman of Zeneca[29]

Today, traditional boundaries between pharmaceutical, biotechnology, agribusiness, food, chemicals, cosmetics and energy sectors are blurring and eroding. Under the "life sciences" banner, transnational firms are using complementary technologies to become dominant actors in all of these industrial sectors. The common denominator is biology. Major enterprises are restructuring to take advantage of the molecular revolution and the complementary use of technologies such as high-throughput screening, combinatorial chemistry, transgenics, bioinformatics and genomics.

The growing commonality between food, medicine and agricultural production is illustrated by the emergence of so-called "molecular farming" (or "pharming") production systems. Life industry corporations aim to manufacture vaccines, drugs, enzymes, plastics and other chemicals in genetically engineered crops and livestock. It is claimed that pharmaceuticals will be commercially grown in genetically engineered plants by the year 2004.[30] In 1998, Monsanto conducted human clinical trials on a cancer treatment that is

produced in corn. Axis Genetics is developing transgenic potatoes for the
development of an oral hepatitis B vaccine. A Texas-based company, ProdiGene,
is developing a natural, dietetic sweetener to be produced in genetically engineered corn that expresses the super-sweet protein Brazzein, derived from
the berries of a West African plant. One kilogram of Brazzein extracted from
a ton of processed corn is equivalent to the sweetness derived from 1,000
kilograms of sugar.[31]

As life industry corporations embrace the biotechnological future, some of
the world's largest chemical corporations have shifted out of commodity pet-
rochemicals into biology – from hydrocarbons to carbohydrates. They are
shedding old-fashioned industrial chemicals and concentrating on agribusiness,
pharmaceuticals and food. Consider the following examples:

- As recently as 1996 Monsanto was the fourth largest chemical company in
  the United States. In a dramatic shift to biotechnology, Monsanto spun off
  its $3 billion chemicals business as a separate company in 1997 (not
  including the company's profitable Roundup herbicide). From 1996 to
  1999 Monsanto spent almost $9 billion acquiring seed and agricultural
  biotechnology companies. In late 1999 Monsanto announced that it would
  merge with drug industry giant Pharmacia & Upjohn to create a new drug
  company, named Pharmacia, with combined annual sales of $17 billion.
  The merged company then spun off its agrochemical, seed and besieged
  biotech business, which kept the name Monsanto.
- In 1998 Hoechst (Germany) spun off Celanese, its big American chemical
  subsidiary, in order to meet its goal of getting out of the chemical industry
  by the end of 2000.[32] In December 1998 the company announced its
  merger with Rhone–Poulenc (France), creating Aventis, at least temporarily
  the world's biggest life sciences company. With combined sales of $20
  billion per annum, Aventis leapfrogged to the world's top-ranking firm in
  sales of pharmaceuticals, agrochemicals and veterinary medicines. The combined research and development budget for Aventis was a staggering $3
  billion – that is equivalent to the budget for the fifteen-year Human
  Genome Project, or, roughly 40 per cent of all funding for agricultural
  research in the private sector.
- Life industry giant Bayer (Germany) is rapidly expanding its life sciences
  operations. In September 1998, for example, Bayer spun off its Agfa subsidiary, and spent $1.2 billion to acquire the diagnostic division of Chiron,
  one of the world's largest biotech companies. Bayer subsequently invested
  $465 million in Millennium Pharmaceuticals, the largest investment to
  date in the field of genomics drug research.
- DuPont, until recently the world's largest chemical producer, has taken
  dramatic steps to bolster its life sciences business. In 1998 DuPont announced that it would divest its petroleum subsidiary, Conoco, the world's
  ninth ranking oil company. The largest-ever initial public offering for a US
  company raised a record $4.4 billion. According to DuPont's chief executive,

Charles Holliday, the sale gave DuPont the war chest it needs to "rapidly accelerate" investment in the life sciences.[33] DuPont wasted no time, spending $2.6 billion to acquire Merck & Co.'s 50 per cent share in their joint venture, DuPont Merck Pharmaceutical, for $2.6 billion. In March 1999 DuPont spent $7.7 billion to acquire the rest of Pioneer Hi-Bred International, the world's largest seed corporation, previously only 20 per cent owned by DuPont.[34]

- The UK-based Zeneca Group PLC and Astra A.B. of Sweden announced the largest-ever European merger in 1998 to form AstraZeneca. The merger transformed two second-tier drug firms into a leading pharmaceutical firm with $14.3 billion in sales. With combined assets in excess of $70 billion, the company was larger than the 1997 gross national product of ninety-three of the world's developing nations.[35] In December 1999 AstraZeneca and Novartis announced they would spin off and merge their agrochemical and seed divisions to create the world's biggest agribusiness corporation – now named "Syngenta."[36]

Some industry analysts have speculated that the announced "de-mergers" by Novartis, AstraZeneca, Pharmacia & Upjohn and Monsanto in late 1999 – that is, the spin-offs of their agricultural input divisions – signalled a major strategic shift and a collapse of the life industry model. It is more likely that the pharmaceutical giants have concluded that that the next half-decade or so of agricultural biotechnology will be fraught with too much risk and controversy. Because of widespread and growing opposition to genetically modified crops and foods, the pharmaceutical corporations have decided to keep the agricultural input side of their interests (biotech seeds and agrochemicals) at arm's length. In the long run, they probably won't drop agricultural biotechnology altogether because the synergies are too lucrative to abandon. As the Rural Advancement Foundation International opined following the Monsanto and Pharmacia merger: "The recent moves by Novartis, AstraZeneca, and Pharmacia-Upjohn are belated – but implausible – attempts to give the Gene Giants safe shelter as the roof caves in on the GM food market."[37]

THE FOOD AND BEVERAGE INDUSTRY: THE MEGA GENE GIANTS

The food and beverage giants are the true titans of the life industry. They are likely to become more visible and dominant players in the next five to ten years. The total retail value of global food sales is estimated at $2 trillion – or six-and-a-half times larger than that for pharmaceuticals. Put another way, the 1997 revenues of the world's largest food and beverage corporation (Nestlé – $45.3 billion) easily surpassed the entire commercial seed industry ($23 billion), the entire agrochemical industry ($31 billion) and the animal health industry ($17 billion). Nestlé's 1997 revenues were more than three times the combined revenues of Hoechst and Rhone–Poulenc, before they merged to create Aventis, the leading pharmaceutical corporation ($13.7 billion).

As genetic engineering and related technologies become more widely used to alter the function and performance of plants, animals and common ingredients, the food and beverage industry is likely to enter into strategic alliances, mergers and acquisitions with seed, biotech and agrochemical and pharmaceutical firms. With a growing emphasis on "output traits," such as oils with longer shelf life, or improved amino acid balance in maize and soybeans, the anticipated value of such alliances increases. Just as chemical and pharmaceutical enterprises have spent billions acquiring seed and biotech firms, acquisition of these same enterprises may prove irresistible to food and beverage transnationals. On the other hand, the pharmaceutical giants generally have higher profit margins, they are science and technology-based and have more cash to plough into major investments. Given current trends, we may soon see huge buyouts and alliances between drug giants and food titans. How long will it be before a Nestlé or Unilever devours a Novartis (or vice versa?), or a ConAgra absorbs a Dow?

With the development of so-called "functional foods" and "nutraceuticals" the lines between food and medicine are blurring, further enticing food processors, agbiotech firms and drug companies to merge complementary interests in food, biotechnology and pharmaceuticals. Companies such as DuPont, Kellogg, ConAgra, Mars, AstraZeneca and others are rushing to engineer foods that claim to enhance health and well-being. According to industry analysts, the sale of foods touting healthy properties is expected soon to reach $29 billion a year, from virtually zero in 1990.[38] (Note that this is higher than the $23 billion global commercial seed market.)

Harvard Professor Dr Ray Goldberg – the man who coined the term "agribusiness" in the 1970s – refers to "agriceuticals" as "the most important economic event in our lifetime."[39] Goldberg points out that the combination of health, science and agribusiness totals $15 trillion, and predicts that the "agriceutical" system will utilize half the assets of the world and supply over half the consumer expenditures in the world.[40]

THE SEED AND AGROCHEMICAL INDUSTRIES

Seeds are software. And we have the seeds.

Alfonso Romo Garza, owner of Empresas La Moderna,
a Mexico-based seed company that controls 25 per cent
of the global vegetable seed market.[41]

Over the past thirty years plant breeding and seed sales have been privatized. For corporations that have combined interests in seed and agrochemicals, the patented seed is the ideal delivery system for a package of proprietary technologies – genes and related inputs. In recent years, many of the world's largest agrochemical and pharmaceutical corporations have spent billions of dollars acquiring seed and biotech companies. DuPont, Monsanto and Novartis are the world's largest seed corporations, together controlling nearly 20 per cent of the global seed trade.

As of September 1999, the top five Gene Giants (AstraZeneca, DuPont, Monsanto, Novartis and Aventis) accounted for nearly two-thirds of the global pesticide market (60 per cent), almost one-quarter (23 per cent) of the commercial seed market, and virtually 100 per cent of the transgenic (genetically modified) seed market. "The Gene Giants' portfolio extends far beyond plant breeding," explains RAFI executive director Pat Mooney. "From plants, to animals, to human genetic material they are fast becoming monopoly monarchs over all the life kingdoms."[42] Five years earlier, none of the top five Gene Giants appeared on the list of leading seed corporations. In fact, three of the top five companies didn't even exist (Zeneca and Astra merged to form AstraZeneca; Rhone–Poulenc and Hoechst became Aventis; Ciba–Geigy and Sandoz became Novartis. DuPont owned only 20 per cent of Pioneer Hi-Bred before March 1999).

Today, the top ten seed companies control over 30 per cent of the $23 billion commercial seed market. But corporate market share is much higher in specific seed sectors and for certain crops. For example:

- 40 per cent of US vegetable seeds come from a single source.[43] The top five vegetable seed companies control 75 per cent of the global vegetable seed market.[44]
- DuPont and Monsanto together control 73 per cent of the US seed corn market.[45]
- Just four companies (Monsanto, Pioneer, Novartis, Dow) control at least 47 per cent of the commercial soybean seed market. An estimated 10 per cent of the market is in public varieties. An estimated 25 per cent of North American soybean seed is farmer-saved, not newly purchased.[46]
- At the end of 1998, a single company, Mississippi-based Delta & Pine Land, controlled over 70 per cent of the US cotton seed market.[47] Delta & Pine Land is perhaps best known for its notorious patent on genetic seed sterilization (a.k.a. Terminator).

A handful of Gene Giants completely dominate the market for bio-engineered seeds. According to estimates compiled by the Sparks Companies, in the world's largest transgenic seed market, the United States, Monsanto's transgenic seeds accounted for 88 per cent of the total transgenic crop area in 1998.[48] AgroEvo (soon to become a subsidiary of Aventis) accounted for 8 per cent, and Novartis's transgenic seeds were planted on just 4 per cent. While leading competitors such as DuPont and American Home Products are expected to commercialize new transgenic crops soon, the market will still be controlled by an elite group of Gene Giants.

THE ANIMAL PHARMACEUTICAL SECTOR: THE GENE GIANTS' ORPHAN INDUSTRY?

With industry-wide revenues of $17 billion, the animal pharmaceutical industry represents only about 5 per cent of the giant human pharmaceutical market. Industry analysts refer to animal health as a "mature" market, with only modest growth (3 per cent plus) in recent years. Nevertheless, the growing synergy between animal veterinary medicine and the human pharmaceutical market is clear. For example, biotech companies are producing human proteins in cows' milk, they are hoping to grow spare-part human organs in pigs, and there is even a rapidly growing market for anti-depressants for dogs.

According to Fountain Agricounsel, the animal health division of most pharmaceutical and agrochemical corporations dilutes the profit margin of the parent company.[49] The animal health industry depends on the economic muscle and R&D budgets of the Gene Giants. Of the top twenty companies, only two are stand-alone businesses with a primary focus on animal health. Analysts predict that the survivors in the animal pharmaceutical sector will include a top tier of four or five mega-size companies with revenues of at least $2 billion (there is only one in that category today) driven by genomics research and technology.

PUBLIC GOODS FOR PRIVATE PROFIT

The vast economic power of life industry corporations, coupled with stagnant budgets for public research, has effectively marginalized the role of public sector researchers − particularly in the agricultural arena. The independent public sector researcher is a vanishing breed. Consider the following examples:

- In November 1998 Novartis and the University of California at Berkeley (Department of Plant and Microbial Biology) signed an unprecedented $25 million, five-year agreement. Although the agreement specifies that Novartis cannot dictate what research will be performed with its money, the company will have first rights to negotiate an exclusive licence on a fraction of all the research developments in the laboratories − whether or not the projects were supported by Novartis funds.[50] Critics charge that the alliance gives a private company unprecedented ability to influence the research agenda at a state-owned university, and it will allow public goods to be appropriated for private profit.
- DuPont and the United Soybean Board (USA) have signed an agreement to form a research partnership which will leverage soybean checkoff dollars and private industry resources. In other words, farmers who voluntarily "tax" and contribute a fraction of their soybean sales for research purposes will see that money used at government laboratories using DuPont technology to identify valuable gene traits. The United Soybean Board, DuPont and university representatives will review proposals and decide who gets funding. Discoveries made by university scientists will become the property of the university, but DuPont has the option to license the technology.[51]

Since 1996 virtually every major seed/agrochemical company has invested in plant genomics research. Driven by the increased efficiency of genomics technology and fierce competition among major agbiotechnology firms, investments in crop genomics accelerated dramatically in 1998.[52] Particularly noteworthy is the very minor participation of public-sector researchers in agricultural genomics. After the Gene Giants and their genomics partners stake patent claims to molecular bits and pieces of commercially important plant genomes, what will be left for the public sector? With patents in hand, the Gene Giants have the legal right to determine who will get access to plant genomics material and at what price.

As more and more researchers have one foot in academia and one in industry, the lure of profits threatens to diminish scientific integrity in the biosciences. Nobel laureate David Baltimore, president of the California Institute of Technology, told *USA Today* that researchers "lost their innocence" with the emergence of biotechnology.[53] "Today, most senior biologists are entangled with one or many companies," Baltimore said. "Does it make people so biased they cannot think properly? I am [on] the board of [biotech company] Amgen, and I would have to think twice before arguing that drug prices are unconscionably high."[54]

HUMAN GENOMICS COMPANIES

Many of the same companies that are employing an industrial, gene-based strategy to predict, understand and manipulate plants and microorganisms are doing the same work with the human genome. Genomics companies aim to decode, map, identify and patent the functional characteristics of "commercially relevant" human genes. These high-tech, entrepreneurial companies were founded on venture capital and the promise of patented products and processes. The majority of genomics companies have no commercial products or profits – only patents. For genomics companies, survival depends on strategic alliances and equity investments from the Gene Giants, or subscription fees to proprietary genome databases.

The furious pace of discovery in the field of genomics is reflected in the growing number of patent claims related to partial gene sequences or ESTs (expressed sequence tags). In 1991 the US Patent and Trademark Office had applications pending on 4,000 EST sequences. In 1996 there were a total of approximately 350,000 EST sequences to be examined, and as of September 1998 there were applications pending on over 500,000 EST sequences.[55] In November 1998 California-based Incyte announced that it had received the first US patent on forty-four ESTs.[56] A year later, Incyte announced that, to date, it "has filed patent applications covering an estimated 50,000 individual human genes. The company was issued 79 new US patents covering full-length genes during the third quarter, bringing its total number of issued and allowed full-length gene patents to 453."[57]

The patenting of partial gene sequences is especially controversial; even the US government's National Institutes of Health object to the patenting of ESTs. How, they ask, can standard patent criteria (novelty, non-obviousness and utility) be met in a case where the function of a partial gene sequence (the protein it encodes) is not even known? Claims on partial gene sequences may also preclude future patenting of a full-length gene containing an already patented sequence. Leroy Hood, head of the University of Washington's department of molecular biology, recently announced that he would terminate research on a possible colon cancer screening test because his work involves a gene that is the subject of an EST patent application filed by Incyte.[58] This example illustrates how gene patents, instead of promoting innovation, are stifling research and hindering competition.

The Human Genome Project, launched in 1990, was conceived as an international, public-sector initiative, a project too massive in scope and too expensive for any single country or company to undertake. With the advent of faster, cheaper sequencing technologies, the race to map the human genome faced increasingly stiff competition from the private sector.[59] In May 1998 a new commercial venture announced that it would start and essentially complete the sequencing of the human genome four years ahead of the US government's target date of 2005. Celera, a joint venture between Perkin-Elmer, the leading manufacturer of sequencing equipment, and the US-based Institute for Genomics Research (TIGR), claimed that its sequencing capacity far exceeds that of all existing genomics laboratories in the world.[60] In October 1999 Celera revealed that it had filed for "preliminary patents" on over 6,500 full or partial human genes – despite Celera's pledge in US Congressional testimony to patent "only" 100–300 human genes.[61]

In January 2000 Celera announced that in excess of 97 per cent of all human genes are represented in its database, giving Celera coverage of 90 per cent of the human genome. Celera describes itself as "the world's largest DNA data factory."[62] Not to be outdone, Incyte's CEO, Roy Whitfield, announced plans to quickly "file a full-length patent on every drug target in the human genome."[63] In June of 2000, scientists and heads of state collaborated in a high-profile announcement of a "working draft" of the human genome, proclaiming it the greatest technological achievement of the new millennium. However, the scientific press criticized it as an "arbitrary milestone," orchestrated to alleviate competition between public and private genome-mapping enterprises, and lingering questions about control and ownership of the human genome were conveniently ignored.[64]

CONCLUSION

The industrialization of gene-based science is touted as the engine that will drive economic development in the twenty-first century. For civil society, it is critically important to understand the technological integration that is now taking place, and to identify the transnational enterprises that are dominant players in every aspect of commercial food, agriculture and health.

Unchecked corporate power coupled with the vanishing role of public-sector research will affect all areas of global health, agriculture and nutrition. Neglect of the public good is inevitable when the research agenda is determined by the private sector in pursuit of corporate profits. Access to food, health and nutrition – once considered a fundamental human right – is now subject to the whims of the free-market system.

At the dawn of the twenty-first century, civil society faces the formidable challenge of demystifying a science that wears a white lab coat and is shrouded in sugar-glazed corporate rhetoric: we are told that biotechnologies will feed the hungry, cure cancer and clean up the environment. The solution, we are told, is in the genes. This reductionist, gene-based approach to food production and human health is dangerous in its simplicity and emotional appeal. When injustice is the root of the problem, technology is not the solution. Our task is to resist monopoly control of life, unmask the corporate Gene Giants, and identify how and why powerful genetic technologies are being used by corporations to restrict choice and undermine democracy.

The heart of the matter is not safety or regulation, but control. As civil society organizations concluded back in 1988, at their first international meeting to address the new biotechnologies in Bogève, France, "Any new technology introduced into a society which is not fundamentally just will exacerbate the disparities between rich and poor."[65] Ultimately, if new biotechnologies are to benefit society, they must be built on informed participation and democratic institutions that are people centred, not profit-centred.

NOTES

1. Krol is quoted in a case study prepared by Professor Jonathan West, Harvard Business School, "E.I. duPont de Nemours and Company," November 19, 1998, no. N-9-699-037.

2. Cary Fowler, Pat Mooney, Eva Lachkvics and Hope Shand, "The Lords of Life: Corporate Control of the New Biosciences," in *The Laws of Life: Another Development and the New Biotechnologies, Development Dialogue* 1–2 (Uppsala: Dag Hammarskjöld Foundation, 1988), p. 191.

3. Pat Mooney, "Private Parts: Privatisation and the Life Industry," in *The Parts of Life: Agricultural Biodiversity, Indigenous Knowledge, and the Role of the Third System, Development Dialogue* (Uppsala: Dag Hammarskjöld Foundation, 1996), p. 134.

4. According to UNCTAD, foreign direct investment is defined as an investment involving management control of a resident entity in one economy by an enterprise resident in another economy.

5. UNCTAD Press Release, "Cross-Border Mergers and Acquisitions Dominate Foreign Direct Investment Flows," November 2, 1998, TAD/INF/2776.

6. Personal communication with Masataka Fujita, International Investment, Transnationals and Technology Flows Branch, UNCTAD, January 15, 1999.

7. The story is told in *The Economist*, January 9, 1999, p. 21.

8. Thomson Financial Securities Data, Press Release, "The World is Not Enough … to Merge; Worldwide Announced M&A Volume Soars to Record $3.1 Trillion in '99," December 22, 1999, http://www.tfsd.com.

9. *Genetic Engineering News*, February 1987, p. 15.

10. Mark Dibner, "Biotechnology in Europe," *Science* 232, June 13, 1986, p. 1367.

11. Peter Drake, "The Biotechnology Industry's Y2K Problem," *Nature Biotechnology* 17, supplement, 1999, pp. BE38–39.

12. Ibid., p. BE38.

13. Ibid.

14. *Nature Biotechnology* 17, February 1999, p. 111.

15. Sean Lance, "Letters to the Industry," in *Biotech 99: Bridging the Gap: Ernst & Young's 13th Biotechnology Industry Annual Report*, Palo Alto, CA: Ernst & Young, 1998, p. 8.

16. Emma Dorey, "Will Investors Return to Biotechnology?" *Nature Biotechnology* 17, February 1999, p. 128.

17. See, for example: Bill Barnhart, "Biotech Stocks are Hot Zone for Mutual Funds," *Chicago Tribune*, February 27, 2000, http://www.chicagotrib.com; and Ronald Rosenberg, "On Wall Street, a Buzz over Biotech," *Boston Globe*, February 18, 2000, p. CO1.

18. Stephanie O'Brien, CBS MarketWatch, "Genomics Stocks Soar: DNA Adds Drama to Biotech Industry," February 20, 2000, http://cbs.marketwatch.com/archive/20000220/news/current/genomics.htx?source=htx/http2_mw.

19. Jürgen Drews MD, "Biotechnology 2000: Pushing Innovation and Reaping the Benefits," Keynote Address at the BIO CEO Conference in New York, on February 21, 2000, p. 1.

20. Ibid., Table 1, "Number of Profitable Companies by Year of Development," p. 2.

21. The Crucible Group, *People, Plants, and Patents*, Ottawa: International Development Research Centre of Canada, 1994, p. 23.

22. Sally I. Hirst, "Biopatents: A Sense of Order," *Trends in Biotechnology* 10, August 1992, pp. 269–70.

23. Myriad Genetics Inc., Corporate Profile, 1995.

24. Tom Wilkie, "Beyond the Genome America Has Been Good at Studying Genes; But what about Studying the Illnesses they're Involved In?" *Daily Davos*, vol. 3, no. 5, January 31, 2000, http://www.dailydavos.com/nwsrv/printed/special/davos/su_vw01.htm.

25. Greg Horstmeier, "Strategic Bedfellows: Consolidation of Seed Companies Heralds Changes in the Food Chain. Where Do Farmers Fit In?" *Farm Journal*, October, 1996. http://www.farmjournal.com.

26. W. Bratic, P. McLane and R. Sterne, "Business Discovers the Value of Patents," *Managing Intellectual Property*, September 1998, p. 72.

27. Novartis, "We Are Novartis," Novartis Communication, Basle, March 1997.

28. Daniel Vasella, CEO of Novartis, quoted in David Pilling, "The Facts of Life: Chemical and Pharmaceutical Companies See their Future in Biological Innovation. *Financial Times*, December 9, 1998, p. 21.

29. Sir David Barnes, chairman of Zeneca is quoted in an article by Clive Cookson and Nikki Tait, "From Corn to Cancer," *Financial Times*, June 2, 1998, p. 17.

30. Joan Olson, "When Crops Save Lives," *Farm Industry News*, vol. 32, no. 6, March 1999.

31. ProdiGene Press Release, "NeKtar Worldwide and ProdiGene to Develop Natural Intense Sweetener in Corn," College Station, TX, April 22, 1998.

32. "Hoechst is Planning to Spin Off Celanese," *New York Times*, November 18, 1998.

33. Deborah Hairston, "Divesting Conoco and Other Crown Jewels," *Chemical Engineering News*, June 1998, p. 46.

34. S. Warren and S. Kilman, "DuPont Co. Lands Huge Biotech Prize," *Wall Street Journal*, March 16, 1999.

35. Figures on the size of national economies comes from the World Bank's *World Development Report 1998/99*, New York: World Bank/Oxford University Press, Table 1, pp. 190–91.

36. Novartis Media Release, "Launch of a Global Leader in Agribusiness, Novartis to Focus on Healthcare," December 2, 1999, http://www.info.novartis.com/media/index.html.

37. Rural Advancement Foundation International, RAFI Genotypes, "Pharma-gedon," December 21, 1999. http://www.rafi.org.

38. J. West, "E.I. duPont de Nemours and Company," Harvard Business School Case Study, N-9–699–037, November 19, 1998, p. 8.

39. Verdant Partners, Company News Release, "Agriceuticals: The Most Important

Economic Event in Our Lifetime," Chicago, IL, December 8, 1999.

40. Ibid.

41. Jonathan Friedland and Scott Kilman "As Geneticists Develop an Appetite for Greens, Mr. Romo Flourishes," *Wall Street Journal*, January 28, 1999, p. A1.

42. Rural Advancement Foundation International (RAFI) News Release, "World Seed Conference: Shrinking Club of Industry Giants," September 3, 1999, available at http://www.rafi.org.

43. Ibid.

44. Lynn Grooms, "With Merger Completed, Harris Moran Focuses on Future," *Seed & Crops Digest*, January 1999.

45. Ann Thayer, "Ag Biotech Food: Risky or Risk Free," *Chemical & Engineering News*, November 1, 1999, p. 17.

46. Ibid.

47. Ibid.

48. Personal communication with Sparks Companies, Inc. on 2 March 1999. Sparks is a Memphis-based consulting firm. Their estimates are based on information provided by seed/biotech companies: US Grains Council Value Enhanced Corn Quality Report, Furman Selz LLC, NatWest Securities. The estimate cited does not include herbicide-tolerant varieties marketed by American Cyanamid and DuPont that are not transgenic.

49. See B. Fountain and D. Thurman, "Animal Health Industry Faces Opportunities, Challenges," *Feedstuffs*, November 9, 1998. Fountain and Thurman are partners in Fountain Agricounsel.

50. "UC Berkeley and Novartis: An Unprecedented Agreement," in *Global Issues in Agricultural Research*, vol. 1, no. 3, January 25, 1999, p. 5.

51. *AgBiotech Reporter*, January 1999, p. 7.

52. M. Ratner, "Competition Drives Agriculture's Genomics Deals," *Nature Biotechnology* 16, September 1998, p. 810.

53. Tim Friend, "It's in the Genes: Scientists Confront Issues," *USA Today*, February 22, 2000, http://www.usatoday.com/life/health/genetics/therapy/lhgth024.htm.

54. Ibid.

55. Personal communication from John Doll, US Patent and Trademark Office, Biotechnology Patent Division, September 22, 1998.

56. Usha Lee McFarling, "The Code War: Biotech Firms Engage in High-stakes Fight over Rights to the Human Blueprint," *San Jose Mercury News*, November 17, 1998.

57. Incyte Press Release, November 22, 1999, *PRNewswire*.

58. McFarling, "The Code War."

59. Nicholas Wade, "It's a Three-Legged Race to Decipher the Human Genome, *New York Times*, June 23, 1998.

60. Lisa Belkin, "Splice Einstein and Sammy Glick. Add a Little Magellan," *New York Times Magazine*, August 23, 1998.

61. Justin Gillis, "Md. Gene Researcher Draws Fire on Filings," *Washington Post*, October 26, 1999, p. E1.

62. Celera Press Release, "Celera Compiles DNA Sequence Covering 90% of the Human Genome." Celera Genomics, Rockville. January 10, 2000.

63. Roy Whitfield, CEO of Incyte, quoted in Kristen Philipkoski, "Incyte Incites Concern," *Wired*, February 17, 2000.

64. See, for example, Colin Macilwain, "World Leaders Heap Praise on Human Genome Landmark," *Nature* 405, June 29, 2000, pp. 983–4. According to *Nature*, only 85 per cent of the genome was sequenced to "draft standard" by that time, and just over 20 per cent was in finished form.

65. Twenty-eight participants from nineteen countries met in Bogève, France, March 7–12, 1987, for a seminar on "The Socioeconomic Impact of New Biotechnologies on Basic Health and Agriculture in the Third World." The Bogève Declaration, which is quoted here, is published in Fowler et al., "The Lords of Life," p. 289.

17

Patents, Ethics and Spin

BETH BURROWS

Theft on the grand scale has always depended on the encouragement of law and public relations ("spin") to flourish. Consider the case of patents.

When Columbus stumbled upon a land new to him, he was carrying "letters patent" from the king and queen of Spain. Those documents made the discovery and the exploitation of a whole "New World" possible, legal and rewarding. The letters were issued for the benefit of Spain by authorities whose right to issue such patents – according to the spin-meisters of another day – came directly from God.

Why bother to create such letters patent? Because – to borrow an insight from Vandana Shiva – what was necessary for the invasion and exploitation of other people's land, what was essential to the colonization, was to have a means of declaring inhabited land "empty" – void of true human beings.[1] With such a legal fiction in place, land could be discovered with impunity and "filled" with a clear conscience. In the moment of discovery, "empty" land could become the queen's (or the king's) and title could later be conferred by sovereign grant.

"Letters patent" served to legitimize theft and the creation of property. In the Old World, they were licences to plunder. Even today, the term is still used to indicate a government grant conferring title to public land. Whether anyone should have the right to bestow such titles is of course a matter of perspective and attitude. Spin.

The process of legitimizing usurpation and colonization – or, to put a different spin on it, the process of legitimizing the transfer of property and the rights to its development – continues today. Today, however, the subject of legitimation is not so much a land rush as a gene rush. The prize is the ownership (and control) of life and the invasion this time, to borrow again from Vandana Shiva, is an invasion of the interior spaces of people, plants, animals and microorganisms.[2]

The new invasion looks remarkably like a second coming of Columbus to some of us. Rather than land-entitling letters patent, there are monopoly-entitling industrial patents on biological materials. Instead of rights descending from God, there are efficiencies contingent on the Market. Standing in for the church and its priests to articulate the blessings for invasion are an assortment of government bureaucrats, and university departments of technology transfer and bioethics.[3] And in place of kings and queens and their ministers, there reign the corporations and their entourage – the World Bank, the International Monetary Fund, the World Trade Organization, NAFTA, APEC, and even the United Nations Convention on Biodiversity.

The association of patents and thievery did not end with Columbus. The tradition continued in the New World. Consider Samuel Slater, for example. The industrial development of the United States derives in a sense from his very skilled act of patent infringement (theft).

In the 1760s, the Englishman Richard Arkwright invented the water-powered spinning frame – a machine that brought cotton spinning out of the home and into the factory and made Britain a world-class power in the manufacture of cloth. To protect this competitive advantage and ensure the market for manufactured cloth in its own colonies, the English parliament enacted a series of restrictive measures, including the prohibition of the export of Arkwright machinery or the emigration of any workers who had been employed in the factories using the Arkwright invention. From 1774 on, those who sent textile machines or their workers abroad from England were subject to fines of £200 and twelve years in jail. That's how serious they were about patent protection.[4]

Samuel Slater had worked for years in the Arkwright mills. In 1790 he left England disguised as a farmer and went to the United States. There, with financing from Moses Brown, he created from memory an entire Arkwright factory and all its equipment. He produced commercial-grade cotton cloth and thereby put the United States on the road to its own Industrial Revolution in manufacturing. His achievement was rewarded and honoured in his lifetime. He became rich and was considered a great American hero. He is still acclaimed today as the father of American manufacturing. Yet clearly Samuel Slater was a patent infringer, an intellectual property thief. He only became a hero to those whom his theft greatly benefited.

Titles then, as now, were matters of legal attitude and public "spin." Even Alexander Hamilton, always on the lookout for federal aid to US industry, once argued that patent infringers – he called them "introducers" – who introduced really useful foreign inventions into the country ought to be granted some kind of benefit in law, much as inventors and authors are benefited by the sections of the US Constitution devoted to the protection of patents, trademarks, copyrights and such.[5]

Attitude and spin have changed with time. Until the push for the inclusion of intellectual property rights – patents, copyright, trademarks and such – in international trade agreements, it was understood, even accepted, albeit with grumbling, that countries did not enforce patent protection until it was in their national interest to do so. Intellectual property protections came very late even in some highly developed places. France, for example, only began to patent drugs in 1958, West Germany in 1968, Japan in 1976, and Switzerland in 1977.

When the young United States pirated the intellectual property of Europe – and Slater wasn't the only infringer – it congratulated itself and saw the theft as evidence of national virility. But by the early 1970s the US was a more mature industrial power which, like Britain before it, was looking for legal means to maintain a competitive advantage.[6] US industry, seeking greater protection for its idea-based products – where it still held the worldwide lead – pushed for inclusion of intellectual property clauses, including standards for patents, in trade agreements. This was a huge change in the way things had worked up until then, and it engendered a fight that is not over yet.[7]

To understand this fight, it helps to remember that even by the 1970s, the use of industrial patents to encourage innovation and to guarantee the sharing of inventions was not universal. Where people made their innovations in groups, bound by an ethic of sharing, or where they considered patents on drugs or agricultural products to be grants of immoral monopolies likely to endanger life, mandatory patents seemed an alien imposition, a scheme whereby outsiders might gain control over local resources and knowledge and industry. Interestingly, both sides in the trade fights since the 1970s – those who wanted intellectual property rules in trade agreements and those who didn't – tried to "spin" the discussion with metaphors for theft.

The US Trade Representative's office and the biotechnology industry explained the need for intellectual property rights in trade agreements with talk of $40–60 billion in losses due to intellectual property piracy; they claimed that the quality of pirated products was lower than the real thing and that the piracy was costing lives; and they blamed all losses on Third World pirates.

Those who criticized the TRIPs (Trade Related Intellectual Property Rights) provisions in GATT (the General Agreement on Tariffs and Trade) or NAFTA (North American Free Trade Agreement) pointed out that the Third World and the indigenous world also suffered losses due to piracy. They noted that many products made in the industrial world, almost all its food crops, and a high percentage of its medicines originate in plant and animal germplasm taken from the developing world. They observed that theft was twofold: first was theft of knowledge of biological material and how to use it, and second was theft of the material itself. Noting that no royalties had been paid for the use of this material, they called the unagreed to, unacknowledged appropriation of the material "biopiracy," and suggested that the World Trade Organization trade rules would likely be interpreted to make continuing theft of genetic material easier for the Industrial World. In counterpoint, spin-meisters from the industrial world retorted that what was claimed to be biopiracy was in reality "bioprospecting" of raw materials.

Parrying the claim that "raw" materials collected in the developing world were "natural" materials and therefore did not qualify as patentable and worthy of industrial-style rewards, the developing world answered that the seeming "natural" materials stolen from them were the result of millennia of study, selection, protection, conservation, development, and refinement by communities of Third World and indigenous peoples, and were therefore no less worthy of recognition and respect and compensation than the products of the industrial world. They also noted that to consider only the inventions of white men in white lab coats to be inventions worthy of recognition and reward is to hold a fundamentally racist view of human creativity: it amounted to once again declaring land "empty" of inhabitants so that it may be claimed for the king and queen of Spain. And, they further pointed out that, by enclosing biological materials in patents, trade agreements were about to transform the rich creative interactions of cultural and biological diversity into a new economics of scarcity.

Finally, Third World critics observed that the patent system of GATT and NAFTA – and likely that of subsequent trade agreements – is a system that benefits highly industrial societies and is not necessarily suitable, desirable or healthy for less industrialized cultures or countries. They recognized – much as the American colonies of England had recognized in the eighteenth century – that no matter how the patent-holders and patent rule-makers chose to spin it, industrial-style patents would not necessarily lead to the transfer of new technologies to or a better life in the developing world but were much more likely to lead to the devastation of local industries, importation of high-cost products by small elites, exportation of (so-called) "raw" materials not protected by patents, and destruction of local and indigenous cultures and cosmologies.

Further, the process that forced some people to adopt others' notions of property and creativity – that thousands of years of ongoing experimentation and production do not deserve the same compensation as a few years of indoor laboratory tinkering – is not only insulting but also very, very costly. To a developing world whose creations may not necessarily result in patent royalties, there was first of all the cost of unrealized profit. Second, there was the cost of added expense: with the extension of patents to living organisms and human body parts and genes, and with the extension of the industrial patenting system to the whole world via various trade agreements, Third World and indigenous communities faced a very legal, sizable, and collectable bill for royalties.

Patents on seeds, for example, could result in (1) farmers denied their traditional rights to save seeds (planting seeds without paying royalties is making an unauthorized copy of a patented product); (2) farmers forced to pay royalties for every seed and farm animal derived from patented stock; and (3) farmers forced to become more dependent on fertilizers and herbicides made by the same companies who collected their traditional seeds in the first place and now sell back the chemically dependent derivatives. The latter concern is heightened by the current direction of research and the increasing ownership of seed companies by agrochemical corporations (see Chapter 16 in this volume). The cost of patents on biologicals used in health care and medicines would be even higher and more horrific. In general, the whole patenting process would lead to greater and greater Third World indebtedness to the industrialized world with little or no recognition of the enormous debt incurred in the other direction.

Thus by the end of the twentieth century sophisticated legal devices called patents were perceived in some places as tools leading to just rewards, and in other places as mechanisms for allowing acts of piracy and imposing crushing costs. One side claimed patents were protection from thieves while the other side remarked that those who demanded patent protection from thieves were once, and continued to be, thieves themselves. With such differing attitudes, the problem became how to know whose yardstick to use and which is the proper spin? In the United States, the Supreme Court decision in *Diamond* v. *Chakrabarty* (see Chapter 21 in this volume) had already greased the way for patenting "everything under the sun made by man." The nations who signed the GATT made patents available for "any invention … in all fields of technology." But court decisions and trade agreements offer no guidance for ethical behaviour when dealing with people for whom everything under the sun is sacred and therefore never to be considered property.[8]

The problem of patent yardsticks also depended on the answer to another question: should the human relationship to the natural world be a commercial relationship? Most might say no and think the relationship should be a matter of ethics. However, in a world governed by trade, ethics may be perceived as a barrier to trade and therefore not an allowable consideration.

Even an agreement intended to deal with the natural world – the UN Convention on Biological Diversity (CBD) – turned out to be a trade agreement and thus engendered the same commercial/ethical dilemmas and the same opportunities for legal spin. The CBD promised conservation and protection of biodiversity and created legal space for the recognition and enforcement of indigenous rights. But the Convention also centred on the use of natural resources (including genetic resources), and the equitable "sharing of benefits," thereby legitimizing a market for owned species and genes (i.e., living resources and their parts) and effectively diminishing most biodiversity to the status of property of the master species. Other beings were not to be seen as honoured fellow members of a greater ecosystem but were to be reduced to the rank of commodities, valuable gene pools, containers to be divided and spliced and owned and priced and sold.

In the rush for genes, theft hardly waited for spin in the case of the CBD. Even before rules of equitable sharing could be worked out, there were attempts to access and patent material collected from Third World and indigenous communities before the CBD came into force and transfers of biological specimens became subject to its rules. There were accessions from public gene banks and botanical gardens containing colonial collections; there were bioprospectors offering inadequate bilateral agreements to communities in which they searched; there were agents freely bioprospecting national parks for corporate clients; and there were companies asking vacationing employees to bring home a spoonful of dirt because it might contain some microorganisms the company could use.[9]

The point of all the biospin, of course, was money. The burden of theft continued to fall so unevenly on Third World and indigenous communities because, in the half-millennium since the king and queen of Spain gave letters patent to Columbus, indigenous and Third World peoples had continued to live where the genetic diversity was richest and to act as its stewards. Having maintained and protected most of the wealth, they were now forced to hold off the thieves.

Once trade agreements or other legal arrangements extended patents to living organisms and their parts, life anywhere could be discovered, transported, owned, manipulated and made worth an investment. When a marketable product could be teased out of what is patentable, the profits could be enormous. Whether the booty was neem from India, or endod from Ethiopia, or the cells of a man from Seattle, or the cheek-scrapings of indigenous peoples, or the biodiversity of an entire Costa Rican rainforest, or one important microorganism from the hot springs of Yellowstone National Park, the value of biodiversity is difficult to exaggerate.[10]

Consider the potential of a few products derived from indigenous knowledge. According to the Rural Advancement Foundation International, neem, a plant that grows mostly in Southeast Asia and produces a kind of natural insecticide, could be worth about $50 million a year.[11] Thaumatin, a natural sweetener derived from a West African plant, might command a sizeable portion of the $2 billion low-calorie sweetener market in the US. And endod, a perennial plant used by Ethiopian women for centuries, has multimillion-dollar profit potential for controlling the zebra mussels that now clog pipes in the Great Lakes.

Modern spin-meisters continue to say the point of patents is not money but human progress. Without the protection of patents, they argue, no scientific (and certainly no medical) progress will be made; no one will risk an investment unless they are guaranteed an eventual monopoly. And yet there are cases where patents stand in the way of human health and progress.[12] And there are all those inventions made in communities throughout the world and throughout all time without the incentives of patents. And there are all those problems created when communities extend the culture of patents to the domain of the living.[13]

SNAPSHOTS FROM BEING THERE

In 1993 I attended a conference in Seattle, Washington, on "The Future of Intellectual Property Protection for Biotechnology in the United States, Europe, and Japan." For three days, eminent speakers discussed patents. On the last day of the conference, one of the panelists bemoaned the situation in

Europe (where at that time it was nearly impossible to obtain a patent on any form of life). The panelist hoped his colleagues in other places would never have to face the situation he faced with "environmentalists and those who would bring ethics and other irrational considerations to the table." Those were his words: "ethics and other irrational considerations." No member of the audience challenged his pairing of "ethics" and "other irrational considerations." Not one lawyer. Not one official. Not one academic. No one.[14]

That same year, a draft proposal for the work of the Human Genome Diversity Project (HGDP) came into my hands.[15] It arrived two days before I was scheduled to go to San Francisco to address a group of indigenous peoples on the intellectual property implications of the (then-proposed) North American Free Trade Agreement and (the then latest version of) the General Agreement on Tariffs and Trade.[16] The HGDP appeared to be a study of high intellectual purpose headed by scientists and scholars with extraordinary credentials. Proposing to collect human tissue from 722 human populations, including indigenous peoples on the verge of extinction, the project aimed to use, to quote one admiring article, "new genetic techniques ... [to solve] the ancient mysteries of mankind's origins and migrations."[17]

Nevertheless, like those who sent me the draft, I found the ethical implications of the HGDP troubling. When I spoke in San Francisco, I used my allotted time to share what I knew of the project. I remember the looks on people's faces and the great silence and sadness that followed my remarks about collecting samples of hair and blood and cheek scrapings. The silence lasted a long time, and when finally someone spoke it was Jeanette Armstrong, an indigenous woman from Canada. She said:

> You people. We thought you folks had taken everything you could. You took our land, you took our homes. You stole our pottery and our songs and our blankets and our designs. You took our language and in some places you even took our children. You snatched at our religion and at our women. You destroyed our history and now, now it seems you come to suck the marrow from our bones.

Other indigenous peoples would later call the HGDP "the Vampire Project" – not because it intended collection of blood samples but because its only apparent interest in the peoples from whom it would collect – some of whom it labelled "Isolates of Historic Interest" and considered peoples on the verge of extinction – was interest in their genetic material.

In 1994 the World Council of Indigenous Peoples issued a pamphlet about the HGDP entitled *Presumed Dead ... but Still Useful as a Human By-product*.[18] The pamphlet charged that the proposed collections of DNA samples "will supposedly help preserve indigenous gene cultures for generations to come ... [but the] real agenda is the future development of pharmaceuticals that will generate huge corporate profits, long after indigenous people have been left to disappear." The pamphlet went on to assert:

> The assumption that indigenous people are doomed adds insult to the indignity of being used as human guinea pigs. The millions of dollars to be spent on the Human

Genome Diversity Project could fund healing and community development for those indigenous peoples that are considered at risk.... The Human Genome Diversity Project has already begun dehumanizing us by labeling us as "Isolates of Historic Interest" (IHI).

While admitting that a "human harvest" of blood samples, cell scrapings and hair roots could have potential scientific value – "The unique characteristics of some genetic samples could support the development of drugs to fight cancer or prevent Alzheimer's disease" – the pamphlet also warned that, "At least two countries, the United States and Japan, allow companies to claim patent rights over human genes and DNA segments. This means that the bloods and living cells of indigenous peoples will become the property of a private research group."

Perhaps no case concerning the patenting of life has been more analysed than the case of *John Moore* v. *the Regents of the University of California*.[19] While Moore was under treatment for hairy cell leukaemia, the doctor supervising his care noticed that Moore's blood cells produced an unusual blood protein that showed promise as an anti-cancer agent. Without Moore's knowledge or consent, the doctor obtained a patent for the university on Moore's cell-line, listing himself and his research assistant as "inventors." In 1984 Moore became aware of the patent and sued the doctor, the university, and the pharmaceutical company to which the doctor and the university had licensed the cell-line.

The case raised novel questions. Did Moore's doctor, an employee of the University of California, wrongfully take and profit from parts of Moore's body? Did Moore deserve a fair share of the profits from the products resulting from his cell-line? What would happen to Science if the court recognized John Moore's property rights? The case took years to be resolved and in 1990 the California Supreme Court ruled that Moore's doctor had breached his "fiduciary duty" to Moore by not revealing his research and financial interest in Moore's cells. Nevertheless, the court denied Moore's claim to ownership of cells removed from his body, explaining:

> Research on human cells plays a critical role in medical research. This is so because researchers are increasingly able to isolate naturally occurring, medically useful, biological substances and to produce useful quantities of such substances through genetic engineering.... The extension of conversion law into this area will hinder research by restricting access to the necessary materials.[20]

Ethicist and lawyer George Annas, commenting on the court decision, noted,

> California's courts decided who could reap profits from a cell line ... as is clear from the text, the majority simply accepts the Chicken Little argument that if John Moore's property interest in his cells is upheld, the biotechnology industry's sky will fall on them and medical progress will suffer a major setback. In this regard the justices seem to have been blinded by science and unable or unwilling to distinguish it from commerce. The court essentially concluded that the biotechnology industry is both wonderful and fragile. Since it is wonderful, we must do our part to foster it; since it is fragile, we must protect it from harm.... In the court's flowery words, recognizing

[Moore's property claim] would threaten to destroy the economic incentive to conduct important medical research.[21]

Finding the decision unsatisfactory, Annas noted that effectively the court had ruled that "only the little people can't sell cells."[22]

Years after the famous court decision, I arranged an interview with John Moore. I began by asking him, "What did it feel like to be patented?" John Moore said an odd thing to me in reply. "Where have you been?" he said. "Where have you been?" I was startled and it took some time for me to understand his meaning. In the years since his case, John had been characterized as everything from a greedy villain out to destroy the progress of medical science to an ungrateful biological freak. No one seemed to remember that others had patented his cell line for profit. Few had noticed his deepest wound:

> "How does it feel to be patented? To learn all of a sudden, I was just a piece of material.... It's so beyond anything you can conceive of. There were so many issues involved.... There was a sense of betrayal.... I was told that, in a dinner conversation with a colleague, [my doctor] had said that, "John Moore is my gold mine." This was certainly true; he had discovered genetic gold in my cells.[23]

In 1996 John Moore appeared before a committee of the United States National Academy of Science. He was there to speak against the Human Genome Diversity Project and he began his testimony to the distinguished panel of scientists saying, "I am known as Patent #4,438,032. Some of you may be familiar with pieces of me in your laboratories." Recounting the story of his cell-line, he reminded the panel:

> Ultimately, everyone was protected and rewarded – the researcher, the physician, the entrepreneur, even Science. Everyone was protected and rewarded ... but the patient, the subject, the source. Yes, me. Didn't I have a right to know? They were exploiting the uniqueness of my genetic material, growing my cells in a Petri dish, patenting my genetic essence. Shouldn't I have been told? Shouldn't someone have explained to me, the one from whom the gold was mined, the person who did not yet know anything about cell lines, genetic codes, blood proteins, GMCSF, patents, contracts, or stock options? Because I knew nothing, was I to be treated as nothing?

Moore continued, "You may at this point be asking yourselves, 'What does this have to do with the HGDP and why is John Moore here telling us his personal story? How does this relate?" Then he concluded:

> Here are my concerns.... I am concerned because the dehumanization of having one's cells conveyed to places and for purposes that one does not know of can be very, very painful. Why should I or any individual or group of individuals have their unique genetic materials borrowed, stolen, or bought for some fraction of their value for some project of others? ... How can anyone else set and get a price on what may be priceless or sacred to someone else? ... I am concerned because even in this country where the rights of the individual are supposed to be protected from the grasp of institutions and certainly from the greed of private corporations and

researchers, they are not.... Do you think a system that could not protect me will protect the rights of peoples and individuals that live in other countries? ... I don't ... I have had a very sad experience. I was violated by a legal system and a set of values in my own country. I know what to expect and I have no reason to believe that the environment which allowed the theft of my genetic material has changed in any positive or material way.[24]

Ironically, at the HGDP hearing there was another John Moore, an anthropologist similar in physical appearance to the patented John Moore. The anthropologist was there to testify on behalf of the HGDP; he was, it turned out, one of the heads of the North American Human Genome Diversity Project.

In November 1996, during an afternoon break at the third Conference of the Parties (COP 3) to the Convention on Biological Diversity, the patented John Moore shared his experiences once again, this time with a group of indigenous peoples. He talked about his sense that the ruling in his case had paved the way for the collection and patenting of indigenous cell-lines. Among those listening to Moore was Colombian senator Lorenzo Muelas, an eloquent campaigner for the return to the indigenous peoples of Colombia of all the genetic materials taken from them. When Moore noted that journalists in the US had branded him a man "against the progress of science," Muelas laughed. Referring to the press reaction in Colombia, he noted, "They said the same thing to me."

In 1997, seventeen years after the *Chakrabarty* decision, a large group of US citizens and public interest organizations wrote to members of the European Parliament, which was on the verge of considering legislation that would extend life patents in Europe:

Once the US Supreme Court decided to allow the patenting of a microorganism ..., other patents for higher life forms followed quickly. Currently, the United States Patent and Trademark Office has granted approximately 35 patents that actually claim animals, including animals that have NOT been genetically engineered. We are now in the midst of massive attempts by patent seekers to divide up the entire biotic community for private interest.

The adverse effects of patenting life daily become more and more apparent (and burdensome) here. Several years ago the California State Supreme Court effectively sacrificed individual civil and property rights for the good of an industry (*Moore v. Regents of the University of California*). Later our leaders thought it wise to pursue patents on the cells of far-off indigenous peoples; when that proved the source of international embarrassment, the patent claims were dropped but the precedent was established. In the meantime we passed technology transfer laws that resulted in great universities corrupting themselves by the pursuit of patents and profits and becoming indentured to industry.

Increasingly, we found that patents constricted innovation and distorted the pursuit of knowledge itself: our peer review system has been put at risk by fear of theft of patentable ideas; universities have sued graduate students over patents resulting from ideas they had while students; graduate students have organized to prevent theft of their intellectual property by their universities and professors; our researchers have postponed publication of medical breakthroughs while they and their corporate sponsors await issuance of patents; and fruitful lines of research have been abandoned altogether to avoid the expense of licensing, the possibility of patent litigation, and the limitations of doing research on the terms of licensing agreements. In the United States researchers boycotted the use of the Oncomouse [the first patented animal], for example, because of the patent holder's licensing agreement sought rights to derivative inventions.

Similarly, farmers have balked at taking up new innovations that carry the burden of licensing agreements that specify in exact and sometimes onerous terms the method of farming that must be used with the patented product. Finally, those who looked on the mire of our patenting regime and despaired of its ability to offer them protection sought to expand our trade secrets law and find solace there. In doing so, they in some cases foreclosed the sharing of innovation altogether. Thus, instead of being a stimulus to creativity, patents on life became the catalyst that narrowed and discouraged research in all directions.

Is this situation what you wish for Europe?[25]

As matters turned out, it was the situation they chose for Europe. In 1998 the European Parliament reversed a decision made three years earlier and extended the possibility of granting patents on life to all the member countries of the European Union. A hard-fought campaign preceded the vote but the "No Patents on Life" proponents, with moral arguments about adverse impacts on the Third World and the erosion of the integrity of life and the health of civil society, proved no match for the better-financed spin doctors. They argued that without proper patent incentives, the jobs, scientists and profits of an uncompetitive European biotechnology industry would drain towards more patent-friendly shores, and then augmented their arguments with crippled children suffering from hereditary diseases, brought in wheelchairs to the halls of parliament. The children wore T-shirts that declared "No Patents, No Cures."[26]

Several years ago, I asked an industry analyst why the public outcry in the case of Diamond v. Chakrabarty was not greater and why the US environmental community did not see an ethical problem. He answered,

> The environmentalists, for one example, can be handled. When we went for life patents, they were kept quiet by the fact that the first patent applied for was for a microorganism that could eat oil. You think that was an accident? What environmentalist was going to get in the way of something that might clean up oil spills? So we obtained the right to own life. Now you're telling me they're going to get upset about the theft of other people's ideas and resources! Get real. With neem, we're giving them natural pesticides. The enviros will never object. They'll probably never even notice they've been handled.... Everybody gets handled.[27]

WHERE ARE WE GOING?

In May 1998, in Bratislava, Slovakia, in an invited presentation to the plenary of the Convention on Biological Diversity, a group calling itself Diverse Women for Diversity stated, "We support Article 8J of this Convention because we recognize that communities have boundaries and rights. And we insist that the sovereignty of communities with respect to their knowledge and resources take precedence over the freedom of outsiders to access and appropriate that knowledge and those resources." They were talking about who has rights to whose commons. They were responding to theft.

By 2000 biopiracy had reached epidemic proportions with cases popping up from Texas to Australia. Everything from the holdings in international seed banks to the biodiversity of national parks seemed ripe for plucking and patenting. At the World Trade Organization in Geneva they were trying to launch a new round of trade negotiations. In Washington DC the Patent and Trademark Office was grinding to a halt under the weight of new patent applications. Genomics companies were sequencing entire genomes in the hope of patenting them. Pharmaceutical companies were asking for patents on everything; one applied for a patent on a bacterium that causes meningitis, thereby raising the possibility of higher costs for treatment if new vaccines were found. And seed companies were eradicating the right of farmers to plant saved seed, justifying their actions with arguments about patent infringement and feeding the world's billions, not seeming to notice that aggressive intellectual property protection was pushing to extinction the very farmers who provided much of the world's food. They did not seem to care that most of the world's small farmers could never afford to purchase patented seed year after year.

Nothing seemed sacred and beyond the reach of the Market. "Hope" was now a word in the official trademark of Monsanto Corporation, one of the biotechnology powerhouses of the world.

NOTES

1. Her analysis was made in the context of a lecture, entitled "Biodiversity and Biopiracy," that she gave January 16, 1996, at the University of Washington in Seattle.
2. Ibid.
3. In the USA, exchange of ideas, information and staff between industry and academia was facilitated by the passage of the Bayh–Dole Act in 1980. The Act allowed universities to obtain patents for discoveries made in federally supported laboratories. Before the Act, the federal government usually held title to discoveries made with the help of public funding. After the Act, universities could transfer to private companies via licence agreements exclusive rights to technologies developed with public moneys.
4. See George S. White, *Memoir of Samuel Slater, the Father of American Manufactures* (1836), Reprints of Economic Classics, New York: Augustus M. Kelley, 1967; and Christopher Simonds, *Samuel Slater's Mill and the Industrial Revolution*, New Jersey: Silver Burdett Press, 1990.
5. Hamilton's recommendations are quoted in White, *Memoir of Samuel Slater*, p. 86.
6. According to a 1995 Pfizer Pharmaceutical advertisement in *The Economist*, by this time "it became clear that tougher global competition lay ahead for the US". Edmund T. Pratt, Jr., Pfizer Forum, "Intellectual Property Rights and International Trade," *The Economist*, May 27, 1995, p. 26.
7. The inclusion of intellectual property rights in international trade agreements was clearly the victory of transnational corporations headquartered in the industrial

world. The corporations bragged openly that they were the ones who pushed Trade Related Intellectual Property Rights (TRIPs) onto the GATT agenda. The 1995 Pfizer advertisement in *The Economist* mentioned in note 6 made clear:

> In conjunction with more than a dozen companies from all the relevant sections of US business, Pfizer and IBM co-founded the Intellectual Property Committee or IPC. The US Trade Representative was impressed and suggested that we increase our effectiveness internationally by joining forced with UNICE, the principal pan-European business group, and its counterpart in Japan, Keidanren.... Working together ... our combined strength enabled us to establish a global private sector network which lay the groundwork for what became "TRIPs".

8. *Diamond v. Chakrabarty*, 447 US 303 (1980).

9. In Europe in 1997, transnational corporations apparently sought to avoid negotiating with Third World and indigenous negotiators by getting their biological samples from European botanical collections containing Third World and indigenous materials collected before the Convention on Biological Diversity came into force. The companies asked towns that held these collections to take small payments in return for guaranteeing that the sampled material was really the property of the local collection. Thus, according to German activist Christine von Weizsäcker, who reported on the incidents at several international meetings related to the CBD, unknowing communities were being asked to condone the historical biopiracy of colonization and to make themselves accessory to a new double-layered biopiracy that violated the spirit of the CBD.

10. To assess the potential value of biodiversity and how "spin" can affect the perception of that value, it helps to examine such recent bioprospecting deals as (a) the arrangement between a pharmaceutical company, Merck, and a non-governmental organization, InBio, in Costa Rica and (b) the contract between Diversa Corporation and the United States National Parks Service. In the InBio deal, according to the Crucible Group's *People, Plants, and Patents*, Ottawa: International Development Research Centre, 1994, p. 11, Merck provided "$1.135 million for 10,000 extracts from biological accessions" and also agreed to a "royalty sharing system if any of the material is commercialized." The royalty rates were not made public but, according to a *RAFI Communiqué* (September/October 1997), they "have been widely reported to be 5% or, more likely, less." In the Diversa deal, royalty rates were also not disclosed publicly and the stipend to the National Parks Service for allowing Diversa to bioprospect Yellowstone National Park was considerably less than that Merck gave InBio. If, on the other hand, the value of "successful" products derived from bioprospected materials is examined, quite another picture emerges. *Thermus aquaticus*, a microorganism taken from Yellowstone several years back, had an enzyme that earns Hoffman LaRoche, the Swiss drug giant that holds its patent, more than $100 million a year, with earnings projected to increase to $1billion a year by 2005.

11. See Rural Advancement Foundation International, *RAFI Communiqué*, "Biopiracy Update: The Inequitable Sharing of Benefits," September/October 1997.

12. See, for example, "American Company to Patent Bacteria," *Guardian*, May 7, 1998; and Seth Shulman, "Cashing in on Medical Knowledge," *Technology Review*, March/April 1998.

13. For an excellent and lengthy discussion of technical patent issues and strategies for dealing with them, see *Signposts to Sui Generis Rights*, resource materials from the international seminar on sui generis right, co-organized by the Thai Network on Community Rights and Biodiversity and Genetic Resources Action International, Bangkok, December 1–6, 1997, Bangkok: BIOTHAI and GRAIN, March, 1998.

14. The remarks were made during at panel discussion at an international conference on "The Future of Intellectual Property Protection for Biotechnology," held October 23, 1993, at the University of Washington School of Law in Seattle.

15. The draft was circulated in 1993 by the Rural Advancement Foundation Inter-

national, a non-governmental organization headquartered in Canada.

16. The event was a conference entitled "Commodification of the Sacred" sponsored by the Cultural Conservancy, a US non-governmental organization headquartered in California.

17. A. Goodheart, "Mapping the Past," *Civilization*, vol. 3, no. 2, 1996, pp. 40–47.

18. *Presumed Dead ... but Still Useful as a Human By-product*, Ottawa: World Council of Indigenous Peoples, 1994.

19. See also B. Burrows, "Second Thoughts about U.S. Patent #4,438,032," *Bulletin of Medical Ethics* 124, 1997, pp. 11–14; and G.J. Annas, "Outrageous Fortune: Selling Other People's Cells," in *Standard of Care: The Law of American Bioethics*, New York: Oxford University Press, 1993, pp. 167–77.

20. *Moore v. Regents of the University of California*, 793P.2d 479, 271, *California Reporter* 146, 1990.

21. Annas, "Outrageous Fortune," p. 172.

22. Ibid. p. 176.

23. J. Moore, "Testimony of John Moore to the Committee on Human Genome Diversity of the National Academy of Sciences, September 16, 1996," an occasional paper of The Edmonds Institute, Edmonds, WA, p. 2.

24. Ibid.

25. J. Mendelson, B. Burrows, E. Flynn et al., Letter to the Members of the European Parliament, Brussels, Belgium, July 4, 1997.

26. The children in wheelchairs incident was widely reported at the time. Tanya Green of the GAIA Foundation (U.K.), confirmed the details in a phone conversation, June 25, 1998.

27. The analyst, originally interviewed in 1994 for an article that later appeared in *Boycott Quarterly*, asked for anonymity.

18

Biotechnology and Indigenous Peoples

VICTORIA TAULI-CORPUZ

In 1993, during the first meeting of the United Nations Commission on Sustainable Development, a group of indigenous representatives met with Rafe Pomerance, former US president of Friends of the Earth and the head of the US government delegation. He patiently answered our questions about biosafety, and his country's refusal to sign on to the Convention on Biodiversity (CBD). But I got worried when he said, "everything within the Convention is negotiable except for one issue, which is intellectual property rights."

I explained that our views diverge from his, from that of transnational corporations, and from Western thinking in general. We simply don't believe that the Western intellectual property rights regime should be imposed on us, nor on the rest of world for that matter. "That is why you need to be part of the global market: to protect your intellectual property rights," he responded. But this is one of the problems: we don't have any control over this global market economy. How can we protect our rights in an arena where we don't have any say over the rules of the game and we are not even acknowledged as key players? It is precisely the market economy which marginalized our indigenous economic systems.

A year later I was on a panel with Andre Langanay, a former committee member of the Human Genome Diversity Project (HGDP), at the "Patents, Genes, and Butterflies" conference in Berne, Switzerland. He was asked to talk about the HGDP and I presented my critique of this project. During the open forum portion he said he couldn't understand what indigenous peoples have against the extraction of their blood if this means that they can contribute to the discovery of new cures for diseases. If he were asked to give his blood in order to help others to get well, he would have no second thoughts about it, he argued.

His statement shows how different our worlds are. He has not gone through the experience of being colonized and having his community militarized because the government or a corporation wants to appropriate his people's lands and resources. Most indigenous peoples have gone through this experience. Much of what we have is being taken away or destroyed in the name of development and progress. The HGDP is still the appropriation of what we have and even of what we are, not just for the sake of science but for more profits.

For those of us whose human rights have been grossly violated, from colonization to the present, it is important that we assert our rights to have control over our own bodies, our territories and resources, and our knowledge and cultures. This is what our opposition to the HGDP is all about. Since the HGDP is one of the biotechnology projects directly impacting on us, it has become a major component of the whole discourse on indigenous peoples and biotechnology.

There are divergent views on the role of biotechnology in bringing about sustainable development. The mainstream view is that this will feed the world, cure diseases once thought to be incurable, clean up the environment, and even increase biodiversity. It is seen as an inevitable development of science and technology and it makes no sense to fight against it. On the other hand, there are many who contest these promises and claims. They also question the soundness and ethics of the science that underpins genetic engineering. Indigenous peoples belong to this latter category.

Corporations engaged in biotechnology are the most ardent proponents of harmonizing intellectual property regimes all over the world. We are worried about how biotechnology and intellectual property rights (IPRs) are being used to undermine further our rights as indigenous peoples. This chapter will attempt to articulate the perspectives we have regarding modern biotechnology, particularly genetic engineering and IPRs. Indigenous peoples do not have a homogeneous view about these. However, there are basic elements that we agree upon. Numerous consultations have been held among indigenous peoples on these issues and this chapter will re-echo some of the views that have emerged from these discussions. I will also share my own views and experiences in dealing with these issues.

THE PROBLEM OF DEFINITION

Biotechnology can be defined as "any technique that utilizes living organisms (or parts of organisms) to make or modify products, to improve plants and animals or to develop micro-organisms for specific purposes."[1] By this definition, biotechnology is as old as humankind. Ancient farmers, women and indigenous peoples have been domesticating and cross-pollinating plants since time immemorial. Cross-breeding and taming of wild animals were also done. Such human interventions have led to the further development of biodiversity, complementing the acts of nature.

Indigenous biotechnologies included fermentation technology to brew beer, wines and other food preparations, and the domestication of wild plants and animals. We, the Igorot people in the Cordillera region of the Philippines, have been fermenting our own *tapey* (rice wine) and *basi* (sugar cane wine) since time immemorial. *Tapey* is made with a native yeast called *bubod*, which is preopared by the women. *Basi* is made with seeds called *gamu* that come from the forest.

There are also a host of cross-breeding efforts by indigenous peoples on animals and plants. Potatoes have been domesticated and bred by Huancapi Indians of the Peruvian Andes. The Igorots have been cultivating and breeding a wide variety of *camote* (sweet potatoes), which were a staple for us before rice was introduced. My father never fails to tell us that as a child, he grew and was nourished by all kinds and colours of *camote*. He never saw rice until he became an adolescent. When rice was introduced different varieties were developed by our people to suit the environmental conditions in our territories. In one village alone there are more than ten varieties of rice seeds planted for different weather and soil conditions. Many varieties of other root crops like cassava and taro were also developed.

Indigenous peoples have also discovered a vast array of medicinal plants through the generations, and are still using many of these. From the late 1970s to the early 1980s I worked with an NGO doing community-based health work, and we did research on the medicinal plants in our region. We published a book that includes not only the medicinal plants but also their uses, which parts of the plant can be used, and the various forms of preparation and administration. With the rush of biopiracy, we have had second thoughts as to whether we should even show this research to others. However, as the book has been published, the information is already in the public domain.

To say that indigenous peoples have contributed significantly to the present body of knowledge possessed by scientists, such as ethnobotanists, ethnopharmacologists, and by agriculturists, foresters and food technologists, is an understatement. The development of these indigenous biotechnologies is continuing. However, the recent moves of biotechnology and agribusiness corporations to appropriate what we have and know will influence whether indigenous knowledge and technologies will continue to flourish.

Today biotechnology is more often associated with the most modern technologies, particularly genetic engineering, new cellular procedures based on the old technology of tissue culture, and embryo transfer. It is this modern biotechnology which poses a major threat to our indigenous values and belief systems, lifestyles, biological diversity, and the last remaining indigenous sustainable resource management systems and socio-politico-economic formations. The philosophical, social, economic, ecological and cultural implications are serious not only for us indigenous peoples but for the whole world. From here on, the term "biotechnology" will refer to these new biotechnologies, particularly genetic engineering.

TECHNOLOGY AND VIOLENCE
AGAINST INDIGENOUS PEOPLES

The capacity of biotechnology to transfer genes within and between species of plants, animals, microorganisms and human beings, to convert living material into new shapes and forms, to redesign and engineer new life forms within a short period of time is unprecedented in human history. This is the big difference between the new biotechnologies and what we previously knew as biotechnology. In the traditional cross-breeding of plants and animals, the reproductive process was not drastically short-circuited. Genetic engineering not only short-circuits the reproductive process, but it creates new life-forms never before seen on the face of the earth.

While it is true that throughout human history people have altered their environment and manipulated living things including themselves, this ability is constrained by nature itself. Plants can be bred with other plants and animals can be bred with closely related species. Genetic engineering, however, can transcend nature's boundaries. It is capable of changing the fundamental structures of living things because genetic material from an animal can be inserted into a plant, and an animal gene can be inserted into a human. The boundaries between species are broken down by genetic engineering. Horizontal gene transfer (transfer of genes between plants, animals and humans), which rarely takes place spontaneously, can now be easily done through genetic engineering.[2]

Microorganisms, plants, animals and human beings, or parts of these, are the main raw materials for the biotechnology industry, just as inanimate, non-renewable matter (mineral ores, oil, petroleum, etc.) were the main raw materials for the Industrial Revolution. The gross exploitation of non-renewable matter is regarded by indigenous peoples as the rape of mother earth. The history of colonization and exploitation of many indigenous peoples in various parts of the world is the story of how the colonizers and corporations got their hands on the rich deposits of minerals and the abundance of forests and forest products found in indigenous peoples' territories. All kinds of methods, legal and illegal, were employed by them to appropriate this natural wealth.

The violence done by the colonizers and even post-colonial governments to our ancestors and ourselves in the process of extracting and appropriating resources make us feel that having this kind of wealth is a curse. If we didn't have these resources maybe we would have been ignored and left to carry on, living our own lives. However, this is not to be because many of our territories are not only rich in minerals but are also biodiversity-rich. Biotechnology can fragment living matter into its smallest components and commodify this process. Now, with the promise of profits in the genetic resources of our bodies, and in plants, animals and microorganisms found in our territories, we are faced with a more insidious and dangerous threat.

VARYING POSITIONS ON BIOTECHNOLOGY

The dilemma for indigenous peoples as they deal with this issue is whether to accept that these developments in science and technology are inevitable. If they are, the only option left for us is to forge the best possible contracts, so we can equitably share benefits derived from these resources and make rules on access which are mutually beneficial. This may be the pragmatic approach, one which follows the advice given by the US delegate mentioned earlier.

A related view is that there is nothing inherently wrong with biotechnology and therefore we should not fight it. The problem does not lie with this science and technology per se, but with who has control over it. If we can have control, then we will be able to use it to our own benefit and to the benefit of humankind. Thus, the strategy should be focused on how to ensure that the transfer of this technology to the Third World and to indigenous peoples can be facilitated. We can lobby governments to exert greater control over the technology on behalf of the people, or build the so-called peoples' government which will do precisely that.

A third view is that biotechnology has its own inherent logic, dynamics and dangers, which will define not only the directions its development will take but also the dominant world-view and individual consciousness. Control over biotechnology is an illusion because you cannot have real control. Its inherent logic will define how you will interpret and organize your research. Therefore the strategy should be to critique biotechnology and oppose its further development. This means exposing and opposing various aspects of biotechnology: the science or world-view which underpins it, its economics, politics and social implications. To protest the patenting of life forms is one aspect of this strategy. Most indigenous peoples support this third position, although there are those who are supportive of the first two views.

Many indigenous peoples' conferences have issued declarations and positions against life patents, calling for a ban on the Human Genome Diversity Project and a moratorium on biopiracy in indigenous peoples' territories.[3] However, we know that in spite of these protests, biopiracy is still taking place, collections of human genetic materials are continuing, and various life forms are still being patented.

While we are not alone in taking these positions, we are the ones who are always accused of being anti-development or anti-progress. We don't apologize for holding these views even though such accusations are levelled at us. At a time when the speed of so-called development or progress is like a runaway train, especially with biotechnology and information technology, it is necessary for some people to suggest stepping on the brakes.

Those of us who have resisted colonization, and whose economies have not been thoroughly eroded by the capitalist market economy, have managed to retain aspects of our pre-colonial cultures. Our cosmologies still revolve around the need to live and relate harmoniously with nature. Our technologies are still rudimentary and not as powerful as those developed in industrialized countries which are capable of redirecting nature and channelling its forces elsewhere.

Indigenous peoples who are in this state of development still maintain an intimate union with nature. Indigenous religion, which is usually a form of animism, in general reflects a reverential attitude towards creation. Even those who were converted to Christianity or Islam or any of the dominant world religions maintain a folk religiosity which combines the dominant religion with indigenous practices.

This is not to say that our own traditions are unchanging in spite of all the developments around us and those brought into our communities. Our cultures are not static. What we have now are results of our survival and resistance strategies to avoid or to cope with the aggressive imposition of the colonizers' ways. While we retain some aspects of our pre-colonial cultures, there has also been an accommodation of the colonizers' cultures.

The alienation between humanity and nature which is characteristic of highly industrialized societies is rarely experienced by indigenous peoples, who still largely rely on nature for their basic survival. Even those who have been introduced to the sophisticated mechanical technology developed since the Industrial Revolution have somehow consciously kept aspects of their ancestors' belief systems and cultures. This can be seen among the indigenous peoples found in industrialized countries. The hunters and fisherfolks among the Inuit in Alaska, Canada and Greenland, for example, do not relate to their prey in the same manner as those who own and manage commercial fish trawlers. They are aware of the need to harvest sustainably to allow for the regeneration of species. They strive to maintain their communities even amidst the strong pressures from the dominant society to assimilate and integrate with the ways of the white people.

The paths of life and spirituality of the Igorots, many of whom are still small-owner tillers, are very much attuned to the agricultural cycle.[4] Community rites and rituals are performed not only during births, weddings and deaths but also during the agricultural seasons of planting, harvesting and weeding. There are rituals to call for the rains to come. The agricultural seasons were determined by the seed varieties we planted and by the climate. For many generations we used indigenous seeds. The introduction of the high-yielding, hybrid seeds of the Green Revolution, however, disrupted the usual periods for community rituals. This is one reason, along with the required chemical inputs, which made many of our farmers go back to the use of indigenous varieties.

CRITIQUE AND IMPLICATIONS OF BIOTECHNOLOGY

The philosophical plane

Biotechnology carries with it a world-view or philosophy that is reductionist and determinist. A living organism is reduced to its smallest component, the gene. The explanation of the way the organism behaves is sought in the genes. This world-view also regards nature as something which should be controlled, dominated, and engineered or re-engineered.

This runs counter to indigenous beliefs, knowledge and practice. The cosmological vision of most indigenous peoples regards nature as divine and a coherent whole, and human beings as a part of nature. Thus, it is imperative that humans should create meaningful solidarity with nature. This is the "web of life" concept, or what is now referred to as the ecosystem approach, which appreciates the relationship and bonds of all aspects of creation. Human beings have to work and live with nature, not seek to control and dominate it. Whether we recognize it or not, we humans are totally dependent on water, air, soil and all life forms, and the destruction or pollution of these will also mean our destruction. The integrity or intrinsic worth of a human being, plant or animal is measured in relation to how it affects and relates with the others.

For indigenous peoples, biodiversity and indigenous knowledge or indigenous science cannot be separated from culture and territoriality. Thus, the genetic determinism of biotechnology conflicts with the holistic world-view of indigenous peoples.

With the invention of technologies that control and re-engineer nature, human beings have succeeded in setting themselves apart from nature. This is what happened after the Industrial Revolution and now with the biotechnology and information revolution. Plants, animals and humans are reduced to their genetic components and their integral wholeness is not important any more. These separate components can be manipulated and engineered at will and for commercial purposes.

The engineering mindset is becoming the norm. Efficiency, not only of machines and human beings but of all living things, is the goal. Since it is life that is being engineered, scientists can act as God. Because profits and economic growth are the most important parameters used to measure development and progress, the adverse environmental, economic, cultural and social impacts of biotechnology are viewed as insignificant.

The way biotechnology further promotes and reinforces the mechanistic, materialistic, reductionist and dualist world-view is a major concern for indigenous peoples. There is an observable, growing intolerance towards other cultures and world-views. Eugenics is promoted along with the universalization of Western standards of beauty and efficiency. Being beautiful means being tall, white, blonde, blue-eyed and slim.

For indigenous peoples to accept the genetic determinist view, they have to alter radically their world-views, their ways of knowing and thinking, and their ways of relating with nature and with each other. Maybe social and natural scientists will say that this is inevitable, because we have to move on with the progress achieved through science and technology. However, with the prevailing environmental, social, economic and cultural crisis, the dominant world-view has lost the moral high ground.

Indigenous peoples, who have not totally surrendered the cosmological vision inherited from their ancestors, and have indeed developed it further, are in a better moral and ethical position. If indigenous peoples keep asserting their own philosophy and their right to believe and practise it, we might someday evolve a different philosophy or perspective that provides a balance between the two extremes.

Ecological and economic implications

The ecological risks of biotechnology have been amply elaborated by NGOs and scientists. Genetically engineered organisms are living beings; if these are released they can mutate, multiply and migrate. Should they have adverse environmental impacts, there is no way to recall or contain them. Since indigenous peoples' territories are the last remaining biodiversity-rich centres, the erosion of this biodiversity could be facilitated by the invasion of more evolutionarily advantaged transgenic plants.

Biotechnology claims that it will be able to clean up the environmental pollution brought about by industrial activities such as mining, oil exploration, and so on. This falls into a typically end-of-the-pipeline kind of pollution management, and we have yet to see this working on a large scale. From the experiences of indigenous peoples, mining and oil drilling operations are still the worst polluters and the most destructive to the land. The track records of mining and oil companies in rehabilitating what they have destroyed in indigenous peoples' territories is very poor, to say the least.

From what we can gather, the efforts of some biotechnologists are being directed towards developing transgenic microorganisms that can eat into mineral ores and isolate precious minerals such as gold. There would be a reduced need for workers and machines to process the ores. The ecological implications of releasing such engineered microbes into the environment, however, is not seriously considered or addressed. Studies on the environmental impacts of the release of transgenic organisms, whether to clean up oil pollution or to ameliorate the pollution of rivers and soils by toxic chemicals used by the mines, are not adequate. The likely consequences are not publicly known, especially by those who are directly affected.

The appropriation of indigenous knowledge on plants and plant uses, along with the destruction of indigenous sustainable resource management and agroforestry practices is also facilitated by biotechnology. Patent applications by scientists, corporations and even governments for medicinal plants used by indigenous peoples since time immemorial are increasing each day. The neem plant and turmeric in India are much used by the tribals. Ayahuasca and quinoa in Latin America, kava in the Pacific, the bitter gourd in the Philippines and Thailand – all are widely used by indigenous peoples.

Quinoa (*Chenopodium quinoa*), for instance, is a high-protein cereal which has long been a staple in the diet of millions of indigenous peoples in the Andean countries of Latin America. It has been cultivated and developed

since pre-Incan times. Two researchers from the University of Colorado received US patent number 5,304,718 in 1994 which gives them exclusive
monopoly control over the male sterile plants of the traditional Bolivian
Apelawa quinoa variety. This crop is exported by Bolivia to the US and
European markets; the value of this export market is US$1 million per year.
The most logical development is that the patent will be taken over by
corporations. The hybrid varieties will be used for wide-scale commercial
production in the US or Europe, and the Bolivian exports will be prevented
from entering the US and European markets. The patent owners will assert
their intellectual property rights.

This will lead to the displacement of thousands of small farmers, most of
whom are indigenous. The other possibility is that lands will fall into the
monopoly control of corporations who own the patents or their subsidiaries
in Bolivia who will produce quinoa using the hybrid commercial varieties.
The genetic erosion of the diverse quinoa varieties developed by indigenous
farmers over centuries will take place.[5]

This process is the most probable course of events for many indigenous
peoples in different parts of the world. This is made possible because of
developments in biotechnology and the legal systems that grant intellectual
property rights to those who are able to innovate in high-technology laboratories. The Trade-Related Intellectual Property Rights (TRIPs) Agreement of
the World Trade Organization has become the standard through which IPR
laws are being harmonized the world over. The contributions of indigenous
peoples in preserving, sustaining and developing biodiversity and resource-
management systems are not recognized and valued by this prevailing system.

THE HUMAN GENOME DIVERSITY PROJECT
AND INDIGENOUS PEOPLES

The ambitious Human Genome Project is a twenty-year effort funded by the
National Institutes of Health (NIH) and the Department of Energy in the
United States of America to the tune of $20 billion. Scientists working on
this belong to an international scientific organization called the Human
Genome Organization (HUGO). The scientists, however, recognized early on
that even if they were able to produce an entire DNA sequence, they would
not still have information on the variation of DNA among humans. They
would like to know the genetic basis of the biodiversity among humans.

So in 1991 they established a committee to develop the Human Genome
Diversity Project (HGDP). "The objectives of the HGDP are to collect,
analyze, and preserve genetic samples from a host of vanishing human populations."[6] It is a massive survey of human genetic diversity. By discovering the
specific DNA differences between populations, they might be able to re-
construct the origins and historical relationships among groups of peoples.
They hope to be able to establish the hereditary basis for differences in human
susceptibility to disease. Researchers have already identified 722 human com-

munities for DNA sampling, and have drafted plans to collect and analyse 10–15,000 samples at a cost of $23–35 million.

They will collect DNA by extracting blood, scraping the inner cheek and collecting hair roots. The collections are called "isolates of historic interest" (IHI). Preservation techniques will be used upon collection, and the researchers will then induce the white blood cells to grow permanently in culture or *in vitro*. This process is referred to as "immortalizing the cell lines". According to HGDP scientists Judith Kidd, Kenneth Kidd, and Kenneth Weiss,

> to ensure permanent samples that can be a resource for many studies, cell lines will be established from individuals in these populations. DNA or the growing cells themselves will then be available to the world research community at no profit or perhaps even at a subsidized cost. Investigators wishing to study questions such as those mentioned here will have access to appropriate material in the cell line "bank".[7]

Leading figures in this project include geneticists and molecular biologists like Luca Cavalli-Sforza from Stanford University, Mary-Claire King of the University of California at Berkeley, Charles Cantor of the US Department of Energy and Kenneth Weiss, a molecular anthropologist at Penn State University, Pennsylvania. In a paper entitled "Call for a Worldwide Survey of Human Genetic Diversity: A Vanishing Opportunity for the Human Genome Project," these researchers said:

> The populations that can tell us the most about our evolutionary past are those that have been isolated for some time, are likely to be linguistically and culturally distinct, and are often surrounded by geographic barriers ... Isolated human populations contain much more informative genetic records than more recent urban ones.[8]

The scientists are aware that their target populations are fast vanishing, so for them time is of the essence. Cavalli-Sforza believes that humans are an endangered species in terms of genetic diversity. He calls the HGDP an "urgent last ditch effort" to collect DNA of vanishing peoples, and is determined to finish the mapping within five to ten years.

CRITIQUE OF THE HGDP

What do we have against this project? The aims of the project look noble and we can grant that the scientists who are involved in it are most sincere in pursuing such aims. However, a look into how the findings of the Human Genome Project are being used would lead one to doubt these noble motives. The working relations between the scientists, the science institutes and the biotechnology corporations such as Genentech and Monsanto are too close for comfort.

It is good that scientists acknowledge that most of the world's human genetic diversity lies with indigenous peoples and that we are endangered; this underscores the urgent need to save this genetic diversity. Indigenous peoples themselves are saying the very same thing. Yet, there is a lack of decisive moves on the part of governments and international bodies to address the genocide and ethnocide of indigenous peoples. To be told that, since indigenous peoples are vanishing fast, there is an urgent need to collect their DNA is adding insult to injury.

In a statement read before the High Level Meeting of the United Nations Commission on Sustainable Development (UN-CSD) in April 1993, I said, "After being subjected to genocide and ethnocide for 500 years, the alternative is for our DNA to be collected and stored. This is just a sophisticated version of how the remains of our ancestors were collected and stored in museums and scientific institutions."[9]

There are many serious concerns to be raised about the project. These revolve around ethical and moral questions. Indigenous peoples' cultural and religious values and rights are being violated by this project. How are the genetic materials and the information going to be used? Who is going to use them and who will benefit from such use? Some of the problems foreseen with the Human Genome Project (HGP) and, subsequently, the Human Genome Diversity Project (HGDP) are the following:

Methods of collecting DNA

Many of the methods employed by corporations to collect genetic materials from indigenous peoples are unethical. One example is the attempt of Hoffman–La Roche to collect the genes of the Aeta people in the Philippines. Since the Aetas became the victims of the eruption of the volcano Mount Pinatubo in 1991, medical missions would periodically visit them. In 1993 Hoffman–La Roche approached the Hawaii-based Aloha Medical Mission, which often visits the Aetas.[10] They tried to link up with this group to collect the genetic materials they need. For people facing calamity, any group that offers charity will be warmly welcomed.

How thoroughly will processes of informed consent be followed, however, considering the time constraints imposed by the proponents on themselves? Will the collectors be thoroughly briefed? It is easy for people from the Department of Health to go to indigenous people's communities and gather blood, cheek tissues and hair roots under the guise of medical missions. Such government agencies can be used to facilitate the collection phase. Health departments do not have a good record of providing health education and services to indigenous peoples, however. In fact, indigenous women have been subjected to forced sterilizations. For such a controversial project there is a strong possibility that the principle of informed consent will not be applied as it should be.

Further, the target populations are those found in remote places, and yet the collected materials, especially the blood, needs to be analysed immediately in a well-equipped laboratory. The proponents are thinking of possible ways in which they can have access to air transportation, such as helicopters. In many Third World countries, where a significant part of the target population is located, it is the military who have helicopters. Such helicopters play a key role in the genocide of indigenous peoples.

The need for sophisticated laboratory equipment to study and preserve the genetic collections means that these collections will usually stay in the developed countries. While the HGDP has proposed to leave duplicate samples of the DNA with the national governments or in regional institutions, the problem of financing such laboratories remains. While the proponents acknowledge that storage laboratories can be in indigenous communities, they still have a rider which says "A condition for establishing such labs ... would have to be that they cooperate on an open basis with investigators interested in the region."[11]

Potential uses of the genetic materials

A new eugenics?

Based on the current uses of genetic materials collected for the Human Genome Project, there is much to worry about. With the discovery of genetic "defects" and "superior" genes, doctors can already proceed with screening "defective" or "superior" embryos and fetuses. The next foreseen step is to abort "defective" fetuses and to clone "superior" ones. Who will determine what bad genes and good genes are?

Is this the first step towards the production of a superior race or super-human beings? Sex determination through amniocentesis is already widely used, especially in countries like India where there is a preference for male babies. With the advances in genetic engineering, where embryos may some-day be manufactured in laboratories, will babies be made to order? The day may come when parents and doctors can create the perfect baby in the laboratory. What will happen if scientists discover the genes that determine the racial characteristics of a fetus? Will parents seek to change the race of their child?

This project is potentially racist, and is actually based on outmoded genetic notions of race. While the proponents claim that the results of the study will erase the basis for discriminating against indigenous peoples, they are not in any position to assert this. The information can be used against indigenous peoples for political purposes. This is like the nuclear bomb. The scientists who created the technology claim it is for peaceful uses. However, when it gets into the hands of those who want to perpetuate their power over the world, political motives overrule the original intent of the research.

Patenting and commercial production of genetic materials

With the additional information and materials which will be gathered from the HGDP, what other possible programmes will be developed? If their aim is to determine the susceptibilities and resistance to diseases, how will such discoveries be used? Will they clone the proteins conferring disease resistances and develop and sell these for profit? The fact that biotechnology corporations are already competing for the control of such materials, and investing in their commercial production and sale, says more than enough.[12]

Patenting is the first step towards the industrial production of inventions or discoveries. Industrial production means the reproduction of millions of identical goods, be these cars, machines, clothes, or whatever. The patenting of life forms will naturally encourage the reproduction of isolated or modified genetic materials, plants, animals and human beings. Scientists are now capable of cloning proteins, microorganisms and even large mammals. The creation of the sheep called Dolly through cloning may have been the first step towards the cloning of human beings.

Craig Venter, a former NIH researcher doing gene mapping and sequencing, has applied for patents on more than two thousand human brain genes. His company has sought ownership over more than 5 per cent of the total number of human genes. Andrew Kimbrell, in his book *The Human Body Shop*, says that

> [should] any one of the genes prove to be extremely valuable, perhaps a key gene for brain cancer research or future therapies to increase I.Q. the researcher ... could then form lucrative licensing agreements with biotechnology companies for exclusive commercial exploitation of the genes.... The entire human genome, the tens of thousands of genes that are our most intimate common heritage would be owned by a handful of companies.[13]

The patent application of the US Department of Commerce for the T-cell line infected with human T-cell lymphotrophic viruses (HTLV) type 1 of a 26-year-old Guaymi woman from Panama was the first attempt to patent genetic materials from indigenous peoples. This application was submitted as early as 1993. International NGOs led by the Rural Advancement Foundation International (RAFI) discovered this application. An international campaign was launched and Isidro Acosta Galindo, the president of the General Congress of the Ngobe-Bugle (Guaymi) wrote to the US Secretary of Commerce demanding that he withdraw the application. The patent claim was denounced by indigenous peoples and NGOs at meetings of the Convention on Biological Diversity and other international gatherings. As a result of this international outcry, the patent application was eventually withdrawn, citing the high cost of pursuing a patent claim.[14]

An indigenous man of the Hagahai people of the highlands of Papua New Guinea had his DNA patented by the United States National Institute of Health on March 14, 1995. This patent covered a cell line containing an unmodified Hagahai DNA. This was also withdrawn under international pressure.

How will genetic materials and genetic information be used?

Indigenous peoples are well known to resist "development" or maldevelopment projects which will destroy their traditional territories. Many indigenous communities are also presently waging armed resistance against the states that are oppressing them. Will genes increasing susceptibility to diseases be used to get rid of belligerent indigenous peoples who are against "development" or "progress"?

If genetic information shows that a certain indigenous group is descended from people from other countries – for instance, that the ancestors of the Igorots come from Southern China – will this be used to deny us our rights to our ancestral lands? What if a group is found to have a genetically high risk of contracting a certain disease? The history of colonization of indigenous peoples shows that biological warfare was often used on them. Smallpox viruses were spread among the resisting Native Americans in North America. Diseases carried by colonial missionaries and soldiers decimated a significant number of Hawaiian natives.

Indigenous peoples have always been discriminated against, and have been portrayed by colonizers as primitive and barbaric. In a world where Western standards and culture are being propagated by media and corporations, the intolerance for diversity is increasing. Will the collection and immortalization of the cell lines of indigenous peoples be a justification for actions that will lead to their final disappearance?

Genetic determinism

It is worrisome to see how DNA or genes are being regarded by scientists. How can sexual orientation and behaviour, for example, be explained by saying that there is a homosexuality or a violence gene? Genes are part of a whole system and an individual is part of a family and society, which are major factors in configuring who she or he is. There is an overestimation of the role played by genes in determining the behaviour and personalities of people.

What could be the possible implications of such conclusions? If the propensity to be a criminal is held to lie with a violence gene, does it follow that this person can be cured through gene therapy? If homosexuality is considered to be produced by a gene and this is regarded as an aberration or a disease, will a time come when gene therapy is applied to "normalize" gays. There is a debate within the gay community over this, with some believing that this supposed discovery will finally diminish the discrimination against gays. Others, however, see this as a reductionist approach. They don't believe that sexual orientation can be explained primarily through genes.

The line of thinking promoted by the HGDP is fraught with dangers. The value of analysing society and better understanding the dynamics between the individual and society will be diminished significantly if we believe that social problems like criminality can be solved by gene therapy, genetic engineering, or by aborting fetuses that are shown to have "criminality genes."

The Human Genome Project and the Human Genome Diversity Project have facilitated the invasion and colonization of the human body by the market economy. Genes are said to be the building blocks of life; thus if life is to be considered sacred, so should the genes be. The effort to map and sequence genes will not just help us learn more about humanity's genetic diversity; it is leading directly towards the commercial exploitation of genes. The patenting of these genetic materials will pass the control over life from nature or God to the patent holders.

RESPONSES TO THE HDGP AND THE PATENTING OF LIFE

The World Council of Churches produced a statement in 1989 calling for a "ban on experiments involving the genetic engineering of the human germline". The outcry of indigenous peoples' groups against the HGDP is another response. Obviously there is a great need to speak out against this sacrilegious treatment of human life. There should be a broad coalition of religious groups, human rights and animal rights groups, the women's movement, indigenous peoples' movements and environmentalists speaking out on these issues. God did not create women and men in his own image only to be reconfigured and commodified by modern society, especially by scientists.

Indigenous peoples have sustained their protests against the HGDP. In June 1993 a conference was held in Aotearoa (New Zealand), from which came the "Mataatua Declaration on the Cultural and Intellectual Property Rights of Indigenous Peoples." This called for a moratorium on the HGDP until such time that its impact has been fully discussed. As early as 1994, I presented a statement at the UN Commission on Sustainable Development asking for a ban on the HGDP. In February 1995 Asian indigenous peoples presented a statement at the European Parliament calling for a halt to the project. During the Fourth World Conference on Women in Beijing, through the leadership of the Asian Indigenous Women's Network, we agreed on the Beijing Declaration of Indigenous Women, which again condemned and called for a ban on the project.

In 1995 seventeen organizations in the Americas signed up to the Declaration of Indigenous Peoples of the Western Hemisphere Regarding the Human Genome Diversity Project. It called on international organizations to protect all life forms from genetic manipulation and destruction. This statement criticized the efforts of Western science "to negate the complexity of any life form by isolating and reducing it to its minute parts ... and [thereby] alter its relationship to the natural order."[15]

Indigenous peoples, especially those who are members of national and international indigenous organizations, networks and alliances, are more or less familiar with these issues. There is still a great need to expand further and deepen the discussions, however. It is also recognized that the whole discussion of biotechnology and biopiracy cannot be tackled without including intellectual property rights and the role of the Trade-Related Aspects of Intellectual Property Rights (TRIPs) agreement of the World Trade Organization (WTO).

This recognition pushed us in the Tebtebba Foundation[16] to organize a workshop of indigenous peoples on Article 27.3.b of the TRIPs Agreement.

This was held in Geneva in July 1999, just before the 16th Session of the UN Working Group on Indigenous Populations. The workshop came up with a statement, "No to Patenting of Life: Indigenous Peoples' Statement on Article 27.3.b of the TRIPs Agreement." This has been sent all over the world through the Internet and at present there are already more than two hundred signatories. Almost all of the major indigenous peoples' organizations and networks from every continent of the world have signed on. We have received very favourable responses to how this statement is being used as a reference for raising awareness on the issue of genetic engineering and the patenting of life.

In Seattle, during the Third Ministerial Meeting of the WTO, a group of indigenous peoples participated in the parallel activities organized by NGOs. We held our own caucus and came up with the "Indigenous Peoples' Seattle Declaration." Again this included a protest against the patenting of life, and is being circulated via the Internet in both English and Spanish versions. There were panel discussions in Seattle on the issue of biotechnology and its implications, and some of us spoke about its impacts on indigenous peoples' communities.

In addition to this, a few of us have participated in the negotiations leading to the adoption of a Biosafety Protocol in the Convention on Biological Diversity. The Biosafety Protocol, which was adopted in January 2000, will regulate the transboundary transfer of genetically modified organisms. The Tebtebba Foundation has worked closely with the Third World Network (an international NGO based in Penang, Malaysia) on this issue. Many indigenous peoples in every part of the world are also taking part in the campaigns against genetically modified organisms and products containing them.

On the national level, there are various efforts on the part of indigenous peoples' organizations to monitor the state of biopiracy and also to lobby for laws which will regulate bioprospecting. In the Philippines, for instance, there is an Executive Order called EO 247, which is supposed to regulate research and bioprospecting in the local communities. This requires prior informed consent before researchers can even set foot in the communities. There are still weaknesses in how this is being implemented, but it has served as a deterrent against the rush of biopirates.

Since 1993 RAFI has consistently maintained "that if any global study of human diversity was to be undertaken, it must be conducted under the umbrella of an intergovernmental organization and with the full, informed consent and participation of indigenous peoples."[17] The HGDP, however, did not welcome this proposal and refused to submit itself to UN supervision.

In spite of these protests, the Human Genome Diversity Project still continues. Those involved are undertaking collections through different channels. In the Philippines, for instance, some professors from the University of the Philippines were given contracts to collect genetic material from indigenous peoples. My daughter, who is a molecular biology student at the university, was asked by her adviser to collect genetic materials from a hundred relatives from her mother's family. She discussed this with me and we agreed that she should not accept this suggestion. We found out later that this professor, who is a molecular biologist, is engaged in gene collections under the auspices of the HGDP.

The UN Commission on Human Rights concludes:

> The HGDP continues, despite the objections of many indigenous peoples. It is arguable that there is a developing awareness and sensitivity to the ethical and legal issues surrounding the collection of the human genome.... It is possible that some of the concerns of indigenous peoples can be addressed through [the] international and local desire to improve consultation with indigenous peoples and through changes in patent law. Some concerns of indigenous peoples, however, cannot be adequately addressed without a complete ban on projects such as the HGDP, and the patenting of human genome.[18]

CONCLUSION

The position of indigenous peoples vis-à-vis biotechnology is still evolving. The common thread in the various positions is the view that life forms should not be patented. If the ownership of patents on life forms is the main incentive for scientists and corporations to invest in biotechnology, it might be a good idea not to allow this. The benevolent motives avowed by scientists who want to contribute to sustainable development should not be tainted by the commercialization or commodification of life.

It is also generally agreed that the harmonization of intellectual property rights regimes to fit the mould of western IPRs is morally and legally indefensible. This is being done to legitimize further the desire of industrialized countries and their transnational corporations to have monopoly control over biotechnology and information technologies. Those who have contributed their centuries-old knowledge to develop and protect the rich biodiversity in their communities will now be accused of biopiracy because the right to this knowledge is going into the hands of the corporations through IPRs.

It should be recognized that indigenous peoples have a right to their intellectual and cultural heritage; this is clearly articulated in the Draft Declaration on the Rights of Indigenous Peoples and other UN standards. This right is being blatantly violated by developments in biotechnology. Even the collection of genetic materials from indigenous peoples' bodies through the HGDP and other similar projects is a violation of the rights and integrity of indigenous peoples.

Indigenous peoples also agree that the protection of biodiversity and cultural diversity cannot be effectively guaranteed if their rights to their ancestral territories are not recognized and respected. Therefore protests against biotechnology cannot be separated from the call for the recognition and respect of the rights of indigenous peoples to their territories and resources and their right to their intellectual and cultural heritage.

The UN Draft Declaration on the Rights of Indigenous Peoples is the emerging standard which should guide states, corporations and society in general on how to deal with indigenous peoples. It was the result of over a decade of intensive dialogues between indigenous peoples, outside experts and government delegations. It is the articulation of the collective values and aspirations of indigenous peoples from the different parts of the world. Indigenous peoples are pushing for the immediate adoption of this before the Decade of Indigenous Peoples ends in 2003.

The march of science and technology will likely proceed in spite of protests from indigenous peoples and NGOs. In the face of the aggressive recolonization of indigenous peoples' territories, bodies and minds which is facilitated by the new science and technologies, it is imperative to support the struggles of indigenous peoples. Whatever gains indigenous peoples will make will also be gains for the whole of humanity and nature.

NOTES

1. US Office of Technology Assessment, "Commercial Biotechnology, An International Analysis," quoted in Henk Hobbelink, *Biotechnology and the Future of World Agriculture*, London: Zed Books, 1991, p. 25.

2. See Mae-Wan Ho, *Genetic Engineering: Dream or Nightmare? The Brave New World of Bad Science and Big Business*, Penang: Third World Network, 1998, pp. 46–7.

3. Examples of these declarations are: Mataatua Declaration on Cultural and Intellectual Property Rights of Indigenous Peoples, June 1993, Aotearoa; National Congress of American Indians, December 3, 1993; Guaymi General Congress, 1994, Panama; Latin and South American Consultation on Indigenous Peoples' Knowledge, 1994, Bolivia; Beijing Declaration of Indigenous Women, 1995, Beijing.

4. See Victoria Tauli-Corpuz, "Reclaiming Earth-based Spirituality, Indigenous Women in the Cordillera," in Rosemary Radford Ruether, ed., *Women Healing Earth: Third World Women on Ecology, Feminism, and Religion*, New York: Orbis Books, 1996, p. 101.

5. GRAIN (Genetic Resources Action International) Briefing Paper, "Patenting, Piracy and Perverted Promises: Patenting Life, the Last Assault on the Commons," Barcelona: GRAIN, 1997, p. 5.

6. Report of the Second Human Genome Diversity Workshop, Penn State University, Pennsylvania, October 29–31, 1992.

7. See J. Kidd, K. Kidd and K. Weiss, "The Human Genome Diversity Initiative," *Human Biology* 65, 1991, pp. 1ff.

8. L. Cavalli-Sforza, A. Wilson, C. Cantor, L. Cook-Deegan and M.C. King, "Call for a Worldwide Survey of Human Genetic Diversity: A Vanishing Opportunity for the Human Genome Project," *Genomics* 11, 1991, p. 490.

9. Victoria Tauli-Corpuz, Statement presented during the 2nd Session of the UN–CSD on behalf of the Cordillera Peoples' Alliance, New York, 1994.

10. The information on the collection of genetic materials from the Aetas was relayed to me by my NGO friends in the Philippines. I was sent copies of the exchange of letters between Dr Philip Camara of the Makati Medical Centre in the Philippines and Elizabeth Trachtenberg of Roche Molecular Systems. The exchange of letters took place between March 1993 and July 1994. A fuller account of this exchange can be seen in the book *The Life Industry: Biodiversity, People, and Profits*, London: Intermediate Technology Publications, 1996.

11. Cavalli-Sforza et al., "Call for a Worldwide Survey."

12. While the official documents on the HGDP deny that what they are doing is for profit, I am positive that eventually some scientist or corporation will apply for patents

on some of the collections. The example of a US corporation called Incyte, which in April 1994 applied for a patent on 40,000 human genes and DNA fragments, is a strong basis for my suspicion. The patent application by the US Department of Commerce for a the T-cell line infected with HTLV-11 virus of a Guaymi woman in Panama, and another application by the same agency for the T-cell line infected with HTLV-1 virus of a Hagahai man in Papua New Guinea, indicate to me that commercialization of these will be a logical next step. Although not directly linked to the Human Genome Diversity Project, these precedents would already indicate the likely future of the genetic collections of the HGDP.

13. See Andrew Kimbrell, *The Human Body Shop: The Engineering and Marketing of Life*, Penang: Third World Network, 1993, p. 46.

14. See Miges Baumann et al., eds, *The Life Industry, Biodiversity, People and Profits*, London: Intermediate Technology Publications, p. 137.

15. "Standard Setting Activities: Evolution of Standards Concerning the Rights of Indigenous Peoples, Human Genome Diversity Research and Indigenous Peoples," United Nations Document E/CN.4/Sub.2/AC.4/1998/4, Geneva: UN Commission on Human Rights, p. 4.

16. Tebtebba Foundation (Indigenous Peoples' International Centre for Policy Research and Education) is an indigenous peoples' NGO whose objective is to help build the capacity of indigenous peoples to fight for their own issues. It does research work, lobbying and advocacy in the national and international arenas, holds training workshops and publishes information.

17. *RAFI Communiqué*, September/October 1997, Ottawa: Rural Advancement Foundation International.

18. United Nations Document, "Standard Setting Activities," p. 11.

19

Shams, Shamans and the Commercialization of Biodiversity

MICHAEL K. DORSEY

At no period of the world has the number of plant-hunters been so great as at present. Will you believe it hundreds of men are engaged in this noble and useful calling? ... They may be found pursuing their avocation in every corner of the world.

Captain Mayne Reid, 1860[1]

Because the "Biorevolution" is becoming a profit-driven business, and as a result of industry's quest for profit, humankind is witnessing the trend towards the commodification of nature – the extension of the commodity logic and form to the natural and cultural resources of biodiversity.

José de Souza Silva, 1994[2]

Bioprospecting – the attempt to identify and eventually commercialize potentially valuable genetic and biochemical resources – is not a new activity; transnational, commercial flows of medicinal plants date back to the sixteenth century.[3] The nineteenth century saw a boom in bioprospecting endeavours, which arguably did not end until the post-World War II period. What is new about the present transnational resurgence in bioprospecting is that it is driven primarily by four interlocking factors: global, market-based economic rationales; rapid and broad technological changes, particularly in biotechnology; a growing interest by pharmaceutical actors to identify their bioprospecting profits with environmental conservation efforts;[4] and efforts to harmonize and standardize global discourses on biodiversity and intellectual property rights regimes.

It is necessary to examine critically current transnational bioprospecting efforts in the tropics for at least two reasons.[5] First, many scholars[6] focus narrowly on the potential of bioprospecting to yield "powerful" new drugs in the fight against diseases and ailments and provide a financial wellspring for myriad conservation schemes.[7] Second, a critical examination of bioprospecting elaborates the neoliberal political-economic context within which prospecting occurs; this enables one to begin to trace its effects on the conservation and protection of biological and cultural diversity and the promotion of particular intellectual property rights (IPR) regimes.

TRANSNATIONAL BIOPROSPECTING (RE)EMERGING

Bioprospecting has its origins in economic botany, often defined as the study of the identification, properties, uses and distribution of economically useful plants.[8] Bioprospecting, however, is a unique subset of economic botany that is done *with the intent* to yield commercially valuable resources. As we shall see, the capitalist profit imperative, especially from the biotechnology industry, is a powerful catalyst and organizer of contemporary bioprospecting projects.

The king of Spain, and other European monarchs, retained botanists and pharmacists to identify, collect, formulate and identify plant medicines for the royal family.[9] The desire to expand personal pharmacopoeias legitimized financing for early exploration projects, especially those to the New World. According to Schultes and Reis, the king of Spain sent his personal physician to live with the Aztecs and study their medicines, less than fifty years after Columbus's first voyages.[10] Indeed it was rare that any ship to or from the New World – or anywhere outside of Europe in the Age of Exploration – did not have a person knowledgeable about plants and potentially capable of exploiting their medicinal properties.

Prior to the mid-1700s, however, plant knowledge was unsystematized. Carolus Linnaeus offered the first systematized plant identification method in the form of the binomial Latin nomenclature (i.e. *Genus, species*) that remains in use today.[11] His students ushered in the modern age of global, commercial plant flows. The Linnaean systematic mapping of plants also correlated with an expanding search for commercially exploitable resources, markets and lands to colonize.[12] This system in essence gave order to otherwise chaotic information and made the task of assigning commercial value that much easier.

The systematicity of Linnaeus's method and his disciples who implemented it around the world legitimized the establishment and growth of state-supported botanical gardens. The gardens, in turn, helped maintain and perpetuate flows in commercially valuable plants. By providing disparately connected botanical gardens with a common discourse which surpassed national boundaries and particular interests, Linnaeus's system spawned and supported transnational plant flows. These gardens and their associated net-works, including botanists and herborists moved species – especially those with medicinal properties or economic value – to the Old World as well as in between the nascent colonies, often by any means necessary. The Royal Botanic Gardens at Kew, in the United Kingdom, for example, sponsored the illegal transfer of many New World plants, most notably *Hevea* (rubber) and *Chinchona* (the "fever bark," capable of curing malaria), back to England as well as to Asia for commercial production.[13]

From the late seventeenth century until the post-World War II period, botanical gardens effectively maintained supplies of much-needed raw materials for medicines and other nascent industrial processes. These flows were closely coordinated with explicitly commercial interests. Arrangements with the overseas trading companies, especially the Swedish East India Company, gave free passage to Linnaeus's students, who began turning up everywhere collecting plants and insects, measuring, annotating, preserving, making drawings, and trying to get it all home intact.[14] Kew prospected *Hevea* to break Brazil's monopoly on rubber. The relationship between commerce and science was a "peculiar one":

> On the one hand commerce was understood as at odds with the disinterestedness of science. On the other, the two were believed to mirror and legitimate each other's aspirations. "A well regulated commerce," said Linnaeus' pupil Anders Sparrman, "as well as navigation in general has its foundation in science ... while this, in return derives support from, and owes its extension to the former."[15]

This "peculiar" relationship still exists today. When I queried officials of the Ecuadorian National Herbarium – which is directed by an American botanist attached to the Missouri Botanical Gardens – about their rationale for seeking support from the transnational pharmaceutical Pfizer, I was asked: "How else will we get the money we need to do our work?"[16]

Several convergent factors in the late 1980s and early 1990s spawned a renaissance in higher plant-related pharmaceutical product research and development, as well as the emergence of a new era of biodiversity prospecting.[17] According to McChesney the renaissance was driven by several factors, most notably the development of new biochemical methodologies:

> advances in bioassay technology; advances in separation and structure elucidation technology; advances in our understanding of biochemical and physiological pathways; the biotechnology revolution; historical success of the approach; loss of practitioners of traditional medicine; loss of biological diversity; loss of chemical diversity; world wide competition.[18]

The decrease in costs, routinization, and increased capacities of some of the new technological methods described above has further driven the bioprospecting renaissance. As Reid, Barber and La Viña note,

> Prior to the 1980s, using test-tube and in vivo assays, a lab could screen 100–1000 samples per week. Now using 96-well microtiter plates and robotics, labs can screen 10,000 samples per week in a broad range of mechanism-based assays.... Where the screening of 10,000 plant extracts would have cost $6 million one decade ago, it can now be accomplished for $150,000. In the next decade, with further miniaturization and high-speed robotics the through-put could grow by another one or two orders of magnitude.[19]

Many pharmaceutical actors have capitalized upon these convergent factors and the associated technological developments. For example, in September 1991 Costa Rica's National Institute of Biodiversity or INBio – a private, non-profit organization – and the US-based pharmaceutical firm, Merck & Co., announced an agreement under which INBio would provide Merck with chemical extracts from wild plants, insects and microorganisms from Costa Rica's tropical forests for Merck's drug screening programme. In exchange, INBio was promised a two-year research and sampling budget of $1,135,000 and royalties on any resulting commercial products.[20] The Merck–INBio agreement, still in effect, represents a synthesis of the aforementioned technological developments, and has been described as a "watershed in the history of biodiversity prospecting."[21] But in many regards the agreement is problematic. Since Costa Rica shares much of its biodiversity with its neighbours, an exclusive royalty only to Costa Rica denies the legacy of neighbouring countries in the conservation of biodiversity. Further, the agreement's secret terms make it difficult to discern and evaluate the accuracy of valuations of Costa Rica's biodiversity.

The INBio–Merck agreement, contract and bioprospecting methodology epitomize how conservation efforts become intertwined with prospecting. Merck paid INBio $1 million in initial capital, and promised royalties on any drugs developed from biological resources prospected by INBio; 10 per cent of the initial fee and 50 per cent of the royalties are to be funnelled back into conservation and biodiversity protection through an arrangement with the Costa Rican government.[22]

Today, it is widely accepted that significant new plant drugs and new methods of producing them will be developed to serve mankind in the twenty-first century. A long list of scientists, primarily botanists and economic botanists support the idea of biodiversity prospecting as a means of revenue generation for pharmaceutical corporations, as well as source for new drug research and development efforts, furthering and establishing conservation programmes and projects, and a means for valuing biological organisms.[23] Here a number of problems arise. Some advocate valuing biological resources through conventional economic analysis.[24] Valuation in this context is essentially utilitarian, anthropocentric and instrumentalist, giving rise to market-based measures of willingness to pay for conservation – or to accept a loss in species diversity – as explicit measures of value.[25] Yet an instrumental notion of value vastly underestimates the significance of myriad cultural values associated with biodiversity.

BIO-MARKETS – BY ANY MEANS NECESSARY

As Stephen King, vice-president of Ethnobotany and Conservation at Shaman Botanicals (formerly Shaman Pharmaceuticals) notes,

> The question is no longer *whether* indigenous and forest dwelling people should benefit from products developed based on their knowledge and forest management technology, *but rather how* to provide these benefits in the most fair and effective method.[26]

The commercialization of biological resources has long been under way in both tropical and temperate biomes. It will continue as new resources are "discovered," expropriated and valued. This acceleration will take place not simply because of the rapid, explosive, radical nature of capital or what William Robbins calls the: "totality of the modernizing forces."[27] Commercialization will advance because it is being posited as a means – if not the sole means – to ensure the protection of the planet. The protection of biodiversity is deemed to depend on the assistance of corporations, especially pharmaceutical firms, which engage in biodiversity prospecting. The relationship between corporations and biodiversity is presented as a cyclical one. Corporations need biodiversity and biodiversity needs them. We are told:

> bioprospecting represents ... [an effective]... "conservation tool" ... [and] ... the bioprospecting industry is dependent on conservation advances, it provides an effective means of bringing critical conservation concerns to the attention of industrial and governmental leaders. They represent one of the most comprehensive conservation approaches to date – providing short- and long-term benefits for both indigenous peoples and national industries.[28]

As von Weizsäcker notes, "there seems to be nothing sinister or ridiculous either in an Institute for Economic Botany in the New York Botanical Garden, or in a sentence by an obviously honest and well-meaning Merck spokesman, 'We have therefore clearly heard the call to come to the rescue of biodiversity by exploring the medical potential of the tropical world.'"[29] Balick opines that "the conservation of biodiversity can be best achieved through establishing linkages with the health-care delivery system in the broadest sense."[30] The problem emerges when we begin to locate and identify the "health-care delivery system." The demands of the commercial delivery system may very well outstrip the potential of tropical forests to supply medicinal resources, thereby undermining conservation. Plants with therapeutic potential are in some cases collected into extinction even faster than their homes – whether in the rainforest or elsewhere – are being destroyed. In one particularly egregious example, the entire adult population of *Maytenus buchananni* – source of the anti-cancer compound maytansine – was harvested when a mission sponsored by the US National Cancer Institute collected 27,215 kilos in Kenya for testing in its drug development programme.[31]

Though the process of commercializing biodiversity *may* have immense potential to contribute to biodiversity conservation, it also may not. As Balick notes, "Finding new drugs has nothing to do with conservation unless [conservation] mechanisms are already in place."[32] Commercialization can be detrimental to conservation, especially in a regulatory vacuum. The use of *Catharanthus roseus*, originally endemic to Madagascar, and sometimes called Madagascar rosy periwinkle, is an especially problematic example.[33]

The rosy periwinkle is often upheld as the plant which (re)ignited current biodiversity prospecting efforts.[34] Extracts of rosy periwinkle enabled Eli Lilly to produce two cancer treatments: vincristine (for paediatric leukaemia) and vinblastine (for Hodgkin's disease). The irony is that Madagascar has realized no financial benefit from Eli Lilly, nor from any of the drug sales. The drugs made from the periwinkle, which was originally found in a severely stressed ecosystem, have not provided any of the valuable incentives for conservation that it might have, despite the fact that in 1985 total sales of vincristine and vinblastine amounted to approximately $100 million, 88 per cent of which was profit![35]

Unquestionably, Eli Lilly reaps the lion's share of benefits from the rosy periwinkle. No money flows back to Madagascar for the drugs produced, and it is unlikely that the drug itself is even available to the poor peasants of Madagascar, should they need it. Thus, there is no reason whatsoever for Malagasy to preserve the rosy periwinkle.

Other examples of how bioprospecting and the subsequent commercialization and marketization of biodiversity failed to contribute to conservation efforts are the cases of *Phytolacca dodecoandra* and *Maytenus buchananni*. *Phytolacca dodecoandra*, the endod berry, was found to be effective against zebra mussels by scientists at the University of Toledo (Ohio). Prior to that time the berry had been used for centuries in Ethiopia as a detergent and a fish intoxicant. By the late 1970s most of the science of endod's molluscicidal properties had been determined.[36] The University of Toledo, however, refuses to donate their process patents on endod to non-profit organizations in Ethiopia. Only in 1995 did university officials offer to sell the endod patent outright for $125,000, or license it for a $50,000 fee plus 2.5 per cent royalties to the Ethiopian scientists who helped the University of Toledo scientists obtain the patent.

Attempts to conserve *Maytenus buchananni*, the source of maytansine in Kenya, were obliterated when the US National Cancer Institute programme harvested the species to extinction. Sustainable harvesting techniques were seemingly disregarded, which effectively thwarted regeneration initiatives.

Shaman Botanicals, formerly Shaman Pharmaceuticals, has been upheld as the "most progressive" of all biodiversity prospecting firms for their efforts to provide compensation packages for the local, indigenous communities with whom they work. Yet the firm epitomizes the maintenance of traditional corporate priorities – in accordance with the law.[37] As the financial services firm Smith Barney notes,

> while Shaman's activities may appeal to certain pools of capital dedicated to socially responsible investment activity, and while Shaman does appear to be an ideal model for environmental and corporate behaviour ... *we have allotted no value to Shaman's enterprise beyond that which we believe to be ultimately realizable in monetary terms.*[38]

Thus even for the most self-promoting socially and environmentally progressive firm, within the confines of the contemporary capitalist context the bottom line can only be the profit line. Corporations say that they may pay a portion of royalties back to collectors in prospected countries, or to the in-country collection organizations, and some do.[39] Yet, collectively, pharmaceutical products based on traditional medicine have returned less than 0.0001 per cent of their profits to the local plant users who assisted research and discovery efforts.[40]

Beyond the desire for profit is the desire to obtain intellectual property rights over the plant-based drugs. One of the biggest problems inhibiting bioprospecting was the difficulty in obtaining adequate patent protection. Certainly, many prospecting firms claim to uphold and secure the intellectual property rights of the communities from which they receive the bio-samples. Nevertheless, at the end of the day indigenous community rights over prospected resources – arguably, their traditionally held and cultivated resources – remain under the strict, if not exclusive, control of the bioprospector, despite talk to the contrary. As information from Shaman Pharmaceuticals noted, their policy is to seek patent protection and enforce all of their intellectual rights.[41]

One of the four key priorities for companies like Shaman, even before they begin to explore for particular plants, is that "there must be a sizable market that warrants the expense of discovery and development."[42] The "sizable markets" that Shaman hopes to attract are in the Northern, developed nations. In the long run of commercial drug development, diseases that plague the people who supply the ethnobotanical knowledge and information to Shaman become irrelevant, or at best superfluous. It is important to emphasize the notion of the long-run, especially for would-be critics (e.g. Shaman employees) of this assertion. Shaman's dual ethnobotanist–physician teams insisted that they were providing immediate reciprocity for the know-how and assistance of local healers.[43] But a bag full of aspirins and other simple remedies that they might provide is a far from fair exchange for information or the actual biological resources which may ultimately be responsible for eliminating a major disease like AIDS or cancer.

In the end local people receive virtually no compensation for the centuries of testing that they have already performed and the associated knowledge (i.e. intellectual property) accumulated. Defenders of these practices may harp back that if a cure for cancer is found everyone will inevitably benefit, eventually. To that I would question what "eventually" really means, and recall the vast number of diseases and ailments that are virtually non-existent in the "developed" parts of the world but ravage the other three-quarters of the planet.

CONCLUSIONS: PROBLEMATIZING
THE SCIENCE OF BIOPROSPECTING

Bioprospecting is a powerful process. By definition it obscures the historical role of indigenous local communities in manipulating and deriving material, social, spiritual and other benefits from biological resources. The language of bioprospecting implies that prior to prospecting, the resources in question lay buried, unknown, unused and without value.[44] To assign value to these resources, so goes the argument, we – scientists and capitalists – must dig them up, hunt them down and/or discover them, like deposits of oil, natural gas or minerals. By definition, then, bioprospecting fails to acknowledge prior use, knowledge and rights over the resources in question. The scientists and policy-makers who define and presumably accept the definition of bio-prospecting frequently (and perhaps quixotically) note the uses of biodiversity by local communities, indigenous peoples, farmers and healers, but more often than not they ultimately fail to account for these economic, social and spiritual values.[45] This conceptualization of value in bioprospecting is a form of violence, a rape. As Vandana Shiva notes, the concept of adding value through the science of bioprospecting hides the removal and destruction of the value within networks of historical and material relations.[46]

Bioprospecting may continue to produce new drugs to address old and new diseases. Yet a question remains: can many of the aforementioned problems – the downsides of the prospecting endeavour – be effectively surmounted to yield the benefits of prospecting? The answer to such a broad question can *never* be a static "yes" or "no." The answer depends on who is asking and who the object of inquiry is. Accordingly the question might be rephrased as: "Who benefits from bioprospecting?" Representatives of Shaman Botanicals present bioprospecting as an "everyone wins" scenario. Local communities reap benefits through benefit-sharing agreements, Shaman makes profits, and people's ailments are served by new drugs. The *benefits* of bioprospecting may indeed be lowest at the local level and highest at the national and global level; while the *costs* are highest locally. Brown's perspective, as noted above, seems to concur with the historical record of resource exploitation in the tropics, which enriches various powerful classes while simultaneously impoverishing those at the margins.[47]

Some argue that "just" benefit sharing mechanisms must be developed between pharmaceutical actors and local communities that utilize biological resources.[48] But what or who defines "just" is not always clear. Who decides the terms of contracts between local communities and bioprospecting actors? What are the specific benefits to be shared and/or exchanged? What is the frequency of exchange of benefits? What are the historical foundations of these mechanisms – that is, to what extent will profit-based benefits be based on a legacy of local community innovation and cultivation of the biological resources in question? Such questions may allow us to understand better, and perhaps someday ensure, "just" benefit-sharing.

Efforts by pharmaceutical companies to secure intellectual property rights (IPR) over myriad biological resources, especially in the form of patents, raise additional, worrisome questions. Patents over biological resources developed by pharmaceutical actors claim to exclude would-be pirates from subsequently unlawful product replication. But what else is excluded – or included – by IPR regimes needs to be investigated as well. Little is known about the methods by which "modern" IPR regimes undermine, or at least alter, longer-standing, community (local or indigenous) rights and already established exclusionary mechanisms over biological resources. Further, to what extent are Western-based IPR regimes biased in favour of corporations that apply for patents by upholding juridical procedures, and thus biased against collective, non-jural (e.g. community decision-making) processes and procedures? Also, if one accepts IPR regimes based on patents, should the patent applications include both developers and providers as contracting parties? If so, how might this take place and what might be ensured by it?

Lastly, and perhaps most importantly, what sort of regulatory regimes (e.g. international, national, and sub-national) are most appropriate to govern and control bioprospecting endeavours? Most of the prescriptions that have been offered are overwhelmingly statist in orientation.[49] The UN Convention on Biological Diversity explicitly gives primacy to "contracting parties" (i.e., nations) in all of its articles.[50] Thus, despite accolades endorsing the need to "respect, preserve, and maintain knowledge, innovations and practices of indigenous and local communities embodying traditional lifestyles relevant for the conservation and sustainable use of biological diversity," the convention's implementation is subject to "national legislation." To what degree will the state intervene on behalf of its constituents and those firms engaged in bioprospecting? If benefit-sharing does take place, what will facilitate inclusion and/or exclusion of sub-national entities? Broadly, political-economic factors which condition bioprospecting must be examined in much more depth.

The aforementioned problems and questions do not represent an exhaustive list. Yet they point out in bold relief that the process of bioprospecting – within a capitalist political-economic framework – is, perhaps, far from likely to net conservation of biodiversity or even its sustained use. Indeed, within such a framework, the "Shaman Botanicals" of the world can only be window-dressing in a world that places profits by any means necessary far above claims of benefiting isolated indigenous communities.

NOTES

1. M. Reid, *The Plant Hunters or Adventures among the Himalaya Mountains*, London: G. Routledge & Sons, 1907.
2. J. Silva, "From Medicinal Plants to Natural Pharmaceuticals," unpublished ms, 1994.
3. See F. Ortiz Crespo, "Dos protobotánicos que se ocuparon de las plantas del nuevo mundo en el siglo XVI," in C. Josse Octubre and M. Rios, eds, *Resúmenes: II Congreso Ecuatoriana de Botánica*, Quito: Pontificia Universidad Católica del Ecuador, 1995. Indeed one may push this date further back to the time of the Sumerians (4000 BC), but this pre-dates the conceptualization of "nation," and hence trans*national* flows.
4. I use the term "pharmaceutical actors" to highlight the fact that myriad entities have a commercial interest in global biodiversity. Thus, "actors" include, at least: independent individuals, often contracted by larger firms; university researchers from myriad departments (e.g., anthropology, botany, biology, zoology, etc.); botanical garden employees; in addition to transnational corporations (TNCs) of various sizes.
5. Bioprospecting is by no means limited to the tropics. These areas, however,

maintain the overwhelming majority of the planet's biodiversity (see J. McNeely et al., *Conserving the World's Biological Diversity.* Gland, Switzerland: IUCN, 1990) and bioprospecting endeavours are most intensely under way in this region.

6. For example see M. Balick, "Ethnobotany, Drug Development and Biodiversity Conservation – Exploring the Linkages," in G.T. Prance, C. Derek, and J. Marsh, eds, *Ethnobotany and the Search for New Drugs: Ciba Foundation Symposium 185*, Chichester: John Wiley & Sons, 1994, pp. 4–17; J.H. Cohen and C.H. Tokheim, *Smith Barney Special Situations Research. Shaman Pharmaceuticals: Ethnobotany, Biodiversity and Drug Discovery* [Location Unknown]: Smith Barney Inc., 1994; A. Cox and M. Balick, "The Ethnobotanical Approach to Drug Discovery," *Scientific American*, vol. 27, no. 6, 1994, pp. 82–7; A. Gentry, "Tropical Forest Biodiversity and the Potential for New Medicinal Plants," in A.D. Kinghorn and M.F. Balandrin, eds, *Human Medicinal Agents from Plants*, Washington, DC: American Chemical Society, 1993, pp. 13–24; S. King, "Establishing Reciprocity: Biodiversity, Conservation and New Models for Cooperation Between Forest-Dwelling Peoples and the Pharmaceutical Industry," in T. Greaves, ed., *Intellectual Property Rights for Indigenous Peoples, A Sourcebook*, Oklahoma City: Society for Applied Anthropology, 1994, pp. 69–82; J.D. McChesney, "Biological and Chemical Diversity and the Search for New Pharmaceuticals and Other Bioactive Natural Products," in Kinghorn and Balandrin, eds, *Human Medicinal Agents from Plants*, pp. 38–47; W. Reid, "Biodiversity Prospecting: Strategies for Sharing Benefits," in V. Sanchez and C. Juma, eds, *Biodiplomacy: Genetic Resources and International Relations*, Nairobi: ACTS Press, 1994, pp. 241–68; A. Sittenfeld and R. Gámez, "Biodiversity Prospecting by Inbio," in W. Reid et al., eds, *Biodiversity Prospecting: Using Genetic Resources For Sustainable Development*, Washington, DC: World Resources Institute, 1993, pp. 69–97.

7. Moreover this literature focuses primarily on the role of pharmaceutical TNCs, and national bureaucratic institutions as proxy representatives for the global "North and South." The role of specific local communities is often obfuscated underneath the heading of "indigenous people" or "rural peoples." Hence, local level impacts (positive or negative) are poorly understood.

8. G.E. Wickens, "What is Economic Botany?" *Economic Botany*, vol. 44, no. 1, 1990, p. 14.

9. S. Ali and S. Liberman "The Biotic Pharmacy: Industry's Involvement in Conserving Natural Resources for Drug Development and Pollution Prevention," paper presented at the conference "Reinventing Environmental Protection," Kennedy School of Government, Harvard University, February 9–10, 1996.

10. R.E. Schultes, and S. von Reis, eds, "Historical Ethnobotany," in R.E. Schultes and S. von Reis, eds, *Enthnobotany: Evolution of a Discipline*, Portland, OR: Dioscorides Press, 1995, pp. 89–91.

11. B.R. Holmstedt, "Historical Perspective and Future of Ethnopharmacology," in Schultes and von Reis, eds, *Enthnobotany*, pp. 320–37.

12. M.L. Pratt, *Imperial Eyes: Travel Writing and Transculturation*, London: Routledge, 1992, p. 30.

13. See L.H. Brockway, "Science and Colonial Expansion: The Role of the British Royal Botanic Gardens," *American Ethnologist*, vol. 6, no. 3, 1979, pp. 449–67. See also: Clements R. Markham, *Peruvian Bark: A Popular Account of the Introduction of Chinchona Cultivation into British India, 1860–1880*, London: John Murray, 1880.

14. Pratt, *Imperial Eyes*, p. 25.

15. Ibid., p. 34.

16. Conversation with Dr David Neil, August 1996. For various reasons the Pfizer deal was never finalized.

17. J.D. McChesney, "Biological and Chemical Diversity and the Search for New Pharmaceuticals and Other Bioactive Natural Products," in Kinghorn and Balandrin, eds, *Human Medicinal Agents from Plants*, pp. 38–47.

18. Ibid., p. 40.

19. W. Reid, C. Barber and A. La Viña, "Translating Genetic Resource Rights into Sustainable Development: Gene Cooperatives, the Biotrade and Lessons from the Philippines," *Plant Genetic Resources Newsletter*, vol. 102, no. 2, 1995.

20. Reid et al., eds, *Biodiversity Prospecting*, p. 7.

21. There is also a large growing network of small-time collectors and suppliers of the large pharmaceutical corporations. See T. Mark, "Indiana Jones, meet Mark Chandler," *Forbes*, vol. 153, no. 11, 1994, pp. 100–103.

22. T. Eisner and E.A. Beiring "Biotic Exploration Fund: Protecting Biodiversity Through Chemical Prospecting," *BioScience*, vol. 44, no. 2, 1994, p. 97.

23. See the references under note 6.

24. See C.M. Peters, A.H. Gentry and R. Mendelsohn, "Valuation of an Amazonian Rainforest," *Nature*, vol. 339, no. 6227, 1989, pp. 655–6; or R. Mendelsohn and M. Balick, "The Value of Undiscovered Pharmaceuticals in Tropical Rainforests," *Economic Botany*, vol. 49, no. 2, 1995, pp. 223–8.

25. K. Brown, "Approaches to Valuing Plant Medicines: The Economics of Culture or the Culture of Economics?" *Biodiversity and Conservation* 3, 1994, pp. 734–50.

26. King, "Establishing Reciprocity," p. 71.

27. W. Robbins, *Colony and Empire: The Capitalist Transformation of the American West*, Lawrence: University Press of Kansas, 1994, p. 3.

28. S.M. Rubin and S.C. Fish, "Biodiversity Prospecting: Using Innovative Contractual Provisions to Foster Ethnobotanical Knowledge, Technology, and Conservation," *Colorado Journal of International Environmental Law and Policy*, vol. 5, no. 1, 1994, p. 25.

29. C. Von Weizsäcker, "Competing Notions of Biodiversity," in W. Sachs, ed., *Global Ecology: A New Arena of Conflict*, London: Zed Books, 1995, p. 122.

30. Balick, "Ethnobotany, Drug Development and Biodiversity Conservation," p. 4.

31. Oldfield, cited in Reid et al., eds, *Biodiversity Prospecting*, pp. 3–4.

32. Personal interview, 28 April 1995.

33. For a limited discussion of the ecology of *C. roseus* see: G. Svoboda, "The role of the alkaloids of *Catharanthus roseus* (L.) G. Don (*Vinca rosea*) and their derivatives in cancer chemotherapy," in *Plants: The Potentials for Extracting Protein, Medicines, and Other Useful Chemicals – Workshop Proceedings*, OTA-BP-F-23. Washington, DC: US Congress, Office of Technology Assessment, 1983, pp. 154–69.

34. Cox and Balick, "The Ethnobotanical Approach to Drug Discovery," p. 82; V.E. Tyler, "Plant Drugs in the Twenty-first Century," *Economic Botany*, vol. 40, no. 3, 1986, pp. 279–88. For a discussion of the political economy of the matter, see C. Joyce, *Earthly Goods: Medicine-Hunting in the Rainforest*, Boston, MA: Little, Brown, 1994, pp. 57–8.

35. See R.J. Huxtable, "The Pharmacology of Extinction," *Journal of Ethno-Pharmacology*, vol. 37, no. 1, 1992, pp. 1–11. In order to furnish initial supply of *C. roseus* the plant was cultivated in a number of farms in India and Madagascar. Yet as Svoboda notes, "because a number of variables could seriously threaten the supply of this life-saving drug, the decision was made to attempt cultivation in the US" ("The role of the alkaloids", p. 33). *C. roseus* now has a pan-tropical distribution and is cultivated as an ornamental in gardens around the world.

36. Madhusree Mukerjee, "The Berry and the Parasite," *Scientific American* 274, April 1996, pp. 22ff.

37. Shaman is portrayed as the "most progressive" inasmuch as they are "attempting to pioneer compensation strategies different from traditional post-commercialization profit sharing." See S. King, T. Carlson and K. Moran, "Biological Diversity, Indigenous Knowledge, Drug Discovery and Intellectual Property Rights," in S.B. Brush and D. Stabinsky, eds, *Valuing Local Knowledge: Indigenous People and Intellectual Property Rights*, Washington DC: Island Press, 1996, pp. 167–85.

38. Cohen and Tokheim, *Smith Barney Special Situations Research*, p. 8.

39. See Rubin and Fish, "Biodiversity Prospecting," p. 30; also, "The Healing Forest

Conservancy: Purpose and Priorities," unpublished ms, Washington DC: The Healing Forest Conservancy.

40. Posey, in Rubin and Fish, "Biodiversity Prospecting," p. 27.

41. Securities and Exchange Commission, *Form 10–K Annual Report Pursuant to Section 13 or 15(d) of the Securities Exchange Act of 1934*, Washington DC: US Securities and Exchange Commission, 1994, p. 13.

42. Cohen and Tokheim, *Smith Barney Special Situations Research*, p. 7.

43. See King, "Establishing Reciprocity," p. 75.

44. V. Shiva, *Biopiracy: The Plunder of Nature and Knowledge*, Boston, MA: South End Press, 1997, p. 73.

45. For an excellent example see Mendelsohn and Balick, "The Value of Undiscovered Pharmaceuticals," p. 16.

46. Shiva, *Biopiracy*, p. 74.

47. Brown, "Approaches to Valuing Plant Medicines."

48. See for example: Reid et al., "Translating Genetic Resource Rights"; J. Mugabe, C. Barber, G. Henne, Lyle Glowka and A. La Viña, "Managing Access to Genetic Resources: Toward Strategies for Benefit-Sharing," in J. Mugabe, N. Clark and W. Reid, eds, *Biopolicy International Series*, no. 17, Nairobi: African Centre for Technology Studies, 1997; K. Ten Kate and S. Laird, "Placing Access and Benefit Sharing in the Commercial Context: A Study of Private Sector Practices and Perspectives, Draft Summary of Results of the First Phase," Paper Presented at the Third Meeting of the Subsidiary Body on Scientific, Technical and Technological Advice of the Convention on Biological Diversity, Montreal, September 1–5, 1997.

49. See Sanchez and Juma, eds, *Biodiplomacy*; Mugabe et al., "Managing Access to Genetic Resources."

50. United Nations Environment Programme, *Convention on Biological Diversity*, New York: United Nations, 1992.

20

Biopiracy:
The Theft of Knowledge
and Resources

VANDANA SHIVA

The poorest two-thirds of humanity live in what can be appropriately called the biodiversity-based economy. As farmers, they select and save their own seeds. As healers, they protect and use medicinal plants. Both the knowledge and the resource are part of an intellectual and biological commons to which the entire community has free access, and there is a long-surviving tradition of free give and-take.

Biopiracy, and patents based on it, are equivalent to enclosing the biological and intellectual commons, while dispossessing the original innovators and users. What was available to them freely and what they have contributed to is converted into a priced commodity, and they will have to pay royalties each time they use it. The duty to exchange and save seed is thus redefined as theft and an intellectual property crime.

The enclosure of the intellectual and biological commons through patents thus creates both material and intellectual poverty for two-thirds of humanity – the poor in the Third World.[1] For the corporations and scientists engaged in biopiracy, patents can become an immediate source of wealth.

Biopiracy includes a threefold theft from traditional users of biodiversity, who constitute the two-thirds marginalized and poor people of the Third World:

- *Resource piracy*, in which the biological and natural resources of communities and the country are freely taken, without recognition or permission, and are used to build up global economies. Examples include the transfer of basmati varieties of rice from India to build the rice economy of the USA, and the free flow of neem seeds from the farms, fields and commons to corporations like W.R. Grace for export; both of these traditional products are being taken out of the reach of the Indian people.
- *Intellectual and cultural piracy*, in which the intellectual and cultural heritage

of communities and the country are freely taken without recognition or permission and are used for claiming intellectual property rights (IPRs) such as patents and trademarks, even though the primary innovation and creativity has not taken place through corporate investment. Examples include the use by US corporations – particularly the Texas-based trading company RiceTec – of the trade name "basmati" for their aromatic rice, or Pepsi's use of the trade name Bikaneri Bhujia, a popular regional snack from the desert regions of western India.

Biopiracy is, in the final analysis, based on a false claim of creativity. It involves the appropriation of the cumulative, collective creativity of traditional societies and projects the theft as an "invention." Thus, for example, the RiceTec company's 1997 patent claim for basmati rice described it as "an instant invention of a novel rice line," even though all RiceTec did was to use Pakistani and Indian basmati rices to derive a basmati strain. Derivation is thus falsely projected as an "instant invention" of total novelty, the source used is appropriated, and the farmers' breeding embodied in it is erased. Biopiracy is based on a racist construction of a fictitious creation boundary according to which creativity is only associated with "white men in white lab coats."

Western systems of knowledge in agriculture and medicine are defined as the only scientific system. Indigenous systems of knowledge are defined as inferior, and in fact as unscientific. For decades, instead of strengthening research on safe and sustainable plant-based pesticides such as neem and pongamia, companies focused exclusively on the development and promotion of hazardous and non-sustainable chemical pesticides such as DDT and Sevin. The use of DDT still causes millions of deaths each year and has contributed to a 12,000-fold increase in the occurrence of pests. The manufacture of Sevin at the Union Carbide Plant in Bhopal led to the disaster which killed thousands and has disabled more than 400,000 people.

Today, in recognition of the ecological failure of the chemical route to pest control, the use of plant-based pesticides is becoming popular in the industrialized world. Corporations that have promoted the use of chemicals are now looking for biological options. In the search for new markets and control over the biodiversity base for the production of biopesticides, transnationals like W.R. Grace have claimed rights to biopesticides based on India's neem tree.[2]

The experience with agrochemicals is replicated in the field of drugs and medicines. Indigenous systems of medicine and the biodiversity of medicinal plants were commonly neglected in our scientific research and health policy, which focused exclusively on the Western allopathic system and on technology transfer from the Western pharmaceutical industry. Indigenous medical systems, which are completely lacking in official support in India, rely on over 7,000 species of medicinal plants and on 15,000 medicines from herbal formulations in different systems. The classical Ayurvedic texts refer to 1,400 plants, Unani texts to 342, the Siddha system to 328. Homeopathy uses 570, of which approximately 100 are Indian plants. The economic value of these medicinal plants to 100 million rural households is immeasurable.

As a result of the increasing public awareness of the side effects of hazard-ous drugs, and the rise of strains resistant to antibiotics, the Western pharma-ceutical industry is increasingly turning to the plant-based systems of Indian and Chinese medicine. Patenting of drugs derived from indigenous systems of medicine has started to reach epidemic proportions. The current value of the world market for medicinal plants from leads given by indigenous and local communities is estimated to be $43 billion.[3]

The effort by Western commercial interests to claim products and in-novations from indigenous knowledge traditions as their "intellectual prop-erty," protected by patents, has emerged as a result of the devaluation, and hence the invisibility, of indigenous systems of knowledge. The reductionist methods of Western science are imposed upon the non-reductionist approaches of indigenous knowledge systems. Since Western-style IPR systems are biased towards Western knowledge systems which reduce biodiversity to its chemical or genetic structures, the indigenous systems get no protection, while the piracy of these systems is protected.

In the absence of a protection system for biodiversity and indigenous knowledge systems, and with the universalization of Western-style IPR regimes, such intellectual and biological piracy will grow. Protecting our bio-logical and intellectual heritage in the age of biopiracy requires the recogni-tion and rejuvenation of our heritage, and the evolution of legal systems for the protection of this heritage in the context of emerging IPR regimes.

TRANSLATION OR CREATION?

Western allopathic systems have no medical cure for jaundice or viral hepa-titis. Jaundice in medical terms is basically the presence of certain signs and symptoms associated with liver dysfunction – for example, yellowness of eyes, nausea, loss of appetite, pain in the liver region, and so on.

Indian systems of medicine – Ayurveda, Unani and Siddha – and folk traditions have used various plants for the treatment of jaundice. *Phyllanthus niruri* is one such medicinal plant used widely in India from north to south, from east to west. It is as much a part of the formal health-care system of Ayurveda as part of local health practices, folk medicine and traditional indig-enous collective knowledge. The plant is called *Bhudharti* in Sanskrit, *Jar amla* in Hindi and *Bhuin amla* in Bengali. It is common throughout the hotter parts of India, growing in fallow land and in shade. An annual herb, 10–30 centimetres high, its leaves are elliptical-oblong, and it flowers and fruits from April to September. The whole plant, its leaves, shoots and roots, are used for treating jaundice. Even though the use of *Phyllanthus niruri* for treatment of jaundice has been an ancient and well-recorded innovation in the Indian systems of medicine, patents are now being applied for this knowledge as if it was a novel invention.

The Fox Chase Cancer Center of Philadelphia has applied to the European Patent Office for the plant's use as a medicament for treating viral hepatitis B.

In spite of the prior knowledge of the use of *P. niruri* as a cure for all forms of hepatitis, including hepatitis B, the Fox Chase Cancer Claim states that "Insofar as is known *P. niruri* had not been proposed for the treatment of viral hepatitis infection prior to the work done by the inventors of the present invention."

In the allopathic system, there is no specific treatment for jaundice. In the case of viral hepatitis, an attempt is made to provide symptomatic treatment by giving glucose, vitamin B complex, along with the avoidance of fatty, fried foods – which cause nausea, vomiting and indigestion in the absence of enough enzymes, as usually produced by a healthy liver. (Genetically engineered hepatitis vaccines have brought mixed results.) In Ayurveda and in traditional systems of medicine, there are products which are known to help in the regeneration of the liver tissue. This treatment is therefore addressed to the root cause of the health problem and not just the symptoms.

By isolating the application of *P. niruri* for treatment of jaundice to one form of infective hepatitis, hepatitis B, and treating this as a novel application, even though medicines derived from *P. niruri* have been used for treating all forms of hepatitis in the traditional systems of medicine, the scientists of the Fox Chase Cancer Center have falsely presented an act of piracy as an act of invention.[4]

THE PIRACY OF NEEM AND OTHER AGRICHEMICALS

Neem, or *Azadirachta indica*, has been used for diverse purposes over centuries in India, including in medicine and in agriculture. Research has shown that neem extracts can influence nearly two hundred species of insect, many of which are resistant to pesticides.[5]

Several extracts of neem have recently been patented by US companies, and many farmers are incensed at what they regard as intellectual piracy. The village neem tree has become a symbol of Indian indigenous knowledge, and of resistance against companies that seek to expropriate this knowledge for their own profit. A number of neem-based commercial products, including pesticides, medicines and cosmetics, have come on the market in recent years, some of them produced in the small-scale sector under the banner of the Khadi and Village Industries Commission (KVIC), others by medium-sized laboratories. However, there has been no attempt to acquire proprietary ownership of formulae, since, under Indian law, agricultural and medicinal products are not patentable.

For centuries the Western world ignored the neem tree and its properties; the practices of Indian peasants and doctors were not deemed worthy of attention by the majority of British, French and Portuguese colonialists. However, in the last few years, growing opposition to chemical products in the West, in particular to pesticides, has led to a sudden enthusiasm for the pharmaceutical properties of neem. In 1971 US timber importer Robert Larson observed the tree's usefulness in India and began importing neem seed to his company headquarters in Wisconsin. Over the next decade he conducted safety and performance tests upon a pesticidal neem extract called Margosan–O and in 1985 received clearance for the product from the US Environmental Protection Agency (EPA). Three years later he sold the patent for the product to the multinational chemical corporation W.R. Grace & Co.

Since 1985 over a dozen US patents have been taken out by US and Japanese firms on formulae for stable neem-based solutions and emulsions and even for neem-based toothpaste. At least four of these patents are owned by W.R. Grace, three by another US company, the Native Plant Institute, and two by the Japanese Terumo Corporation. Having garnered their patents and with the prospect of a licence from the EPA, Grace set about manufacturing and commercializing their product by establishing a base in India. The company approached several Indian manufacturers with proposals to buy up their technology or to convince them to stop producing value-added products and instead supply the company with raw material. Eventually W.R. Grace managed to arrange a joint venture with a firm called P.J. Margo Pvt. Ltd. They have set up a plant in India which processes up to twenty tons of seed a day. They are also setting up a network of neem seed suppliers, to ensure a constant supply of the seeds at a reliable price.

Grace's aggressive interest in Indian neem production has provoked a chorus of objections from Indian scientists, farmers and political activists, who assert that multinational companies have no right to expropriate the fruit of centuries of indigenous experimentation and several decades of Indian scientific research. This has stimulated a bitter transcontinental debate about the ethics of intellectual property and patent rights.

Western scientists, bolstered by a 1993 report by the US Congressional Research Service (CRS), argue that any synthetic form of the neem-derived compound azadirachtin should be patentable, because the synthetic form is not technically a product of nature. However, neither azadirachtin nor any of the other active principles of neem has yet been synthesized in laboratories. The existing patents apply only to methods of extracting the natural chemical in the form of a stable emulsion or solution, methods which are simply an extension of the traditional processes used for millennia for making neem-based products. The biologically active polar chemicals can be extracted using technology already available to villages in developing countries.[6]

W.R. Grace's claim of novelty exists mainly in the context of the West's ignorance. Over the two thousand years that neem-based biopesticides and medicines have been used in India, many complex processes were developed to make them available for specific use, though the active ingredients were not given Latinized "scientific" names. The reluctance of Indian scientists to patent their inventions, thus leaving their work vulnerable to piracy, may in part derive from a recognition that the bulk of the work had already been accomplished by generations of anonymous experimenters. This debt has yet to be acknowledged by the US patentors and their apologists.

One reviewer of W.R. Grace's US neem patent (US Patent no. 5,124,349) commented, "I find it incomprehensible that W.R. Grace could have been granted a patent ... claiming novelty on a process the whole world has known for years."[7] Solvents mentioned in the patent and solvents similar to them have been used on neem seeds and described in publications before the company's application.

BIOPROSPECTING AND "BENEFIT SHARING"

Biopiracy is often camouflaged in the language of "bioprospecting" and "benefit sharing." However, most discussions on access and benefit sharing narrow the interpretation of access as access to global commercial interests to Third World biodiversity. They interpret benefit sharing not as the benefits the farmers have shared freely with each other and with the gene banks from where the seed corporations freely take the genetic wealth of the farmers. "Benefit sharing" is reduced to the corporations providing a small percentage of their profits to a biodiversity fund or a farmers' rights fund.

Discussions of benefit sharing have been based on proposals to treat farmers' rights as merely a fund supported by voluntary payments/taxes of the order of 1 per cent. Meanwhile, the potential market for the seed industry is estimated at $7.5 billion. This is a benefit forcefully taken from the farmers by the seed industry, since farmers have to give up farmers' rights of free exchange, free sale, and free saving to provide markets protected by markets. The low-cost, royalty-free seed economy for the farmer is thus substituted by a high-cost, royalty-burdened economy controlled by the seed industry. Instead of focusing on the $7.5 billion benefit provided by farmers to the seed industry, all discussions on benefit sharing have focused on the remote possibility of a 1 per cent fund based on voluntary contributions. This nonexistent fund is then claimed to address the "farmers' rights."[8]

Protection of such theft is achieved through intellectual property laws. Any benefit sharing model which defends IPRs is in essence a "crumb sharing" model established to hide the theft of the loaf.[9] Equitable benefit sharing implies that the loaf itself should stay in the hands of indigenous and local communities. As the original innovators and custodians of agricultural biodiversity and as providers of food security, they have a right to more than the crumbs.

NOTES

1. Vandana Shiva, Afsar Jafri, Gitanjali Bedi and Radha Holla-Bhar, "The Enclosure and Recovery of Commons," New Delhi: Research Foundation for Science, Technology and Ecology (RFSTE), 1997.

2. Vandana Shiva and Radha Holla Bhar, "Intellectual Piracy and the Neem Patents," New Delhi: RFSTE, 1993.

3. F. Grifo and J. Rosenthal, *Biodiversity and Human Health*, Washington DC: Island Press, 1997.

4. Mira Shiva and Vandana Shiva, "Patents on Phyllanthus Niruri: The Plant for

Indigenous Medicinal Cure for Jaundice," New Delhi: RFSTE, 1995.

5. Vandana Shiva, "Protecting our Biological and Intellectual Heritage in the Age of Biopiracy," New Delhi: RFSTE, 1996.

6. Shiva and Bhar, "Intellectual Piracy and The Neem Patents"; K. Vijayalakshmi, K.S. Radha and Vandana Shiva, "Neem: A User's Manual," Madras: Center for Indigenous Knowledge Systems and New Delhi: RFSTE, 1995.

7. Quoted in Vandana Shiva, "Captive Minds, Captive Lives," New Delhi: RFSTE, 1995. In May of 2000 the European Patent Office responded positively to an international petition urging that they revoke W.R. Grace's patent for a neem-derived fungicide in Europe. It is widely hoped that this historic victory for indigenous communities against false claims of innovation and novelty by corporations and colonial science will mark a turning point in the struggle against biopiracy.

8. Vandana Shiva, "Farmers' Rights: Access and Benefit Sharing," *Bija Newsletter* no. 17/18, November 1996.

9. David Hathaway at IUCN meeting on Biodiversity and TRIPs, March 19, 1998.

21

Exclusive Rights, Enclosure and the Patenting of Life

KIMBERLY A. WILSON

Once upon a time, no one had the right to exclude others from using natural resources. No one could claim to have a patent, or exclusive rights for a plant, animal species, or human gene. Today, Americans encounter patented objects, the registered trademark of a brand name, and copyrighted material on a daily basis. Whether on a billboard, the pages of a book, our clothes, or a disposable cup lid, patents and trademarks pervade US culture. Our contact with patented objects becomes even more intimate as the food we eat, the plants we grow, the livestock we raise, and the genes in our bodies become patentable "inventions." The extension of patent protection to nearly every area of life alters social relationships and human relationships to nature.

The English Enclosure Acts of the 1500s and 1600s marked the beginning of the assignment of exclusive rights to the natural world. Land, waterways, mineral deposits, crops and animal herds all became forms of property to be bought, sold and traded. "Letters Patent" gave English lords ownership rights to "newly discovered" lands, and early industry in England used exclusive rights to establish monopolies and patents on mechanical inventions. The Enclosure Acts changed the face of England's rural society by eliminating the peasants' right to hold land in common. The Acts simultaneously created and transferred property rights over common land to royalty and upper-class estate owners. The effects of the Enclosure Acts are well documented. "Enclosure concentrated wealth. It ruined small farming families and drove them into towns; it raised prices; it intensified labour and encouraged luxury. Above all it destroyed equality."[1]

The New England colonies included individuals and families who had been displaced by the enclosure movement in England. These early settlers were reluctant to concentrate power in the hands of a few. Exclusive rights were addressed in the Body of Liberties, the code of laws adopted by the General Court of Massachusetts in 1641, which specifically outlawed monopolies except where the colony could benefit. The Body of Liberties states, "There shall be no monopolies granted or allowed among us, but of such new inventions as are profitable to the country, and that for a short time."[2] The path the colonies chose, then, was one of moderation. Exclusive rights were granted, but American colonies were not yet committed to a formal patent system.

The need to establish economic independence from England, while maintaining English standards of living, made the granting of exclusive rights a familiar and simple method to secure economic growth and produce necessary goods for the entire community. Colonies began actively to encourage monopolistic practices. The colonies took a "dramatic departure from common law principles"[3] and provided privileges to monopolists that far exceeded the rights of the citizen. The needs of industry came to define the needs of the community, as the desire for surplus production led courts to support the extension of monopoly rights to both essential and non-essential industries. As the states unified under the constitution, a uniform system to grant exclusive rights was created.

The Constitution of the United States, Article 1, Section 8, empowers Congress "To promote the Progress of Science and useful Arts, by securing for limited Times to Authors and Inventors the exclusive Right to their respective Writings and Discoveries."[4] Continued legislative attention led to the Patent Act of 1793, which defined inventions eligible to receive patent protection. The Act, authored by Thomas Jefferson, defined its subject as "any new and useful art, machine, manufacture, or composition of matter, or any new or useful improvement [thereof]."[5] Jefferson did not mention organic or naturally occurring substances, plants, animals, or any other life form as eligible for patent protection.

The ideology of progress celebrated the use of patents to ensure the growth of national industry. What some colonists had feared was manifested in the 1920s and 1930s, only this time it was the corporation, not the individual, which had been granted too much power:

> The growth of the antitrust movement led to an increasing focus on patents, which were viewed as important weapons in the suffocating arsenal of big business. The exclusive nature of the patent grant, coupled with the actual market power that some patents conferred on their holders, seemed closely related to many of the monopolist's oppressive practices. Expert consultants to Congress criticized the role of patents and called for radical reforms such as the compulsory licensing of all patents to anyone who wanted to use them. The central idea behind the anti-patent movement was that the rights of powerful corporations had come to dominate the interests of the community.[6]

The anti-patent sentiments within the antitrust movement were attacked as unpatriotic because they were anti-industry, anti-progress and anti-technology. The balance between public good and private rewards was brought to a crisis point and several sections of existing patent laws were revised.

A 1939 revision to Title 35 of the Patent Act followed the antitrust move-ment and allowed patents to be granted for new and useful information rather than only for finished products.[7] It is at this point in history that patent rights were expanded to include ideas, processes and methods, leading to another form of individual property called "intellectual property." Since the 1940s, the term "intellectual property" has been used to refer specifically to the protec-tion of ideas, processes and methods, and also to refer to the system of exclusive rights in a larger sense, including patents, trademarks, copyright and trade secrets. Several different types of intellectual property are now available for organic materials. They include Plant Variety Protection Certificates, plant patents, and "composition of matter" patents which are issued on genetically engineered life forms and the genetic material of animals and humans.

The United States first treated plants as property in 1931, when a patent was given to Henry F. Bosenberg for the "Climbing or Trailing Rose."[8] The law states that,

> Whoever invents or discovers and asexually reproduces any distinct and new variety of plant, including cultivated sports, mutants, hybrids, and newly found seedlings other than tuber propagated plant or a plant found in an uncultivated state, may obtain a patent therefore, subject to the conditions and requirements of title.[9]

The appearance of plants in patent history marks the first of a series of events in which nature and natural products were enclosed by the law and made into property.

The 1970 Plant Variety Protection Act (PVPA) expanded patent protection to certain sexually reproduced plants, in the hope that plant breeders would innovate to solve recurring crop problems and thus stabilize the nation's food supply. The Act sought to encourage crop diversity by offering Plant Variety Protection Certificates, which, like patents, give the owner exclusive rights to the new plant breed for twenty-two years. The result of the Act, more than twenty-five years later, has been extreme consolidation of the seed industry, so that just a few multinational companies own the majority of PVP Certificates.

The patenting of life was extended beyond the plant kingdom beginning in 1972, when Ananda Mohan Chakrabarty filed a patent application that contained thirty-six claims outlining his use of four different plasmids. Chakra-barty found that these plasmids worked in different combinations to increase the oil-degrading properties of certain bacteria. As a microbiologist doing research on microorganisms capable of breaking down crude oil, he had al-ready made agreements to license the patent to his employer, the General Electric Company. The patent examiner accepted the claims in which Chakrabarty used the process of genetic engineering to contain oil spills, but rejected those in which Chakrabarty claimed rights to the actual bacteria. The patent examiner's "decision rested on two grounds: (1) that micro-organisms are 'products of nature,' and (2) that as living things they are unpatentable subject matter under Title 35, US Code, section 101."[10] The patenting of living organisms was intentionally omitted in the US Code and the examiner's rejection based on the Code was then standard practice.

General Electric appealed the patent examiner's decision to the Patent Office Board of Appeals. Despite the importance of this decision to the industrial development of commercial products from nature, the Board affirmed the examiner's decision that the organism was alive, and, as a life form, unpatentable under section 101 of Title 35.

General Electric and Chakrabarty, unsatisfied with the Board's decision, appealed to the Court of Customs and Patent Appeals. In a remarkable turn-around, the Court ruled in their favour, allowing Chakrabarty to receive a patent on a living organism. The Court declared that there was no distinction between animate and inanimate inventions, stating that for the purpose of patent law, "the fact that micro-organisms are alive is without legal significance."[11] Chakrabarty and the General Electric Company were granted Patent #4,259,444.

In 1978, the Commissioner of Patents and Trademarks sought *certiorari*[12] for the *Chakrabarty* case, due to the incongruity of the Court of Customs and Patent Appeals decision, its direct contradiction to existing law in Title 35 and US Patent and Trademark Office policy and practice. The Supreme Court declined to review the case. After a second appeal from the Commissioner, the Supreme Court agreed to review the Chakrabarty case in 1980.

In a close 5–4 decision in *Chakrabarty*, the Supreme Court majority argued that "the relevant distinction was not between living and inanimate things, but between products of nature, whether living or not, and human made inventions."[13] According to this line of reasoning, a scientist could create a new, functional animal by combining cells, genes or even body parts from other animals and receive a patent for this "invention" because (1) this particular type of animal is new and does not exist in nature; (2) this particular type of animal could not exist without human innovation; and (3) this new type of animal has utility (i.e., capacity for labour, organ transplantation, pharmaceutical and medical testing, etc.).

Although the decision to allow Chakrabarty's patent contradicted existing practices, the justices explained their authority to set the direction of the US Patent Office, stating "that Congress, not the courts, must define the limits of patentability; but it is equally true that once Congress has spoken it is 'the province and duty of the judicial department to say what the law is.'"[14] Justice Burger, writing for the majority, asserted that the Court was "without competence to entertain these [patentability] arguments,"[15] that the issue of patenting life forms was one of great importance which needed legislative attention and depthful examination. With no guidance from Congress, however, they jumped ahead of policy-makers to alter human relationships to nature drastically by allowing life forms to be patented.

Three other Supreme Court justices joined Justice Brennan's dissent in *Diamond* v. *Chakrabarty*. Justice Brennan stated, "I read the Court to admit that the popular conception, even among advocates of agricultural patents, was that living organisms were unpatentable."[16] Since there was no legislative direction on the topic of genetic engineering, the dissenting Justices felt it was wrong to "extend patent protection further than Congress has provided."[17]

The 1970 Plant Variety Protection Act excluded bacteria from patentable subject matter. This legislation suggests that Congress specifically chose not to extend patent protection to bacteria or other life forms. The Act states, "The breeder of any novel variety of sexually reproduced plant (other than fungi, bacteria, or first generation hybrids) who has so reproduced the variety, or his successor, in interest, shall be entitled to plant variety protection therefor…"[18] Justice Brennan highlighted this legislation in his dissent, stating, "the 1970 Act clearly indicates that Congress has included bacteria within the focus of its legislative concern, but not within the scope of patent protection."[19]

Brennan's dissent ended with these words, "As I have shown, the court's decision does not follow the unavoidable implications of the statute. Rather, it extends the patent system to cover living material even though Congress plainly has legislated in the belief that section 101 does not encompass living organisms. It is the role of Congress, not this court, to broaden or narrow the reach of patent laws. This is especially true where, as here, the composition sought to be patented uniquely implicates matters of public concern."[20]

The *Chakrabarty* case has been cited by scientists, lawyers and geneticists as opening the door for patent applications on all kinds of life forms. The ability of individuals and companies to attract capital through patent ownership has led to a huge increase in the number of life-form patent applications, and has helped create a multi-billion-dollar biotechnology industry funded almost entirely by venture capitalists. Genetic research, genetic engineering, cloning and genomics (human genetics and biopharmaceuticals) make up a highly concentrated industry that is reshaping concepts of legal ownership, natural resources, disease, health and identity.

In 1987 the *Chakrabarty* decision was used again by the Patent Office to expand the legal definition of "man-made." This time it was all "multicellular living organisms, including animals"[21] that became eligible for patent protection. A year later, the world's first patent on a mammal was issued for a genetically engineered mouse, widely known as the Oncomouse, due to its induced genetic propensity to be stricken with cancer. Approved by the US PTO in April 1988, the patent covers *all* non-human "onco-animals." Patents are given for any transgenic animal whose genetic composition, by definition, includes genes selected from other animals or animal species by non-traditional breeding methods. Considered "inventions" by the US Patent Office, these new breeds may vary only slightly from their "natural" predecessors.

In 1988, the Transgenic Animal Patent Reform Act was introduced in Congress to address these concerns. The Act called for a two-year moratorium on animal patenting, while Congress studied the issue. After heavy lobbying by the biotechnology industry, the Act failed in the Senate and the concerns of small farmers and others became legislative history.[22]

Brennan's dissent and the amicus briefs filed in the *Chakrabarty* case made clear that "progress" in the field of genetic engineering is a matter of public concern, one which citizens and elected officials should monitor and direct. US legislation can prevent the granting of patents for materials or products that are undesirable, dangerous and threatening to world security. The Atomic Energy Act of 1954 specifically "excludes the patenting of inventions useful solely in the utilization of special nuclear material or atomic energy for atomic weapons." These legislative safeguards become more necessary as the line between "natural" and "man-made" blurs.

The US patent system is a tool by which multinational corporations, investing heavily in research and development, can seek returns on their investments by claiming exclusive rights to new resources. The ownership that patents grant the biotechnology industry encourages a concentration of wealth and control over biological resources unparalleled in US history. The social impacts of the so-called "biotechnology revolution" are as far-reaching as the effects of the Enclosure Acts on England's rural society.

By allowing human, plant and animal life to be commodified, the US patent system has moved life forms from the natural realm into the industrial realm, where they become property and products of human innovation. The reduction of people to "tools" and raw materials and the reduction of people to genetic data similarly transform identities and social roles. For example, Kevin Kinsella, CEO of a leading genomic company, says "Within a few years you will be able to hold up a compact disc and say: 'That's me.' The 3 billion bits of data that make up a human genome could be stored on a single disc."[23] When everything can be bought and sold, what is sacred? A shift in the relationship between the law and technology is needed so that new ideologies of progress can emerge reflecting human values and protecting the integrity of the natural world.

NOTES

1. J.M. Neeson, *Commoners: Common Right, Enclosure and Social Change in England, 1700–1820*, Cambridge: Cambridge University Press, 1993, p. 24.

2. Fred Warshofsky, *The Patent Wars: The Battle to Own the World's Technology*, New York: Wiley & Sons, 1994, p. 32.

3. Morton Horwitz, *The Transformation of American Law 1780–1860*, Cambridge, MA: Harvard University Press, 1977, p. 40.

4. The Constitution of the United States of America, Article 1, Sec. 8.

5. Patent Act of February 21, 1793, 1,1 Stat. 319.

6. Robert P. Merges, *Patent Law & Policy: Cases & Materials*, Charlottesville, VA: Mitchie, 1997, p. 8.

7. US Code, Title 35 (Patents), Chapter 10, Sec. 101.

8. US Patent and Trademark Office, *Revolutionary Ideas: Patents and Progress in America*, Washington DC: US Government Printing Office, 1976, p. 28.

9. 35 USC Sec. 161 Patents for Plants, Amended September 3, 1954, 68 Stat. 1190.

10. *Diamond v. Chakrabarty*, 100 S.Ct. 2204, 2206.

11. *Diamond* v. *Chakrabarty*, 100 S.Ct. 2204, 2204.

12. *Certiorari* is a process by which a superior court, in this case the Supreme Court of the United States, can re-examine the decision of a lower court, like the Court of Customs and Patent Appeals.

13. *Diamond* v. *Chakrabarty*, 100 S.Ct. 2204, 2209.

14. Ibid., 2210.

15. Ibid., 2211.

16. Ibid., 2213.

17. Ibid., 2213.

18. Title 7, US Code, Section 2402(a), emphasis added.

19. *Diamond* v. *Chakrabarty*, 100 S.Ct. 2204, 2214.

20. Ibid., 2214.

21. US Patent and Trademark Office, "Animals – Patentability," April 7, 1987.

22. Corporate Counsel, "When Law and Nature Collide," *National Law Journal*, June 18, 1997.

23. "Genes and T-shirts," *The Economist*, January 4, 1997.

22

Paving the Way for Biotechnology: Federal Regulations and Industry PR

JENNIFER FERRARA

The evolution of genetic engineering from a laboratory science to a means for creating commercial products happened very fast – within a single decade. The US government saw the commercialization of biotechnology coming and deliberately chose a path that has amounted to non-regulation. Genetic engineering breaks through natural barriers of reproduction and alters the processes of plant and animal breeding, but agribusiness corporations were wary that burdensome regulations would hinder new discoveries and therefore the technology's commercial development. The federal government took up industry's cause: instead of establishing strict, precautionary regulations that gave priority to public and environmental health, agencies patched together an inadequate regulatory system that relied on risk assessment, industry science and corporate voluntarism.

The United States was in the heat of a high-tech economic race with Japan in the 1980s, and lawmakers saw genetic engineering as the new technology that would allow the US to maintain its position as the world's "leader" in export agriculture. The federal government would erect no law that might reduce America's competitiveness in the future world market for bioengineered products.

The first government body to establish guidelines for biotechnology research was the National Institutes of Health (NIH) in 1976.[1] Since the NIH is a funder of research and not a regulatory body, it could formulate guidelines but it had no power to enforce them. From the beginning, the NIH guidelines relied on the scientific community's and industry's self-regulation, starting a trend that continues today. As corporations became more involved in genetic engineering, NIH guidelines made accommodations for field tests and the mass production of genetically engineered organisms. In 1977 and 1978, sixteen bills to regulate genetic research were introduced in the US Congress. None was passed, and the NIH guidelines – which dealt primarily with medical and pharmaceutical research and did not take a precautionary approach – remained the sole regulatory mechanism for biotechnology research.

In the early 1980s, agribusiness corporations were beginning to develop genetically engineered plants, animal drugs and livestock, but no system was in place to regulate the development, sale or use of these products.[2] This was the era of the aggressively deregulatory Reagan administration, which developed the framework by which genetically engineered products, including food, are "regulated" today. Industrial profit, not public safety, was the administration's top priority. Officials at the Office of Management and Budget, the State and Commerce departments, and the White House Office of Science and Technology Policy wanted to ensure that the administration did nothing to "stifle" the development of biotechnology or to send the "wrong" message to Wall Street.[3] When George Bush, Sr took office in 1988, the President's Council on Competitiveness, chaired by vice-president Dan Quayle, joined the biotechnology industry in opposing strong regulations and close oversight by federal agencies.[4]

The main policy to emerge from that period was the "biotechnology regulatory framework," codified in 1986.[5] It was founded on the corporate generated assertion that genetic engineering was just an extension of traditional plant and animal breeding, and that bioengineered products did not differ fundamentally from non-engineered organisms.[6] The Reagan administration determined that existing federal agencies could regulate genetically engineered products sufficiently, and gave them overlapping regulatory authority.[7] The Food and Drug Administration (FDA) would regulate genetically engineered organisms in food and drugs, the United States Department of Agriculture (USDA) would regulate genetically engineered crop plants and animals, the Environmental Protection Agency (EPA) would regulate genetically engineered organisms released into the environment for pest control, and the NIH would look at organisms that could affect public health. In determining that existing agencies could do the job of regulating these products, the administration avoided passing new, more stringent federal laws or establishing a new regulatory agency devoted to the task.

The policy left gaping communication gaps between agencies, considerable regulatory ground uncovered, and much confusion over who would regulate what.[8] But, most importantly, the regulations were founded on the false premiss that engineered organisms used for food and agricultural products are no different to non-engineered, conventional products.[9] We know that genetic engineering deletes essential proteins, adds entirely new ones, and can modify genetic characteristics in entirely unexpected ways. Still, as long as the new genes come from an approved food source, the US government treats new or altered genes in engineered foods as natural, not novel, additives. In most cases, regulators are not required to take a precautionary approach when evaluating new genetically engineered food products; products are considered safe until proven otherwise.

Nearly a decade later, it appeared that the federal government was still playing catch-up in establishing working biotechnology safety regulations. The Union of Concerned Scientists (UCS), which continues to monitor the biotechnology industry and the federal regulatory system closely, was pointing out big holes in the so-called framework.[10] "Fundamentally, it does not contain sufficient statutory authority to oversee all of the products and activities entailed in genetic engineering," wrote UCS scientists in February of 1994. "Where authority does exist, there are problems with implementing regulations and policies." For example, a 1992 FDA policy exempted corporations from having to test most genetically engineered foods for safety or get FDA approval before the foods are put on the market.[11] Unless the corporation itself determines that "sufficient safety questions exist,"[12] they are only required to undergo voluntary, private "consultations" with the agency before marketing their product.[13]

It is not unusual for agribusiness corporations such as Monsanto to manipulate the limited safety regulations that exist. To establish safety standards for new products, federal agencies rely on studies performed by the very companies that are trying to get their products on the market. Studies to determine the long-term health consequences of new products are not always required. Over the years, many corporations have submitted fraudulent test results showing that their products are safe, or they have simply withheld information or studies indicating otherwise. Because the federal government protects corporate safety studies as trade secrets, they are not available for public scrutiny. By sheltering corporations in this way, federal agencies hold corporations' pursuit of profits above the public's right to health and a safe environment.

Laws governing biotechnology continue to favour agribusiness and biotechnology companies, but as the industry has developed the corporate agenda with respect to regulation has taken ironic twists. Initially, the lack of a cautious regulatory approach enabled small biotechnology companies to develop and market new products at a rapid pace. Throughout the 1990s, larger agribusiness corporations like Monsanto and Novartis were buying up these small companies, while developing their own expansive biotechnology research and marketing operations. From their position at the top of the industry in the US, Monsanto and other large corporations have favoured some regulations, but only when they serve corporate marketing purposes. Regulations that require companies to submit a plethora of costly scientific data to regulatory agencies, for example, discourage competition from smaller biotechnology and seed companies while giving the public the illusion that new biotechnology products undergo rigorous safety evaluations and are therefore safe.

In 1995, for example, Monsanto lobbied against a provision in the EPA funding bill that would have prevented the EPA from regulating agricultural plants bioengineered to contain the toxin from the bacterium *Bacillus thuringiensis* (Bt).[14] Genetically engineered foods were just about to hit the market, and Monsanto was fully aware that almost any EPA regulations for Bt plants would publicly sanction the genetically engineered products and possibly help defuse resistance from public interest environmental groups. Corporations would only be able to get their Bt products to market if they had extensive money and resources to jump through all the regulatory hoops, however pro forma. With the competition out of the way, the market would belong to Monsanto and just a few other agribusiness giants.

FDA SCANDALS AND REVOLVING DOORS

To understand better how genetically engineered foods and the associated safety hazards were unleashed onto the American public, let us take a look at the story of the first mass-marketed bioengineered food product, Monsanto's recombinant Bovine Growth Hormone (rBGH). This artificial hormone has been linked to cancer in humans and serious health problems in cows, including udder infections and reproductive problems (see Chapter 4 in this volume). The development and approval of rBGH was rife with scandal and protest, but the right combination of government backing, corporate science, and heavily funded corporate public relations schemes paved the way for the first major release of a genetically engineered food into the US food supply. The FDA and Monsanto hid important information about safety concerns, masked disturbing conflicts of interest, and stifled those who were asking the "wrong" questions and telling the truth about rBGH.

The FDA declared rBGH milk safe for human consumption before important information about how rBGH milk might affect human health was ever available.[15] When critical information about how rBGH raised the levels of insulin–like growth factor (IGF-1) in milk[16] and the possible link between IGF-1 and human cancer began to emerge,[17] the FDA was already apparently in too deep to change its mind or raise more questions about the drug's effects on human health. Instead, the agency relied almost exclusively on data generated by Monsanto – and highly criticized by independent scientists – to justify a decision it appeared to have made years earlier.[18]

In 1991 a veterinary pathologist at the University of Vermont, where Monsanto spent nearly half a million dollars to fund test trials of rBGH, leaked information about severe health problems affecting rBGH-treated cows, including mastitis and deformed births.[19] The UVM scientist heading the research had already made numerous public statements to state lawmakers and the press, and had released a preliminary report indicating that rBGH-treated cows suffered no abnormal health problems compared to untreated cows.[20] The US General Accounting Office (GAO), the research arm of the Congress, was called in to investigate. During the investigation, the FDA stalled in providing the GAO with original Monsanto test data,[21] and the GAO was unable to obtain critical data from UVM and Monsanto.[22] The GAO terminated its investigation, concerned that Monsanto had the time to manipulate the questionable data and that any further investigation would be fruitless.[23]

Even FDA insiders have criticized the agency for its spotty review of the drug, but the FDA has dismissed these concerns and fired at least one official who blew the whistle on the FDA's corrupt drug approval process.Veterinarian Dr Richard Burroughs reviewed animal drug applications at the FDA's Center for Veterinary Medicine from 1979 until he was fired in 1989.[24] In 1985 Burroughs headed the FDA's review of rBGH and remained directly involved in the review process for almost five years. Burroughs wrote the original protocols for animal safety studies and reviewed the data that rBGH developers, including Monsanto, submitted as they carried out safety studies.

A 1991 article in *Eating Well* magazine quotes Burroughs describing a change in the FDA beginning in the mid-1980s: "There seemed to be a trend in the place toward approval at any price. It went from a university-like setting where there was independent scientific review to an atmosphere of 'approve, approve, approve.'"[25] According to Burroughs, the FDA was totally unprepared to review rBGH, the first genetically engineered animal drug to go through the approval process; rBGH was beyond the scope of most FDA employees' knowledge. But rather than admit incompetence, the FDA, according to Burroughs "decided to cover up inappropriate studies and decisions," and agency officials "suppressed and manipulated data to cover up their own ignorance and incompetence."[26] Burroughs was pressured by corporate representatives who wanted the agency to ease strict safety testing protocols, and he saw corporations drop sick cows from rBGH test trials and manipulate data in other ways to make health and safety problems disappear.

Burroughs challenged the agency's lenience and its changing role from guardian of public health to protector of corporate profits. He criticized the FDA and its handling of rBGH in statements to congressional investigators, in testimony to state legislatures, and in briefings to the press.[27] Inside the FDA, he rejected a number of corporate-sponsored safety studies as insufficient and was prevented by his superiors from investigating data submitted by industry revealing possible health problems caused by rBGH. Though Burroughs had a record at the FDA showing eight straight years of good performance, he began receiving poor performance reports, for which he claims he was set up. Finally, in November 1989, he was fired for "incompetence."

Not only did the FDA fail to act upon evidence that rBGH was not safe, the agency actually promoted Monsanto's product before and after the drug's approval. In doing so, the FDA took on the impossible double role of regulator and promoter of bioengineered foods. Dr Michael Hansen of the Consumers Union notes that the FDA acted as an rBGH advocate by issuing news releases promoting rBGH, making public statements praising the drug, and writing promotional pieces about rBGH in the agency's publication, *FDA Consumer*.[28] In an apparent attempt to quell public controversy over rBGH, two FDA researchers published a paper in the journal *Science* in 1990 to show that rBGH was safe for consumers.[29] Gerald Guest, director for FDA's Center for Veterinary Medicine, told *Science*, "We'd like to get our side of the story out, to show why we're comfortable with the safety. We'd like for people to know that it's a thoughtful process, and we want it to be open and credible."[30]

Guest was apparently engaged in some wishful thinking. Dr Samuel Epstein criticized the FDA for acting "as a booster or advocate for an animal drug that hasn't yet been approved."[31] Epstein and others faulted the FDA for including portions of unpublished studies about rBGH in the *Science* article, but not making the full studies available for independent review.[32]

The FDA's pro-rBGH activities make more sense in light of known conflicts of interest between the FDA and the Monsanto corporation.[33] Michael R. Taylor, the FDA's deputy commissioner for policy, wrote the FDA's rBGH labelling guidelines. The guidelines, announced in February 1994, virtually prohibited dairy corporations from making any real distinction between products produced with and without rBGH.[34] To prefent rBGH-milk being "stigmatized" in the marketplace, the FDA announced that labels on non-rBGH products must state that there is no difference between rBGH and the naturally occurring hormone. In March 1994 Taylor was publicly exposed as a former lawyer for Monsanto. While working for Monsanto, Taylor had prepared a memo for the company discussing whether or not it would be constitutional for states to erect labelling laws concerning rBGH dairy products.[35] In other words, Taylor helped Monsanto figure out whether or not the corporation could sue states or companies that wanted to tell the public that their products were free of Monsanto's drug.

Taylor wasn't the only FDA official involved in rBGH policy who had worked for Monsanto. Margaret Miller, deputy director of the FDA's Office of New Animal Drugs, was a former Monsanto research scientist who had worked on Monsanto's rBGH safety studies up until 1989. Suzanne Sechen was a primary reviewer for rBGH in the Office of New Animal Drugs between 1988 and 1990. Before coming to the FDA, she had done research for several Monsanto-funded rBGH studies as a graduate student at Cornell University. Her advisor at Cornell was one of Monsanto's university consultants and a known rBGH promoter. Remarkably, the GAO determined in a 1994 investigation that these officials' former association with the Monsanto corporation did not pose a conflict of interest. But for those concerned about the health and environmental hazards of genetic engineering, the revolving door between the biotechnology industry and federal regulating agencies is cause for concern.

THE ROLE OF PUBLIC RELATIONS

The International Dairy Foods Association and the National Milk Producers Federation, two leading dairy industry trade groups, formed a public relations arm called the Dairy Coalition in anticipation of public controversy over the approval of rBGH.[36] Early on, the industry polled US citizens about their opinion on rBGH and whom they would trust for information on food safety.[37] Though people did not want rBGH dairy products and did not trust information on genetically engineered products provided by government agencies or corporations, they could be convinced that the drug was safe by university scientists and carefully presented pro-rBGH "educational" material.

In addition to doing its own "public education," the Coalition suppresses controversial media stories about rBGH.[38] For example, at a Washington press conference in January 1996, Food & Water, Inc., a national non-profit organization working to end corporate control of the food supply, along with the Cancer Prevention Coalition, directed by Dr Samuel Epstein, released Epstein's study demonstrating possible links between IGF-1 and breast and colon cancers. The Dairy Coalition immediately went to work. In a February internal memo, the Coalition congratulated itself on convincing editors at the *New York Times*, the *Wall Street Journal*, and the *Washington Post* to dismiss the story.[39] Editors at *USA Today* and the *Boston Globe* who slipped through the Coalition's fingers were subjected to heated meetings with dairy industry representatives, letters questioning the newspapers' integrity, and reams of pro-rBGH material.

Monsanto also used strong-arm tactics to keep a Florida television station from airing information about the possible connection between rBGH and cancer.[40] Fox television station WTVT in Tampa, Florida, hired two award-winning television journalists, Steve Wilson and Jane Akre, to do a series on rBGH in the fall of 1996. The series was heavily advertised and scheduled to air on February 24, 1997, but after receiving two letters from Monsanto lawyers the station pulled the series.

Monsanto lawyers told the station that the Monsanto corporation would suffer "enormous damage," and the station might be subject to "dire consequences" if it ran the series in its original form. Fox lawyers tried to change the series and offered to pay the journalists to keep quiet about the censorship. The journalists refused, and the station fired the journalists and watered down the story, omitting all references to cancer. Wilson and Akre filed a lawsuit charging that the station violated its licence from the Federal Communications Commission by demanding they include known falsehoods in their report.[41]

The public relations schemes used to introduce rBGH and keep it on the market are part of a bigger trend within the food industry. As John Stauber and Sheldon Rampton describe in their book, *Mad Cow USA*, the food industry has been operating for over a decade in a "crisis management" mode.[42] As the public becomes more aware of the dangers of industrial food-producing technologies like toxic pesticides, antibiotics and rBGH, the food industry is spending hundreds of millions of dollars on public-relations campaigns to quell health and environmental concerns. As these authors put it, corporations are waging "an all-out war for the hearts and minds of consumers." In this war, the "good guys" are innocent corporations, the "enemies" are public interest groups, independent scientists and journalists, and the weapons are paid industry scientists, lobbyists and politicians.

NOTES

1. Jack Doyle, *Altered Harvest*, New York: Viking Penguin, 1985, pp. 244–6.
2. Ibid., p. 247.

3. Ibid., pp. 249–50.

4. Michael Fox, *Superpigs and Wondercorn: The Brave New World of Biotechnology and Where it All May Lead*, New York: Lyons & Burford, 1992, p. 11.

5. "The Clinton Administration and the Biotechnology Framework," *The Gene Exchange*, Washington DC: Union of Concerned Scientists, February 1994, pp. 6–7, 11.

6. Fox, *Superpigs and Wondercorn*, p. 35.

7. Ibid., p. 42.

8. Ibid., p. 37; also "The Clinton Administration."

9. Erik Millstone, Eric Brunner and Sue Mayer, "Beyond 'Substantial Equivalence'," *Nature* 401, October 7, 1999, pp. 525–6.

10. "The Clinton Administration," 5.

11. Kristin Dawkins, *Gene Wars: The Politics of Biotechnology*, New York: Seven Stories Press, 1997, p. 33.

12. Ibid.

13. "FDA Implements Informal Food Safety Reviews: Allows Food Uses of Seven Genetically Engineered Crops," *The Gene Exchange*, Washington, DC: Union of Concerned Scientists, December 1994, p. 3.

14. Rachel Burnstein, "Paid Protection," *Mother Jones*, January–February 1997, p. 42.

15. Michael K. Hansen, *Biotechnology and Milk: Benefit or Threat?* Mount Vernon, NY: Consumers Union of the United States, 1990, pp. 22.

16. Samuel S. Epstein, "Unlabeled Milk from Cows Treated with Biosynthetic Growth Hormones: A Case of Regulatory Abdication," *International Journal of Health Services*, vol. 26, no. 1, 1996, pp. 173–85.

17. Peter Montague, "Milk Safety," *Rachel's Environment & Health Weekly* 454, Annapolis, MD: Environmental Research Foundation, August 10, 1995.

18. Epstein, "Unlabeled Milk"; Joel Bleifuss, "Mucking with Milk," *In These Times*, January 10, 1994, pp. 12–13.

19. Andrew Christiansen, *Recombinant Bovine Growth Hormone: Alarming Tests, Unfounded Approval*, Montpelier, VT: Rural Vermont, July 1995, p. 8.

20. Ibid., pp. 7, 12.

21. Ibid., p. 14.

22. Ibid., p. 16.

23. Ibid., pp. 19–21.

24. Craig Canine, "Hear No Evil," *Eating Well*, July/August 1991, pp. 41–7.

25. Ibid., p. 41.

26. Ibid., p. 43.

27. Hansen, *Biotechnology and Milk*, pp. 19–20.

28. Ibid., p. 23.

29. Judith C. Juskevich and C. Greg Guyer, "Bovine Growth Hormone: Human Food Safety Evaluation," *Science* 249, August 24, 1990, pp. 875–84.

30. "FDA Publishes Bovine Growth Hormone Data," *Science* 249, August 24, 1990, p. 852.

31. Ibid.

32. Epstein, "Unlabeled Milk."

33. Christiansen, *Recombinant Bovine Growth Hormone*, p. 23; Jim Ridgeway, "Robocow: How Tomorrow's Farming is Poisoning Today's Milk," *Village Voice*, March 14, 1995.

34. Keith Schneider, "F.D.A. Warns the Diary Industry Not to Label Milk Hormone-Free," *New York Times*, February 8, 1994.

35. Ridgeway, "Robocow."

36. Jennifer Ferrara, "Monsanto and the BGH Blues," *Food & Water Journal*, vol. 5, no. 2, Spring 1996, pp. 12–13, 15.

37. Colleen M. Sauber, "How Will Consumers Respond to BST?" *Dairy Herd*

Management, April 1989, pp. 18–20, 22, 24.

38. Ferrara, "Monsanto and the BGH Blues."

39. Ibid.

40. Peter Montague, "Milk, rBGH and Cancer," *Rachel's Environment & Health Weekly* 593, April 9, 1998.

41. Wilson and Akre have posted all the key documents in their lawsuit, including a transcript of their original series, at http://www.foxbghsuit.com. In August 2000 a Florida jury awarded Akre $425,000 in damages, accepting her claim (but not Wilson's) that her firing was in retaliation for her threat to petition the Federal Communications Commission regarding the suppressed series.

42. Sheldon Rampton and John Stauber, *Mad Cow U.S.A.: Could the Nightmare Happen Here?*, Monroe, ME: Common Courage Press, 1997, p. 4.

23

Biotechnology and the New World Order

MITCHEL COHEN

The globalization of capitalism has a new weapon, about which people around the world know very little – the colonization of our genes. Genetic engineering is the ideal technology for corporatizing whole new areas of nature. Thus, it is an essential component of the new globalization of capital. It conquers those parts of life that have thus far stood outside of its domain: the inner workings of the living cell.

Some see this as "science," which we are taught (incorrectly) is inherently good and "free from politics." In actuality, there is no such thing as a "neutral" science. Science and technology – as our way of seeing the world around us, and as tools for manipulating and "developing" it – are dripping with politics. Far from being "neutral," biotechnology and genetic engineering are, to companies like Monsanto and Novartis (and their paid apologists), a new engine for the accumulation of capital and huge profits, regardless of the toll it takes on the world around us.

As September 1998 edged into autumn, the new issue of *The Ecologist* was sent to press. Twelve thousand copies of this well-respected British journal were being readied for distribution when, suddenly, activist email and phone lines buzzed with the news: *The Ecologist*'s printer, who had been under contract with the magazine for many years, had suddenly shredded the entire print run, without notifying the editors.

The shredded edition examined, in depth, the machinations of the Monsanto Company, the powerful multinational conglomerate with billions of dollars invested in patenting the genetic sequences of living organisms and in genetically engineering agriculture and pharmaceutical products for private profit. "We were afraid of being sued by Monsanto for libel," the printer said, explaining why he destroyed the print run despite the fact that Prince Charles himself had written an article in that very issue about the importance of organic farming and against biotechnology in agriculture. Fortunately, *The Ecologist* managed to find another printer, who reprinted the issue in its entirety, with reprints numbering over 400,000 copies.[1]

What happened to *The Ecologist* is just the tip of the civil liberties iceberg. We are seeing a transformation of law, nature and society right before our eyes. Genetic engineering – a form of biotechnology – and the new laws and mechanisms designed to protect, develop and impose it on the natural world at the expense of democratic rights most people hold as sacred, is a key to the successful globalization of capital and its New World Order. Genetic engineering is as fundamental to the consolidation of the new stage of capitalist accumulation as the steam engine was to the development of mercantile capitalism, the mechanical loom to the creation of factories, and the technology of chattel slavery to early capitalist agriculture.

In the 1700s, publicly used lands in Europe were suddenly "privatized" by wealthy individuals and companies. In England, as elsewhere, laws were enacted that allowed lands that were owned by no one and which had been used in common for centuries to be expropriated by the emerging capitalist class. These became known as the Enclosure Acts. In Germany, those who accused women of being witches to be burned at the stake were, in many cases, allowed to confiscate the property of those they had accused.

All of this "privatization," this *theft*, was given legalistic cover in new laws, as public lands and early machinery were becoming privatized and reshaped by the needs of capital. Over time, those "enclosures" came to receive social acceptance and sanction in law. And so, throughout Europe, the taking of dead wood for heat and cooking by peasants became a criminal offence. By 1842, 85 per cent of all prosecutions in the Rhineland, as in much of Europe, dealt with a new crime: "the theft of wood." Corporations – but not workers – were soon freed, under the new laws, to strip all of the trees on public lands – whole mountains! – with impunity.

Today, the same process is happening again. It goes by various names, "structural adjustment" and "neoliberalism" being the two most widely known around the world. Treaties such as the North American Free Trade Agreement (NAFTA), the General Agreement on Tariffs and Trade (GATT), and the proposed Multilateral Agreement on Investments (MAI), and the imperious bodies they legislate into being, such as the World Trade Organization (WTO), International Monetary Fund (IMF) and World Bank, as well as the US Agency for International Development (USAID), are political mechanisms brought into being to codify "neoliberalism" through the mechanism of "intellectual property rights" of corporations over the sovereignty of existing nation-states and the basic rights of every country's citizenry.

With genetic colonialism, corporations are winning rights legally which they had always taken extra-legally. Biotech corporations are demanding human rights for corporations, while at the same time supporting legislation curtailing the rights of actual people. And through such institutions as the World Trade Organization and the proposed Free Trade Agreement of the Americas

(FTAA), corporations are able to lay claim to their novel and newly won legal status as virtual (and ideal) citizens of the world, at the expense of the general population and the sovereignty of nation-states themselves.

Under the World Trade Organization, international corporations are permitted to sue nation-states for daring to enforce environmental regulations, claiming that such regulations infringe on the corporations' "right" to free trade and private enterprise. The European Union, for instance, prohibits the use of hormones in raising beef cattle. The hormones, implanted as pellets in the ears of the cattle, make them grow bigger and faster. At least one of the six compounds in question is a proven carcinogen. Because most US beef producers use the hormone, the European market has been largely off limits to them. The US government challenged Europe's exclusion of American beef before the World Trade Organization, and won. A WTO panel ruled that there was not enough scientific evidence against the hormones for them to be prohibited. The US government responded by imposing an impossibly high tariff on cheese imported from France, among other products. French farmers responded by attacking American corporations there, particularly McDonald's. French farmer and activist José Bové was given a huge welcome by WTO protesters in Seattle in November of 1999, as the French delegation to the WTO led a protest of thousands at a downtown McDonald's.

Today, in the US and in Europe, we are experiencing the first wave of legislation to consolidate and institutionalize new political configurations of power which, in general, undermine the ideal of the democratic nation-state. Of course, capitalism has always privileged corporate profits over peoples' needs. This is nothing new. For example, in 1999 US vice-president Al Gore ordered the Environmental Protection Agency to *slow down* its implementation of stricter standards for agricultural pesticides. But the new technologies require even greater measures – a qualitatively different paradigm – in which to exploit nature and labour to their full capacity.

Attacks on the sovereignty of people and the sacred "wholeness" of nature are vastly magnified by the new technology. Freedom of speech, the inalienable right to address one's government demanding redress of grievances, is under sustained attack in the US in a way that is fundamentally different than ever attempted before. Criticism of corporate policies or technologies is increasingly portrayed as libel or slander – as if a corporation, which is not a living person, can be slandered! So-called "food disparagement" laws now exist in thirteen states; they proclaim the "rights" of corporations on a vast scale once presumed to be the hallmark of the citizenry, not private businesses.

Public criticism of corporate policies is now held to be criminal activity subject to imprisonment. Chiquita Banana recently threatened a lawsuit against a newspaper which ran an important story documenting the company's domination of Central America's economies. Chiquita officials did not take issue with the truth of the story itself but with the methods the reporters used to gain their information – they had ingeniously found a way to tap into Chiquita's voice-mail system without authorization. Although the reporters uncovered all sorts of nefarious schemes Chiquita was plotting, such as the forced removal of peasants in Haiti and the conversion of their lands into vast plantations of export crops, the *Cincinnati Enquirer*'s publishers buckled under the pressure that millions of dollars can buy. The newspaper agreed to pull the story; it issued a retraction (despite the fact that the story was *true*), paid Chiquita an unprecedented $10 million, and fired its reporters.[2]

Corporations are now regularly filing Strategic Lawsuits Against Political Participation (SLAPP suits) against individuals protesting the destruction of forests or leafleting about high levels of hormones and antibiotics in meat, often not because they expect to win such lawsuits, but to tie up the far less affluent activists in court until they submit to industry's edicts. One such recent suit was brought by the giant Beverly Enterprises – the nation's largest for-profit nursing home chain – against Kate Bronfenbrenner, a professor and scholar at Cornell University.[3] Bronfenbrenner had testified before a congressional panel about the company's union-busting tactics. So the nursing home chain decided to sue her, claiming that her testimony "defamed" it (as if "it" was a person that could sue over such things; and as though it was okay to intimidate a person from telling the truth to the US Congress). Many condemned the suit as an attack on free speech and academic freedom and rallied to her defence. Nevertheless, it took an enormous and expensive legal effort and public campaign before Beverly Enterprises' lawsuit against Bronfenbrenner was dismissed in June 1998.

Monsanto was able to get away with strong-arming small dairies that refused to allow farmers to inject their cows with the hormone rBGH. Immediately after the genetically engineered drug was approved for commercial use in the United States, several companies were sued to keep them from labelling their dairy products "rBGH free." Despite failing to win a single round in the courts, Monsanto has nevertheless been able to create enough economic and political intimidation to win economically what it cannot win in the courts.

The company has also gone after the media to keep the public from finding out about the links between injections of cows with its product and cancer in humans. Fox TV in Tampa, Florida, fired two award-winning reporters under pressure from Monsanto, as they were about to air the story of rBGH (see Chapter 22 in this volume). That is why the shredding of *The Ecologist* came as no surprise to activists in the US.

The first targets of a lawsuit under the new state "food disparagement" laws were Howard Lyman (of the Humane Society of the United States, now president of EarthSave International), and talk-show host Oprah Winfrey. Lyman and Winfrey were sued for publicizing on national television in the US the dangers associated with Britain's epidemic of mad cow disease.[4] A former rancher, Lyman explained how the beef industry fed its cows "rendered" cattle protein: ground-up carcasses of other animals. Winfrey said she would never eat another hamburger. A cattle rancher representing the industry sued Winfrey and Lyman for libel for "criticizing a food or agricultural product without scientific basis" under a food disparagement law passed in Texas. The lawsuit stated that Lyman's warning about mad cow disease "goes beyond all possible bounds of decency and is utterly intolerable in a civilized community," even though it is true.[5]

Under previous laws, the food industry bore the burden of proof. To win a libel case, it had to prove that its critics were deliberately and knowingly circulating false information. Under the new standard, however, it doesn't matter that Lyman believes in his statements, or even that he can produce distinguished scientists to support his conclusions.

"Agricultural disparagement statutes represent a legislative attempt to insulate an economic sector from criticism, and, in this respect, they may be strikingly successful in chilling the speech of anyone concerned about the food we eat," observes David Bederman, Associate Professor of Law at Emory University Law School.

> The freedom of speech, always precious, becomes ever more so as the agricultural industries use previously untried methods as varied as exotic pesticides, growth hormones, radiation, and genetic engineering on our food supply. Scientists and consumer advocates must be able to express their legitimate concerns. The agricultural disparagement statutes quell just that type of speech. At bottom, any restriction on speech about the quality and safety of our food is dangerous, undemocratic, and unconstitutional.[6]

Fortunately, in this case, Lyman and Winfrey prevailed, albeit on a technicality. That was one crucial but still limited victory against corporate libel lawsuits that are upholding the "rights" of corporations as if they are people, and bringing "the formidable powers of government and industry together for the purpose of suppressing the views of people with complaints against the system."[7]

What will most likely turn out to be so striking about the present period in years to come, however, is not only the geographic extent of capitalism's reach – commonly known as "globalization," as shaped by the International Monetary Fund (IMF), World Bank and World Trade Organization – but its plundering of new dimensions of being, the Nature *within*.

A spectre is haunting the planet – the spectre of biological devastation and ecological catastrophe. The ecosystems sustaining life are being ravaged. Many familiar organisms – butterflies, frogs, whole species – are in sudden danger of being wiped out, and mechanisms for propagation – even seeds – are coming under the private ownership and control of a few very large pharmaceutical corporations.

Civil liberties take a back seat to the exigencies of the biotech industry. All the good things that human beings have achieved, and all the beauty of the world around us – the once magnificent old growth forests, pristine drinking water, healthy soils, seas teeming with fish, indeed the sanctity of life itself as manifest in our genetic codes – are being grabbed, privatized and pillaged by corporate, technological and political powers and legitimized by new laws in a shameless orgy of material profit.

With changes to the gene pool verging on irreversibility and biotechnology becoming ensconced as essential to this new era of capitalism, the enforcement of so-called intellectual property rights has political ramifications far beyond the biology of a few individual organisms. The biotech industry is hammering structures of power and domination into new configurations, so that our political institutions parallel, intersect and serve the needs of bio-technological corporations. These require new "power formations" on an international level.

The new technologies constitute modes of production and reproduction that intersect capitalist relations of exploitation, shaping and, ultimately, dominating our approach to science, art and even so-called "pure research." Science has its part to play; researchers are victims, but also perpetrators of the dominant determinist paradigm looming over our lives. By the end of 1999, at least six people had died as a result of experimental (and unapproved) "gene therapy," performed by scientists who had ties to corporations seeking to sell genetic technology to the health care industry.[8] But these deaths were covered up. Because of their investments, the scientists were allowed to conceal the results of their medical experiments by writing them up *not* as deaths, but as "proprietary business information."

Catastrophe is, literally, blowing in the wind. The biotech industry is charging ahead full speed, knocking aside all who dare to question its apparent willingness to sacrifice our lives and the environment in its rush for profits.

Genes, like every other entity, are context dependent; nucleic acids do *not* constitute the "blueprint of life." They are *not* determining agents but part of complex dialectical interactions involving DNA and genes, genes and traits, traits and behaviour, and behaviour and overlapping and mutually defining systems of capitalism and patriarchy. These enter and shape every level, from the way we observe the interaction of molecules (reductionistically, deterministically) to the way we conceive of and research genes, chromosomes and cells, to the more familiar conditions of alienation we experience and re-create as human beings.

With the new microbiology fast becoming the dominant framework for examining life, the *doing* of scientific work itself has become more and more atomized, fragmented, broken down into specialized disciplines and sub-disciplines: not just Biology, Chemistry, Physics, Ecology, but Molecular Biology, Evolutionary Genetics, Cytology, and Developmental Embryology, overwhelming us with its plethora of disconnected parts. Can there be any appreciation of the whole, of the complex interaction between the whole and the parts? Such appreciation is becoming more difficult as scientific work and thought feed on and reproduce patterns of exploitation, linearity and domination, even when one does not mean to do so. The more fragmented our focus, the more "deterministic," in this culture, we become. The strictly deterministic, quasi-religious cause-and-effect model of DNA as blueprint still predominates. Too many scientists just substitute "genes" for "God" as the ultimate determining force. Why are ultimate determining forces needed, anyway?

My biologist friends offer one of two rejoinders to my sweeping statements: "That's not true," or "We already know that, what else is new?" They offer examples where genetic engineering has done some good: the healing of heart tissue after a heart attack, an emergency injection of engineered insulin, and so forth. "If *we* controlled this technology we'd put it to work for the public good," they suggest. But such arguments turn out to be rationalizations for scientists working within the biotech juggernaut, which always tries to sell itself with a humane face so as to ease the qualms not only of the population at large but of its own scientists, trapped as they are in what has become an increasingly commercial and reductionist approach to research. Contrary to the red herring statements issued by the industry, radical ecologists do not propose to deny sick people relief of suffering obtained from, for instance, protease inhibitors – however temporary that relief will turn out to be – nor other such treatments, sometimes genetically engineered, for AIDS and other diseases. But we do attempt to provoke society as a whole to address why people are sick to begin with and to delegitimize the industrial framework as the dominant paradigm for doing scientific research. Why are scientists researching what they are researching to begin with, let alone in the ways they are doing it? The arguments of the biotech industry and its apologists turn out to be little more than sophistry allowing them to protect their investments and bet the world against their anticipated profits.

Scientists, researchers and technicians are, they say, engaged in biotech development "for the good of humanity." Let me offer, then, some modest proposals for the common good:

• Ban all genetic engineering of agriculture, plants, pesticides and foods.
• Abolish the private patenting of genetic sequences as "intellectual property."
• Take private profit out of research and development of health-related drugs.
• Require all bioengineered products, and those derived from them, to be clearly labelled.

We need not merely to question but to *challenge* authority and the privatization so central to the new technologies. Changing the world, standing up to Monsanto, Novartis, Aventis, DuPont, Eli Lilly and the like requires taking risks, sometimes very serious and personal ones, so that we can begin to determine our own destinies.

But in a world where the concept of "self" is littered with industrial genes (and who really knows any longer where those genes have been?), what is the meaning of self-determination? What self is doing the determining? If there is to be any hope at all in literally reclaiming our "selves," let alone our world, creating a new society and saving the planet, one can no more take hold of capitalist technology and wield it for the public good than one could the apparatus of the state, for inherent in the technology of genetic engineering, as in the state, are all the relations of exploitation, domination and power over others, and over Nature, that we need to overthrow. These relations inevitably reassert themselves unless we dismantle both the technology and the state altogether, along with the system of capitalism in which we live. Only in the course of doing all of that can we re-envision the world we hope to live in and take the kinds of action needed to bring it about.

NOTES

1. *The Ecologist*, vol. 28, no. 5, September/October 1998. Copies are available from *The Ecologist*, Unit 18, Chelsea Wharf, 15 Lots Road, London SW10 0QJ, or ecologist@ gn.apc.org.

2. Anna M. Busch and Larry Burns, "Chiquita Coverup: Money and Power Beats Integrity and Truth," Council on Hemispheric Affairs, reprinted in the *Albion Monitor*, July 20, 1998, at http://www.monitor.net; also, *Democracy Now* Pacifica radio network broadcast, July 7, 1998, at http://www.democracynow.org; Associated Press, "Chiquita Satisfied with Settlement with the Cincinnati Enquirer," July 4, 1998, via *Nando Times News*, http://www.nando.com.

3. *Democracy Now*, Pacifica radio network broadcast, April 6, 1998, http://www. democracynow.org; Canadian Broadcasting Company Radio, "As It Happens," April 18, 1998, http://www.cbc.ca; Ben Dobbin, Associated Press, in *Nando Times Business Archives* 1998, www.nando.com; American Association of University Professors, press release (undated), 1998. See also Kate Bronfenbrenner, "A Statement on the Victory over Beverly Enterprise," http://www.rci.rutgers.edu.

4. Sheldon Rampton and John Stauber, "Mad Cow Disease: Industrial Farming Comes Home to Roost," *Covert Action Quarterly*, Fall 1997.

5. Petition by Paul F. Engler and Cactus Feeders, Inc. against Oprah Winfrey, Harpo Productions, Howard Lyman and Cannon Communications, US District Court, Texas Northern District, May 28, 1996.

6. David J. Bederman, quoted in Rampton and Stauber, "Mad Cow Disease," p. 61.

7. Ibid.

8. Sheryl Gay Stohlberg, "Death Raises Questions about Gene Therapy's Safety," *New York Times*, November 4, 1999.

PART IV

The Worldwide Resistance
to Genetic Engineering

In recent years the worldwide resistance to genetic engineering has become one of the fastest-growing social movements in a generation. Along with its closely linked counterpart, the movement against corporate globalization, this resistance has inspired many thousands, perhaps millions, of people to question the technological choices being made in our name, and challenge the global institutions that seek to impose this dangerous and ethically troubling technology on people and the earth.

Opposition to genetic engineering has taken on as many diverse forms as there are communities of people that have come to appreciate this technology's potentially devastating impacts on their lives and local ecosystems. It is a movement with a deep commitment to the integrity of life on earth, a determination to uncover the links among related issues, a profound sense of urgency, and often an outrageous sense of humour. In every part of the world where people are affected by the aggressive promotion of biotechnology and the corporate ownership of life, people have found unique and often very colourful ways to dramatize the issues and put a stop to this industry's outrages.

Throughout the British Isles, well over a hundred experimental plots of genetically engineered maize, sugar beet and oilseed rape have been cut down or pulled out of the ground by local activists. Sometimes it happens in the dark of night, sometimes in broad daylight, accompanied by public festivals, processions, music, costumes and the planting of trees and flowers. Nearly all the large supermarket chains in Britain have pledged to keep genetically engineered food out of their store brand products, and advertise this pledge widely. In Germany, activists have set up protest camps alongside, and sometimes directly on top of, the biotechnology industry's experimental plots. Such camps have received widespread support from residents of the surrounding towns and villages, who would gratefully share local foods, music and community traditions.

In France, radical farmers have been in the forefront of opposition to genetic engineering and to the threat of transnational corporate dominance over food. In 1998 members of the French Peasant Confederation (Confédération Paysanne) entered a Novartis warehouse containing 5 tons of genetically engineered maize and destroyed the crop by spraying it with water hoses and fire extinguishers. A year later, members of the same organization focused on McDonald's as a symbol of their opposition to threatened US trade sanctions. McDonald's fast-food shops in southern France were blockaded with tractor-loads of rotten fruit and manure, and sometimes filled with flocks of live chickens and turkeys.[1] Greenpeace campaigners in Switzerland cut down several tons of Novartis's Bt corn in one well-publicized action and deposited it at the company's hazardous waste incinerator in Basel. Since 1996 Greenpeace has blocked numerous shiploads of biotech crops from the US from entering seaports all across northern Europe.

In India, hundreds of thousands of farmers have demonstrated against transnational corporate ownership of seeds, and a large coalition of groups launched a nationwide "Monsanto Quit India" campaign in 1998, on the anniversary of the day Gandhi urged the British finally to "Quit India." In a separate incident, farmers burned several test plots of Bt cotton and declared their intention to "Cremate Monsanto." In the central Indian state of Andhra Pradesh, local farmers uprooted a Bt cotton crop, and then successfully petitioned their state government to ban further trials and uproot several remaining test plots. The Indian Supreme Court responded favourably to a petition to consider whether the planting of genetically engineered crops violates fundamental constitutional rights.

Genetic engineering has been a matter of intense public controversy in Italy, Austria, Luxembourg, the Netherlands, Australia, New Zealand, Japan, Brazil, Mexico and much of Southeast Asia as well. Many European countries have taken measures to prevent the commercial growing of genetically engineered crops.[2] Only threats of trade war and intense political pressure from the highest levels of the US government have prevented several European countries from completely banning imports of engineered crops, instead going along with a European Union policy that makes approvals of new genetically engineered crop varieties as politically difficult as possible.[3]

In the United States, opposition to genetic engineering also has deep roots and a long history. An examination of this history gives the lie to industry claims that the American public has chosen to accept genetic engineering quietly. Rather, companies like Monsanto succeeded for much of the 1980s and 1990s in keeping biotech controversies out of the mainstream press, aside from a few exceptional cases. As late as 1999 half of those surveyed in the US were not aware that products of genetic engineering were currently for sale. Repeated surveys have shown that opposition to genetically engineered foods is only muted when people are kept unaware of their existence. By the end of the 1990s, the political tide had begun to turn, and the combination of large activist gatherings, clandestine field actions, corporate campaigns and efforts to regulate engineered foods in individual US states was beginning to create the outlines of a truly nationwide movement.

The dramatic events in Seattle in the fall of 1999, when tens of thousands of people gathered to protest and obstruct the ministerial meeting of the World Trade Organization, breathed new life into all forms of social resistance in the United States. An entire generation of activists had never before experienced a rapidly growing movement, never seen people in the streets have a significant and decisive effect on the powerful institutions that so often appear politically invincible. While the police response was itself quite unprecedented – particularly the use of rubber bullets and a vast array of chemical weapons against nonviolent protesters in the United States – the courage and determination of many thousands of people ultimately enabled dissenting officials from scores of nations to obstruct the launch of a destructive new round of international trade talks. In Europe, the successful effort to stigmatize Monsanto, once the most aggressive and apparently invincible proponent of genetic engineering, sent shock waves through the world's financial markets and made agricultural biotechnology an investment to be avoided.[4] These successes are sure to have a lasting impact on the character of successive generations of actions against the dual threat of genetic engineering and corporate globalism.

The movement against genetic engineering is also noteworthy for its efforts to highlight the alternatives to this costly and unsustainable technology. People opposed to genetic engineering support food cooperatives, organic farmers and experiments to improve people's access to more locally grown food. The GMO debates have encouraged changes in personal diets towards less processed food and more local, organic alternatives. There has been a parallel rise in interest in alternative healing regimes, preventive medicine, and a new synthesis of Eastern and Western approaches to healing.[5] People in the industrialized world are creating new ties of solidarity with people in the South, especially indigenous communities, and many organizations have emerged to create fair-trade relationships between consumer cooperatives in the north and traditional agrarian communities worldwide.

The contributors to this section offer many inspiring tales of determined resistance to genetic engineering in all its forms. I begin by recounting the story of the opposition to genetic engineering in the United States, from the earliest origins of this technology in the 1970s to today's movement against genetically altered foods. Jim Thomas follows with his own story of the "genetic resistance" in Britain, which has probably seen the most sustained, visible movement against GMOs of any in the world. It is a movement sparked by direct action, lifted by humour and celebration, and sustained by a virtually unanimous consumer rejection of genetically engineered products. Next, Vandana Shiva tells the parallel story of India's growing farmers' movement against globalized agriculture and the resistance to genetic engineering. The power of the seed as both a means of sustenance and a central cultural symbol has helped nurture a movement that reaches people at all levels of society.

Many people around the world have asked why Europeans seem so much more unified and vocal in their opposition to genetic engineering than Americans. Thomas Schweiger sets out to answer this question, based on his own experiences as an activist in Austria and a lobbyist against GMOs in the European Parliament, recounting the origins of the current anti-GM movement in continental Europe along the way. Steve Emmott, a long-time policy advisor to the Green Members of the European Parliament, describes the numerous ups and downs of the ten year campaign against the European Patent Directive, the EU initiative designed to impose the continent-wide patenting of life. Then Lucy Sharratt of the Canadian Sierra Club describes how a mix of grassroots opposition, persistent lobbying of public officials, and the revelations of sceptical government scientists led to a nationwide ban on the use of rBGH in Canada. A diverse group of farmers, consumer activists and conscientious public officials kept this issue in the headlines for several years, and withstood the sometimes overwhelming pressure brought to bear by corporate interests based in the United States.

The central role of alternative, community-based institutions in opposing genetically engineered food is described by Robin Seydel of the La Montañita Food Cooperative in New Mexico. In a time when many once-alternative institutions in the US have succumbed to the pressures of the marketplace and turned away from public controversy, Seydel shows how one cooperative, committed to public education and supporting local farmers, has greatly enhanced its reputation by taking centre stage in the struggle against genetic engineering. Finally, Chaia Heller of the Institute for Social Ecology situates the GMO debate in the context of larger economic and social trends and suggests a radical new way forward. Biotechnology, she explains, typifies a qualitatively new stage in the development of capitalism, a service-based economy that is vastly different from the dazzling, prosperous "information society" promised by the global mass media. Heller proposes a redoubled opposition to genetic engineering and corporate globalism that embodies a radical new view of democracy, a redefinition of citizenship and a resurgence of communities of people acting to reclaim power over our lives and our society.

As we have seen throughout this book, genetic engineering, cloning and other biotechnologies have radical implications for our health, the environment and the future of life on earth. Today's engineered foods and medicines are setting the stage for the creation of a literally "redesigned nature," increasingly subject to corporate control and ownership. These technologies have inspired creative, radical action, as well as radical thinking about the nature of social and economic institutions, the relationship between human societies and the natural world, and the possibilities for a genuinely ecological alternative. In this concluding section, we will see how today's activists are working to expose the consequences of biotechnology, while simultaneously seeking to illuminate a path towards a more humane, truly ecological society.

NOTES

1. This campaign was initiated by Roquefort cheese producers in the south of France, after the US imposed new tariffs in retaliation against Europe's refusal to accept imports of hormone-injected beef.

2. Thomas Schweiger, *The Current Legal Situation within the EU Concerning the Marketing and Commercial Growing of GMOs*, Brussels: Greenpeace International, April 1999.

3. Bill Lambrecht, "World Recoils at Monsanto's Brave New Crops," *St. Louis Post-Dispatch*, December 27, 1998. The Clinton administration's most active biotech advocates included vice-president Gore and national security advisor Sandy Berger, according to Lambrecht.

4. See the September/October 1998 special issue of *The Ecologist*, entitled "The Monsanto Files." This issue survived attempted censorship and ultimately sold over 400,000 copies worldwide, translated into several languages.

5. A series of insightful articles on the politics and practice of alternative medicine is available at http://www.tibetanmedicine.com.

24

Resisting the Engineering of Life

BRIAN TOKAR

For more than a quarter of a century – since the first successful attempts at splicing and recombining DNA in the laboratory – people knowledgeable about genetics, ecology, agricultural science and numerous related subjects have voiced concerns about the social and environmental consequences of genetic engineering. But for much of the American public – and people around the world – biotechnology seemed to be just another new idea, only recently emerged from the annals of science fiction. Compared to such pressing environmental concerns as the disappearance of living species and habitat, the destruction of forests and rivers, and the chemical poisoning of our air and water, biotechnology seemed to be a relatively distant concern, one that most people could safely put on the back burner.

The world looks very different today. The rapid commercialization of genetically engineered crops, the stunning pace of developments in animal cloning, human "gene therapy," and the effects of bioprospecting and gene patenting on traditional cultures are awakening people from their complacency. Public actions against the biotechnology industry are spreading throughout the world. They have taken on as many different forms as the diverse peoples and societies that are threatened by the biotechnology industry's effects on the earth's biological and cultural diversity. These actions have begun to inspire a new wave of activism against biotechnology in the United States.

For several years, officials of biotechnology companies like Monsanto have told people in Europe and Asia that farmers and consumers in the United States willingly accept genetic engineered products. While every survey of public opinion in recent years has revealed overwhelming scepticism towards genetic engineering, especially in agriculture,[1] there have only been a few of the high-profile public actions and campaigns we have seen elsewhere. Activists from other countries often ask, "Why is there not yet such a widespread and highly visible movement against genetic engineering in the United States?"

While public opposition to genetic engineering in the US is just beginning to attain the high profile of efforts against nuclear power, forest destruction, and toxic chemical pollution, developments in biotechnology have been opposed and resisted in the country ever since the technologies of genetic manipulation were first developed by scientists. In fact, the first words of caution about the possible dangers of genetic engineering came from the molecular biologists themselves. It all began in 1973 when two researchers at Stanford University succeeded in transferring a hybrid plasmid – a ring of DNA including a foreign, spliced-in bacterial gene – into a particular strain of E. coli bacteria. Almost immediately, proposals emerged for a plethora of previously impossible experiments: cloning cancer genes, merging and cloning viruses, exchanging genes among different species of animals and plants, and numerous others.

In the winter of 1975, over a hundred internationally respected molecular biologists met near Monterey, on the central California coast, to discuss the future of recombinant DNA technology. Wishing to head off the possibility that their own experiments or those of their colleagues might present a public hazard, and fearful that Congress might otherwise soon act to regulate their experiments, the scientists endorsed a call for national research guidelines to contain potentially hazardous experiments. The more hazardous an experiment might be, the more special protection would be required for laboratory facilities. The scale of planned experiments was to be limited, and intentional releases of genetically modified organisms into the open environment would be prohibited.[2]

As soon as these guidelines were enacted by the National Institutes of Health (NIH), the main federal agency that funds biological research in the United States, research universities across the country announced plans to build specialized containment laboratories for potentially hazardous gene-splicing experiments. In university cities from Cambridge, Massachusetts, to San Diego, California, people alarmed about the dangers of these facilities actively opposed their construction. Scores of hearings, public forums, letter-writing campaigns and demonstrations were organized for people to air their opposition. The effort in Cambridge brought together figures such as Nobel prizewinning biologist George Wald and the city's populist mayor Al Velucci, along with other critical scientists and community activists. But most biologists, even those who had just recently raised concerns about the hazards of gene-splicing, were now called upon to defend facilities that they needed in order to continue their research. While scientists such as Wald, and members of the activist group Science for the People, raised scientific and ethical questions in opposition to the labs, others condemned lab opponents as misinformed Luddites spreading an irrational fear of science.

In many communities, the issue became a lightning rod for residents' long-simmering anger over the local political and economic power of large, wealthy universities located in economically troubled urban neighbourhoods. Several cities, including Cambridge, established citizen review commissions to oversee gene-splicing experiments, and local regulations were passed in nearly a dozen cities overall.[3] Meanwhile, the NIH guidelines helped forestall any congressional action against genetic engineering. Although several minor scandals involving the possible release of dangerous materials were reported in the scientific press, the public debate was muted and the guidelines were progressively weakened over time.[4] Gene-splicing rapidly became the technology of choice in an ever-widening range of biological disciplines, and corporations large and small were beginning to invest heavily in the new technology. "The promise of economic payoff," reports historian of science Pnina Abir-Am, "led to a quick dismantling of the remaining regulations."[5]

By 1983 the NIH had withdrawn its list of prohibited experiments, done away with containment requirements for all but the most severely pathogenic organisms, and moved to replace the ban on intentional releases of engineered organisms with a comprehensive review process. A Stanford University scientist had been approved to field-test a lysine-enhanced corn variety in 1980, and a Cornell experiment in 1982 involved tomato and tobacco plants containing bacterial genes for antibiotic resistance. But the first widespread public opposition to the deliberate release of genetically engineered organisms came in California, after scientists at the state university gained approval to field-test a variety of soil bacteria that were genetically altered to inhibit frost formation on the leaves and flowers of plants.

Scientists had learned that the formation of ice crystals on plants depends on the ability of particular soil bacteria to act as catalysts for ice formation. By deleting the gene for a key "ice nucleation" protein, and then releasing populations of so-called "ice-minus" bacteria onto a field, the Berkeley-based researchers suggested that the naturally occurring bacteria would be unable to perform their icing role, and crops could be protected from freezing. Jeremy Rifkin's Foundation for Economic Trends filed a suit against the NIH, arguing that the ecological consequences of this unprecedented experiment had not been considered, and that an Environmental Impact Statement was needed. Echoing the concerns of several well-known ecologists and atmospheric scientists, the suit highlighted the possibility that ice-minus bacteria could alter the wintering cycles of native plants, or perhaps interfere with the formation of ice crystals in the upper atmosphere that are necessary for the development of clouds.[6] The particular bacteria being studied were also known to be pathogenic to some plants, a quality which might be enhanced or otherwise affected by genetic manipulation. In May of 1984 a federal judge halted the planned experiment and ordered an Environmental Impact Statement for any future releases.[7]

While the University of California worked to satisfy federal requirements, a private company closely linked to the University initiated its own petition to test ice-minus bacteria. As a private company, not reliant upon NIH funding, Advanced Genetic Sciences (AGS) was able to bypass NIH oversight and apply directly to the Environmental Protection Agency (EPA) for a release of frost-inhibiting bacteria in agriculturally rich Monterey County. A local Christmas tree farmer named Glenn Church petitioned county officials to intervene, citing possible effects on the area's farm economy.[8] County residents opposed to the test wrote letters and filled hearing rooms; they gathered support statewide, and internationally as well. A telegram from the Green delegates in the West German parliament (Bundestag) helped convince Monterey County officials first to postpone the test, and then to amend the county's land use plan so as effectively to prohibit releases of genetically engineered organisms.[9]

In 1986 AGS made headlines in the San Francisco Bay Area when an employee leaked the news of an illegal release of ice-minus bacteria into trees planted on the roof of the company's Oakland headquarters; the protests that followed gained nationwide media attention.[10] Further information leaks from AGS laboratories revealed that employees had been suffering from persistent allergic reactions and sinus troubles, possibly associated with these bacteria. Still, despite repeated legal interventions and intense local controversy, AGS gained approval to release its frost-inhibiting bacteria in the East Bay town of Brentwood in April of 1987.

It was a chilly April night, the beginning of northern California's long dry season. Through well-groomed apple and pear orchards and newly planted corn fields, a small band of eco-saboteurs crept towards their target. Ahead was a field of strawberries, just beginning to show their spring blossoms, but this was no ordinary strawberry field. It was surrounded by a high chain-link fence with barbed wire, a trailer full of hired guards, and a symmetrical array of towers with Petri dishes, sensors and other modern biological test equipment. The next morning at 7 a.m., Advanced Genetic Sciences was scheduled to conduct the first authorized open-air test of genetically altered bacteria, and demonstrate whether these bacteria would indeed inhibit the formation of frost crystals on the leaves and flowers of the plants. Before scores of reporters, and television cameras from as far away as Tokyo, AGS was ready to inaugurate officially the Age of Biotechnology.

It must have been quite a shock to the assembled corporate and government officials when dawn finally arose on the morning of April 24. Over two thousand strawberry plants, more than 80 per cent of the total, had been pulled out of the ground during the night. Almost none had any blossoms left. The perpetrators had come and gone while the security guards were asleep, and had disappeared into the night.[11]

Public relations, not science, was the order of the day, however. As reporters began to arrive on the scene, AGS hurriedly stuffed the uprooted plants into the ground and scientists dressed in full-body hazardous materials suits, as ordered by EPA, began spraying their bacteria. The next day's world headlines reflected the company's thoroughly dubious claim that their experiment was a success. Industry spokespeople trumpeted their promises of bigger and better miracles ahead.

Just a year later, however, the situation looked very different. Two more tests of the anti-frost bacteria were carried out in California and both were effectively sabotaged by test opponents. Local opposition continued to grow, among both suburban dwellers and the area's Mexican farmworker communities. The local press revealed, contrary to company claims, that escaped bacteria from the first test were detected well beyond the test plot.[12] When plants were put in laboratory freezers – natural frost is rare in this part of California at any time of year – they still froze within a degree or two of 32°F. AGS research director John Bedbrook told the *New York Times*, "If we have to go through a huge amount of effort to educate [*sic*] every community, the cost is going to be beyond us."[13] As the 1989 growing season began, AGS withdrew their application to continue field tests of ice-minus bacteria in California.

Printed statements from those involved in the first-ever direct actions against genetic engineered organisms reflected their committed local support, and a wide-ranging political outlook. Prior to the second Brentwood test, opponents prepared leaflets in both English and Spanish and held a public march and caravan from the local school to the test site. When the University of California began testing ice-minus bacteria in Tulelake, near the Oregon border, local residents reportedly helped eco-saboteurs find the carefully hidden test site. A group claiming responsibility for pulling up thousands of ice-minus-sprayed potato plants in Tulelake issued a communiqué stating in part:

> The genetic engineering industry is only the most recent example of this civilization's drive to subjugate nature to its own ends. This world-view has resulted in unprecedented attacks against the ecosystems we depend on for life. We need to evolve beyond the worldview that pits humanity against nature, and which is a product of the conjunction of patriarchy and capitalism.

Test opponents of all political views knew that the release of anti-frost bacteria was only the beginning. These organisms clearly were not being released into the environment because farmers wanted them, and they would probably never be a profitable commercial product, given the exorbitant costs of production. The bacteria were being released, many concluded, because they appeared to have fewer obvious pitfalls than many of the other genetically engineered life forms that were being developed in laboratories around the world. Once these bacteria were released, the industry hoped it would have free rein for the future.

The Foundation on Economic Trends, which had filed suit to halt the ice-minus field tests in California, continued its series of pioneering lawsuits against the biotechnology industry and its cohorts in government. They sued the Department of Agriculture for encouraging the development of genetically engineered farm animals, and obtained a permanent injunction against the Defense Department's plan to test biological aerosols at the Pentagon's Dugway Proving Ground in Utah. The Foundation's lawyers challenged the first genetically engineered rabies vaccine (which was to be tested on wildlife), the development of patented mice with the AIDS virus spliced into their chromosomes, and the first attempts at human gene therapy.[14]

Still, despite these interventions, and those of several national environmental groups, the Reagan and Bush administrations effectively dismantled the long-range effort by a number of federal agencies to develop a comprehensive regulatory framework for products of genetic engineering. This left a patchwork of regulations, each developed for a different purpose, which are applied to genetically engineered organisms on a case-by-case basis. One EPA office regulates pesticides, another toxic substances; the Department of Agriculture (USDA) reviews field tests, while the FDA is mandated to evaluate food safety.

This multiplicity of hurdles may have held up approval of a few products; however, many others have simply fallen through the regulatory cracks. For example, the EPA regulates pesticides, including biopesticides such as Bt, but the FDA is barred from offering information about pesticides on food labels. Thus food safety officials claim they are unable to require labelling of Bt crops, even though every plant cell produces active bacterial toxin.[15] In 1989, when twenty-seven people died of a rare blood disease associated with a particular genetically engineered source of the amino acid tryptophan, the FDA suddenly pulled all tryptophan supplements from the market.[16] There was never a thorough investigation of the source of the suspect batches – in this case, the Japanese company Showa Denko. If a substance is considered safe when extracted from natural sources, the agency simply assumes that genetically engineered varieties are safe as well; if a genetically engineered product causes problems, the problem must be with the product itself, not the manner in which it was produced.

During the presidency of the elder George Bush, the controversial Council on Competitiveness, operated out of vice-president Dan Quayle's office, reduced the biotechnology industry's regulatory accountability even further. Under the leaky umbrellas of "performance-based standards" and "substantial equivalence," products of biotechnology were deemed in 1991 to require no special scrutiny beyond what is required for non-genetically engineered varieties of the same product. The inherent differences between genetically engineered and conventional varieties of various food crops, drugs, and so on, were declared irrelevant, except for special cases, such as substances already known to be allergenic. The USDA's requirement for field test permits for engineered crops was replaced by a simple, after-the-fact notification. Genetically engineered drugs, food products and microbes would now be approved without any consideration of the special risks and uncertainties of gene manipulation.[17] This is the period when biotech industry representatives would have us believe that the initial "public debate" on genetic engineering happened in the United States; in reality only the very closest observers knew anything of this consistent pattern of regulatory neglect and abandonment.

The US government's abdication of meaningful regulatory responsibility for products of biotechnology soon opened the floodgates for the entry of genetically engineered ingredients into our food supply. The first such product was genetically engineered rennet for the production of cheese. Jeremy Rifkin and other biotechnology opponents perceived that it would be difficult to raise a public outcry against this product, which is not present in food in measurable quantities. Rennet, essentially a mixture of digestive enzymes, is traditionally extracted from one of the lower stomachs of young calves. Rifkin and others were concerned that producing enzymes from vats of genetically engineered bacteria might be seen by many environmentalists and animal rights advocates as preferable to extracting rennet from calves.

Bovine Growth Hormone was a different story. Farmers and food safety advocates immediately realized that this genetically engineered hormone (produced by bacteria genetically "enhanced" with cow DNA) represented a potential threat to both milk producers and consumers. While the human health consequences of synthetic (recombinant) BGH were just beginning to be studied, farm activists realized that the promised increase in milk production could have dire consequences for an already depressed family farm sector.[18] One study predicted that a third of all dairy farmers in the United States could be forced out of business in five years if the hormone were to be widely used.

This concern was dramatized by the efforts of a lone Wisconsin dairy farmer named John Kinsman. Kinsman discovered that the student union at the University of Wisconsin in Madison was serving ice cream made with milk from cows that were being injected with experimental rBGH. He travelled to Madison in mid-winter and began picketing the union with a sign explaining what the students were unknowingly consuming. Kinsman's one-person campaign helped spark a nationwide alliance of farmers and citizens that would delay by several years the government's approval for commercial use of the hormone. Jeremy Rifkin's Foundation on Economic Trends announced plans for a nationwide alliance of farmers, consumers and animal rights advocates to prevent the acceptance of rBGH by the dairy industry.

In 1989 Richard Burroughs, an FDA veterinarian, was fired for raising questions about the manipulation of experimental data on the effects of BGH injections on milk cows (see Chapter 22 of this volume). Then, a disgruntled staff scientist at the University of Vermont (UVM) went public with data that revealed serious problems with that institution's Monsanto-funded BGH research. Just as the FDA's approval of rBGH appeared imminent, the Vermont revelations helped force a public reopening of the case.[19]

As the story goes, veterinary pathologist Marla Lyng was refused permission to perform autopsies on certain aborted calf fetuses at the UVM experimental farm in Burlington. A few abnormal fetuses may have been buried surreptitiously. Lyng lost her job at the end of 1989, and then released copies of the farm's cow health records to state legislators while her firing was under appeal. One state representative with a dairy-farming background was given the raw data to re-analyse and was appalled by the results. In addition to nutritional problems and frequent udder infections (mastitis), BGH-injected cows showed a high incidence of reproductive problems, from retained placentas and uterine infections to high rates of spontaneous abortions and severely deformed offspring.[20] Even daughters of rBGH-treated cows had aborted fetuses with bizarre genetic defects: fluid-filled holes in the head, improper bone development, and even double pelvises supporting extra legs. The UVM findings also confirmed that medical problems in cows injected with rBGH required higher than normal doses of antibiotics.[21] These revelations sparked an investigation by the US Congress's General Accounting Office, which helped delay the approval of rBGH by highlighting "potential biases" in Monsanto's research methods and decrying the company's systematic withholding of data.[22]

Monsanto's recombinant Bovine Growth Hormone was approved by the FDA for commercial sale beginning in 1994, however; and widespread reports of serious health problems in injected cows soon followed.[23] Instead of addressing the causes of farmers' complaints about rBGH, Monsanto went on the offensive, threatening to sue small dairy companies that advertised their products as free of the artificial hormone.[24] Opponents of rBGH stepped up their efforts. The Minnesota-based Pure Food Campaign, originally an offshoot of the Foundation for Economic Trends, coordinated demonstrations in major cities across the country, including numerous high-profile public milk-dumpings by farmers and consumers. Media coverage was impressive in the first months of rBGH use, and the reporting was often uncharacteristically sympathetic to the opponents. The Vermont state legislature passed the first mandatory labelling bill for rBGH-tainted dairy products in March of 1994, and Maine, Wisconsin and Minnesota attempted similar legislation. Over a hundred school districts from Vermont to Los Angeles passed resolutions against rBGH products in their cafeterias, and a lawsuit was filed against the FDA to expose the rampant conflicts of interest among the staff responsible for rBGH approval.

The landmark Vermont labelling law proved short-lived, however, and the story offers important lessons for continuing efforts to label genetically engineered foods in the US. Over a year after the law's passage, there were still no labels. Activists became increasingly mired in legislative rule-making and hesitated to confront a Democratic governor who was conspicuously trying to play both sides. There were endless legislative attempts to weaken the law and, finally, a lawsuit by leading trade associations representing the dairy industry, grocers, food processors and several others.

Instead of going on the offensive against the industry's blatant attempts to sabotage rBGH labelling, staff members for Vermont farm and consumer groups sought compromise. Lacking confidence that the industry lawsuit could be beaten in the courts, and reluctant to raise the stakes politically, they quietly supported plans to modify the labelling rules and make them less costly to manufacturers. The onus for labelling would fall upon individual grocers rather than the manufacturers, eventually allowing industry lobbyists to portray rBGH labelling as an undue burden on small, rural shopkeepers. The state's Republican Attorney General at first hesitated to defend the law in court, then agreed to do so only on the grounds of consumer preference, rather than the necessities of public health. This opened the door to a federal appeals court ultimately striking down Vermont's labelling law on the grounds that it violated the companies' constitutional right to refuse to speak.[25] By the end of the saga, many supporters had tired of the long, defensive battle to defend a law that, in the end, only indirectly offered consumers the information they wanted.

Local and national efforts continued to focus on major dairy processors that supported the use of rBGH by farmers. A few companies agreed to go rBGH-free, while others dug in their heels in defence of this first widely available product of genetic engineering.[26] Several national organizations focused their efforts on a federal rBGH-labelling bill, sponsored by Vermont Representative Bernie Sanders, although the industry's powerful friends in Washington were clearly going to succeed in preventing such a bill from ever reaching the House floor.[27]

While political initiatives against rBGH foundered, consumer opposition remained strong. Whole milk dealers throughout the Northeast continued to risk lawsuits from Monsanto and label their products as free of synthetic hormones. Ben and Jerry's Ice Cream successfully challenged an Illinois state law banning "rBGH-free" labels. Organic milk, which was extremely hard to find even in dairy-producing regions in the early 1990s, became much more widely available. Farmers' continuing troubles with Monsanto's hormone, combined with the vocal concerns of dairy consumers, have effectively derailed the company's plans to make rBGH injections a routine practice for dairy farmers across the US.

A similar mix of consumer pressure and simple product failure ended the short career of the first genetically engineered vegetable to be approved for sale in the United States, the so-called Flavr-Savr tomato. Developed by the Davis, California, biotechnology company Calgene, these tomatoes were genetically altered to ripen more slowly and thus appear fresher on store shelves; the company promised a shelf life of several weeks. First, a threatened boycott drove Campbell Soup to announce that it would not use the product, despite the company's considerable investment in its development. Efforts to test-market the tomatoes in California and the Chicago area foundered, as news reports raised questions about the product's much-touted flavour. Activists raised concerns about the safety of the engineered tomatoes, rival companies challenged Calgene's patent, and Calgene found that their tomatoes were easily bruised by mechanical harvesting and packing.[28] By 1995 production was suspended both in Florida and in Mexico. In 1996 Monsanto bought Calgene, along with Florida's leading tomato packing company, Gargiulo L.P., but even this new corporate alliance was unable to salvage the first genetically engineered tomatoes.

Meanwhile, genetically engineered varieties of other basic food crops were being developed and tested at an astonishing pace. In 1990, two scientists at the National Wildlife Federation, Drs Margaret Mellon and Jane Rissler began closely monitoring applications to the USDA and EPA for field tests of genetically engineered plants and microorganisms. By the spring of 1991 they had identified 149 such tests in thirty states (plus Puerto Rico); a year later the number of tests had more than tripled. By the time Mellon and Rissler shifted their project to the Union of Concerned Scientists during the winter of 1993–4, there had been over a thousand field tests, and by the end of 1994 there were well over two thousand in all but seven states.[29] Mellon and Rissler intervened against government moves to weaken regulations for genetically engineered organisms, petitioned federal agencies to consider the health risks of engineered foods, and carried out the first comprehensive analysis of the potential ecological risks of genetically engineered crops.[30]

Other organizations, from the Consumers Union and the Environmental Defense Fund to a wide variety of grassroots state and regional groups, also questioned the adequacy of health and safety research, challenged inadequate enforcement and worked to educate consumers. But the overwhelming influence of the biotechnology industry at the highest levels of the Clinton/Gore administration severely limited these groups' ability to make change within official channels. One of the early signals of Clinton's loyalty to the biotechnology industry was the release of an "interpretive statement" on the UN Convention on Biological Diversity. The statement sought to eliminate the requirement for companies to share research on biological resources with the people from whose lands those resources are derived.[31] The ensuing controversy helped set the stage for the administration's consistently obstructionist role in the negotiations that ultimately led to the Cartagena Biosafety Protocol.[32]

During the same period, biotechnology industry funds were beginning to alter research priorities drastically at universities across the United States. This became a matter of intense controversy on many campuses and in their surrounding communities. Students protested the construction of new biotech research facilities at Ohio University, the University of Pennsylvania, and several University of California campuses, among others. In Burlington, Vermont, the city sponsored a series of public forums in anticipation of the University of Vermont's new biotechnology centre, and unruly protests forced the university to cancel a gala outdoor opening ceremony for the facility. In 1998, public opposition and NIH scrutiny helped scale back a proposed $300 million research agreement between Sandoz (now Novartis) and the Scripps Research Institute, near San Diego; a similar deal between Novartis and the University of California at Berkeley, which offered research funds and access to corporate databases in exchange for first rights to exploitable discoveries, sparked city-wide controversy as well.[33]

In New York City, officials of Columbia University and a number of biotech companies announced plans to build a major biotechnology research park in Harlem, on the site of the famous Audubon Ballroom. The Ballroom had played an important part in the lives of countless African-American activists and artists, and was the site of Malcolm X's assassination; the project became the centre of a struggle pitting Harlem-based activists, Green Party members, and many others against much of the city's business and political establishment.[34] In Philadelphia, activists opposed the demolition of a historic university building for a biotechnology facility funded largely by the Defense Department. Activists in San Francisco, New York and other major cities challenged public officials from across the political spectrum who touted biotechnology as a futuristic new industry that would revive their cities' then-faltering economies.

Several prominent scientists also voiced criticisms of the biotechnology industry's claims, and decried its insidious influence on scientific practice and research agendas. The Boston-based Council for Responsible Genetics, which emerged from the original 1970s' debates over the containment of engineered organisms, articulated a wide-ranging critique of the emerging genetic determinism in biological research and medical practice.[35] The group spoke out in defence of those who faced discrimination due to the findings of genetic screening tests, critiqued pseudo-scientific claims of a genetic basis for intelligence and "criminality," and helped raise public understanding of the continuing risks of genetic experimentation. Critical scientists and other US activists and NGOs also played an important role in international debates around biosafety, intellectual property and bioprospecting, exposing the behind-the-scenes machinations of the biotechnology industry to increasing public scrutiny.

The large-scale commercialization of genetically engineered foods began in the United States in the fall of 1996, as approved varieties of engineered soybeans, corn, potatoes and squash were harvested by growers and shipped without notice to supermarkets and food-processing plants nationwide. The Pure Food Campaign and other groups organized demonstrations across the country, eventually calling for a series of Global Days of Action against genetically engineered foods. The first such event, in April of 1997, featured demonstrations and public gatherings in nineteen US cities, as well as seventeen European countries, India, the Philippines, Malaysia, Japan, Canada, Australia, New Zealand, Brazil and Ethiopia. Six months later, there were protests and educational events during a two-week period in some fifty cities worldwide.[36]

Greenpeace undertook two dramatic early actions against genetically engineered crops in the US. In October of 1996, Greenpeace activists used a milk-based dye to paint a 100-foot X on a field of Roundup-resistant soybeans in Iowa. The paint was visible from the air, and the action attracted significant media attention. A month later, Greenpeace's inflatable boats sought to block a ship containing genetically engineered soybeans from leaving Cargill's grain facility on the Mississippi River, outside of New Orleans.[37] But this soon appeared to be the peak of Greenpeace's involvement in the issue on this side of the Atlantic. In the summer of 1997, the organization closed all of its campaign offices in the US, and their involvement with biotechnology was limited for many years to issue updates, lobbying in Washington, and support for a few small, local demonstrations.

An important shift in US activism against genetic engineering occurred in 1998 when members of the Gateway Green Alliance, based in Monsanto's home town of St Louis, organized a major international activist conference. Dubbed the "First Grassroots Gathering on Biodevastation: Genetic Engineering," the July 1998 gathering was the first in the US to emphasize grassroots activism and public education over policy analysis and behind-the-scenes interventions. It was co-sponsored by over thirty groups, and attracted participants from all across the country, as well as from Canada, the UK, Ireland, Mexico, India and Japan. A coalition of Japanese farmer and consumer groups brought a delegation of more than twenty representatives.

The gathering's closing plenary approved a "Biodevastation Declaration," which had been drafted by a participant from the Council of Canadians, and offered a comprehensive outlook on the problems of biotechnology. The two hundred participants also agreed by acclamation to several specific demands: a ban on rBGH and the genetic engineering of plants and animals; an end to the patenting of DNA sequences, the Human Genome Diversity Project, the National Violence Initiative Project, and the development of Terminator seed technology; a halt to the testing of experimental drugs on prisoners, Native Americans and others; and the encouragement of organic farming, local food systems, home-scale gardening and ecosystem restoration, as alternatives to corporate monocrop agriculture. A colourful demonstration outside Monsanto's world headquarters in a suburb just outside St Louis featured puppets, skits and speeches by international activists.[38]

The year 1998 also saw an impressive public outcry by consumers of organic foods, who demanded that genetically engineered products be excluded from the US government's labelling rules for organic foods. When the White House budget office (OMB) insisted that the USDA consider allowing products of genetic engineering to be labeled organic, more than 275,000 people wrote cards, letters and testimony objecting to this and other perceived violations of widely accepted organic practices. Agriculture Secretary Dan Glickman, who two years earlier had traveled through Europe threatening a trade war if countries tried to ban genetically engineered imports from the United States, soon promised that engineered foods would not be labeled organic by the USDA.[39]

Following the St Louis gathering, biotech opponents across the United States sought to strengthen local and regional networks, and find ways to raise the profile of their issues in communities nationwide. They pledged to step up public actions against biotechnology companies, seed suppliers and food processors, as well as universities and government agencies with close ties to the biotech industry. Over the next year or so, regional grassroots groups formed under the banner of Resistance Against Genetic Engineering (RAGE) in the Northeast, Pacific Northwest, San Francisco Bay Area, mid-South and upper Midwest,[40] while more traditional NGOs coalesced as the Genetic

Engineering Action Network (GEAN), and advocates for direct action against genetic engineering created the Bioengineering Action Network (BAN).

A follow-up "Biodevastation" conference was hosted by Vandana Shiva's organization in New Delhi, India, in March of 1999, and a third one, in Seattle two months later, featured a lively demonstration at the annual convention of the Biotechnology Industry Organization (BIO). "Biodevastation 2000," which preceded that year's BIO convention in Boston, culminated in a rally and parade by some four thousand people. This was by far the largest show of opposition to genetic engineering in the United States, and one of the largest protests specifically focused on biotechnology anywhere in the world to date.[41]

Farmers again came to the forefront of opposition to genetic engineering in the USA after large grain processors such as Archer Daniels Midland began to lower prices for genetically engineered crops, towards the end of the 1999 growing season. Gary Goldberg, CEO of the American Corn Growers Association, urged members to "consider alternatives," and called GMOs an "albatross around the neck of farmers."[42] In November of 1999, a coalition of more than thirty farm groups, including the corn growers and the National Family Farm Coalition, issued a public declaration, warning that farmers who plant genetically altered seed are risking their own livelihood and the future of the entire farm economy.[43] Pressure from farmers and gardeners nationwide compelled over fifty US and Canadian seed companies to sign a pledge promising "not knowingly" to offer genetically engineered varieties.[44] Food companies from Gerber Baby Foods to the Whole Foods chain of natural foods supermarkets signed similar pledges amidst growing consumer pressure.

Other groups focused on various legal avenues to oppose genetic engineering. Groups such as the Center for Food Safety, Alliance for Bio-Integrity, and the Foundation on Economic Trends sued several government agencies, including the FDA (for allowing unlabelled genetically engineered foods onto the market), the EPA (for ignoring the dangers of pesticidal crops such as Bt corn) and the Department of Justice (for allowing the biotechnology industry to acquire monopoly power in the commercial seed industry). In November of 1999, a bill calling for the labelling of genetically engineered foods was introduced in the US House of Representatives, and a similar bill was introduced in the Senate in February 2000.

The year 1999 also saw the re-emergence of direct actions against genetic engineering in the United States. Approximately twenty separate clandestine uprootings of genetically engineered crops were reported that summer and fall throughout the US and Canada.[45] Most were in California, where activists targeted genetic engineering research around the University of California campuses in Berkeley and Davis. When activists cut down half an acre of herbicide-tolerant corn at the University of Maine, it attracted statewide and national media attention; three actions in California, Minnesota and Vermont focused on large commercial crops rather than university research stations.[46]

While specific tactics and organizing methods vary widely, US activists increasingly agree on the need for a more systemic critique of biotechnology in all its forms. While some biotech opponents insist that pressing for food labelling legislation is the most realistic approach, given the limitations of mainstream political discourse in the United States, others see this as a half-way measure at best. Campaigns focusing chiefly on labelling, without a deeper analysis of the institutional roots of genetic engineering, offer the biotech industry many possible outs. Labeling does little to address the environmental hazards of engineered crops, the needs of farmers, or the problems of less affluent families, whose food choices are often quite limited. Some elements of the biotechnology industry clearly seek a compromise on labelling in an attempt to assuage consumer fears, while releasing a minimum of real information. Grassroots activists have voiced dismay that US NGOs might agree to an inadequate compromise on labelling, just as people are becoming aware of the wider implications of the new genetic technologies.

The opposition in Europe, India and elsewhere has not only exposed the profound underlying hazards of genetic engineering and other biotechnologies, but also tapped into a deeply ingrained scepticism toward views of people and the rest of nature as objects to be manipulated and controlled. In Europe, the spectre of Nazi eugenics hangs over discussions of genetic engineering and cloning; in India, the seed is a powerful cultural symbol and its manipulation and appropriation by capital is an abomination.

In the USA there are considerable cultural, as well as political, obstacles to developing a wider movement against genetic engineering. We are forced to confront pervasive cultural myths of untrammelled consumerism and a popular science-fiction outlook toward the future. But we also share with our neighbours an ethical commitment to democracy, freedom and love for the land, however manipulated these ideas have been in the past. Biotechnology seeks to rally the public behind its promises of future miracles and dazzle them with its technical wonders. Our movement, if it is to succeed, must peel away the science-fiction blinders, deepen our understanding of the integrity of the natural world, challenge the structures of power in society, and present a clear vision of a more ecologically and humanly sustainable way of life.

NOTES

1. Barnaby Feder, "Biotech Firm to Advocate Labels on Genetically Altered Products," *New York Times*, February 24, 1997; James Walsh, "Brave New Farm," *Time*, January 11, 1999, p. 87.

2. Sheldon Krimsky, *Biotechnics and Society: The Rise of Industrial Genetics*, New York: Praeger, 1991, pp. 100–101; see also James D. Watson and John Tooze, *The DNA Story: A Documentary History of Gene Cloning*, San Francisco: W.H. Freeman, 1981.

3. *Biotechnics and Society*, p. 102.

4. For example, Nicholas Wade, "DNA: Chapter of Accidents at San Diego," *Science* 209, September 5, 1980, pp. 1101–2; Charles Marwick, "Genetic Engineers in the Sinbin..." *New Scientist*, September 11, 1980, p. 764; Richard A. Knox, "US Halts One

DNA Experiment at Harvard," *Boston Globe*, December 16, 1977, p. 1.

5. Pnina G. Abir-Am, "'New' Trends in the History of Molecular Biology," *Historical Studies in the Physical and Biological Sciences*, vol. 26, no. 1, p. 176.

6. Reginald Rhein, "'Ice-minus' Bacteria Gain EPA Approval," *Chemical Week*, November 27, 1985, pp. 98–9.

7. Krimsky, *Biotechnics and Society*, pp. 117–29; Colin Norman, "Judge Halts Gene-splicing Experiment," *Science* 224, pp. 962–3, June 1, 1984; also Rhein, "'Ice-minus' Bacteria Gain EPA Approval."

8. Jonathan C. Drake, "Manmade Mutant Microbes Invade the Central Coast, *Santa Cruz Express*, February 13, 1986.

9. Philip J. Hilts, "Microbe Test Hits New Snag," *Washington Post*, January 17, 1986.

10. Krimsky, *Biotechnics and Society*, pp. 123–4; "Halting Designer Bacteria," *Newsweek*, February 10, 1986, p. 8.

11. "Homo Fragaria" (pseud.), "The Strawberry Liberation Front," *Earth First! Journal*, June 1987, p. 1; Elliot Diringer, "Vandals Fail to Prevent Cloned-bacteria Spray," *San Francisco Chronicle*, April 25, 1987.

12. Amy Axt Hanson, "Second Phase of Frostban Test Started," *Antioch Daily Ledger*, April 29, 1987; on the predictability of this, see Peter Aleshire, "Altered Bacteria Spread Very Easily, Expert Says," *Oakland Tribune*, April 6, 1986.

13. Keith Schneider, "Biotechnology Lags Despite Success," *New York Times*, January 18, 1988.

14. Krimsky, *Biotechnics and Society*, pp. 121–3; Jeffrey L. Fox, "Gene Test Redux, Eugenics Advisory Proposed, *Bio/technology* 7, March 1989, pp. 205–6.

15. Michael Pollan, "Playing God in the Garden," *New York Times Magazine*, October 25, 1998, p. 51.

16. See the introduction to Part II of this volume.

17. Christopher Anderson, "US Science's 'Stealth Agency'," *Nature* 353, September 19, 1991, p. 198; Christine Triano, *All the Vice President's Men*, Washington, DC: OMB Watch, September 1991, pp. 28–9.

18. See, for example, Keith Schneider, "Gene-altered Farm Drug Starts Battle in Milk States," *New York Times*, April 29, 1989, p. 1.

19. This account is partially excerpted from Brian Tokar, "The False Promise of Biotechnology," *Z Magazine*, February 1992, pp. 27–32.

20. Andrew Christiansen, *Recombinant Bovine Growth Hormone: Alarming Tests, Unfounded Approval*, Montpelier, VT: Rural Vermont, July 1995, pp. 8–10.

21. Many years of research on the effects of rBGH is summarized in D.S. Kronfeld, "Health Management of Dairy Herds Treated with Bovine Somatotropin," *Journal of the American Veterinary Medical Association*, vol. 204, no. 1, January 1994, pp. 116–130; Samuel S. Epstein, "Unlabeled Milk from Cows Treated with Biosynthetic Growth Hormones: A Case of Regulatory Abdication," *International Journal of Health Services*, vol. 26, no. 1, 1996, pp. 173–85; Samuel S. Epstein, "BST: The Public Health Hazards," *The Ecologist*, vol. 19, no. 5, September/October 1989, pp. 191–5.

22. US General Accounting Office, "FDA's Review of Recombinant Bovine Growth Hormone," Document no. GAO/PEMD-92-26, Washington, DC: US GAO.

23. Mark Kastel, *Down on the Farm: The Real BGH Story*, Montpelier, VT: Rural Vermont, Fall 1995; Brian Tokar, "Biotechnology: The Debate Heats Up," *Z Magazine*, June 1995, pp. 49–55.

24. Diane Gershon, "Monsanto Sues over BST," *Nature* 368, March 31, 1994, p. 384.

25. Nick Marro, "Court Strikes Down BGH Label Law," *Barre-Montpelier Times-Argus*, August 9, 1996, p. 1.

26. See Tokar, "Biotechnology: The Debate Heats Up."

27. Michael Colby, "Activist Malpractice," *Safe Food News*, Fall 1994, p. 4.

28. Scott McMurray, "New Calgene Tomato Might Have Tasted Just as Good With-

out Genetic Alteration," *Wall Street Journal*, January 12, 1993; "Stomach Erosions in Genetic Tomato Animal Study Discounted," *Food Chemical News*, March 15, 1993; Anne Gonzales, "Production in Mexico Halted," *Packer*, January 30, 1995; Ralph T. King, Jr., "Low-Tech Woe Slows Calgene Super Tomato," *Wall Street Journal*, April 11, 1995; Ralph T. King, Jr., "Expert Calls Calgene Research On Gene-Altering Method Flawed," *Wall Street Journal*, April 24, 1995, John Unrein, "Biotech is Pulled in Florida," *Packer*, November 6, 1995.

29. The states with no field tests at the end of 1994 were Nevada, New Mexico, Wyoming, Utah, Vermont, New Hampshire and Rhode Island. Figures are from Mellon and Rissler's publication, *The Gene Exchange*, archived at http://www.ucsusa.org.

30. Jane Rissler and Margaret Mellon, *Perils Amidst the Promise: Ecological Risks of Transgenic Crops in a Global Market*, Washington DC: Union of Concerned Scientists, 1993, revised and republished as *The Ecological Risks of Engineered Crops*, Cambridge, AM: MIT Press, 1996.

31. See Brian Tokar, "Environmentalism, Clinton Style," *Z Magazine*, October 1993, pp. 34–5.

32. For an account of these proceedings, see Beth Burrows, "Resurrecting the Ugly American," *Food & Water Journal*, Spring 1999, pp. 32–5.

33. Elizabeth Wilson, "Berkeley and Novartis Strike a Deal," *Chemical and Engineering News*, December 14, 1998, p. 41.

34. Peggy Dye, "Harlem Faces the Vulture Culture, *Z Magazine*, February 1992, pp. 55–9, and "The Night the Grassroots Got Away, *Village Voice*, September 18, 1990, p. 9.

35. Jon Beckwith, "Thinking of Biology: A Historical View of Social Responsibility in Genetics," *BioScience*, vol. 43, no. 5, May 1993, pp. 327–33; Stuart A. Newman, "Idealist Biology," *Perspectives on Biology and Medicine*, vol. 31, no. 3, Spring 1988, pp. 353–68; Ruth Hubbard and Elijah Wald, *Exploding the Gene Myth*, Boston, MA: Beacon Press, 1997.

36. Ronnie Cummins, "Activists in 16 Nations Carry Out Successful Global Days of Action," *Food Bytes* 3, November 4, 1997, at http://www.purefood.org.

37. "Greenpeace Quarantines Genetically Altered Monsanto 'X-field' in Iowa," Greenpeace USA press release, October 10, 1996; "Greenpeace Stops Genetically Engineered Soybeans Destined for Europe on Mississippi River," Greenpeace press release, November 19, 1996.

38. Freida Morris, "'Monsanto: You Have Shamed Us'," *The Ecologist*, vol. 28, no. 5, September/October 1998, pp. 304–5; proceedings of the first Biodevastation gathering are available in the US Greens' quarterly journal *Synthesis/Regeneration* 18 and 19, Winter and Spring 1999.

39. "USDA Bows to Public Pressure on 'Organic' Standards," EnviroNews Service, May 8, 1998.

40. The acronym "RAGE" was first used by feminists opposed to reproductive technology and genetic engineering in the mid-1980s (FINRRAGE: Feminist International Network of Resistance to Reproductive and Genetic Engineering, founded in 1985), adapted by social ecologist Zoë Erwin for her MA thesis "The Revolution Against Genetic Engineering," (Goddard College, 1996), and renewed by the New England caucus at the first Biodevastation gathering in St Louis to form NERAGE (New England, later NorthEast, Resistance Against Genetic Engineering). The other RAGE groups as of February 2000 are (in order of their creation), BayRAGE (San Francisco), NWRAGE (Portland, Oregon), Down South RAGE (Memphis) and GrainRAGE (Minneapolis).

41. See Brian Tokar, "Resistance against Genetic Engineering," *Z Magazine*, June 2000, available at http://www.zmag.org; documents form Biodevastation 2000 are available at http://www.biodev.org.

42. American Corn Growers Association press release, "Corn Growers Call on Farmers to Consider Alternatives to Planting GMOs if Questions are Not Answered," August

25, 1999, at http://www.acga.org.

43. "Family Farmers Warn: If Your Next Crop is GMO, it May Be Your Last," Farm Aid press release, November 23, 1999, posted to electronic list, biotech_activists@iatp.org.

44. Kimberly Wilson, "Safe Seeds," *GeneWatch* (Cambridge, MA), vol. 13, no. 1, February 2000, p. 16.

45. Reports and communiqués from many of these actions are archived at http://www.tao.ca/~ban/ar.htm.

46. Jeff Tuttle, "Vandals Hit Modified Corn at UM," *Bangor Daily News*, August 20, 1999, p. B1; "Taking it to the Fields: A Tale of Corn Sabotage," *Food & Water Journal*, Fall 1999, pp. 14–24. Advocates of clandestine actions published a twenty-page illustrated *Nighttime Gardener's Guide*, available from nighttimegardeners@angelfire.com.

25

Princes, Aliens, Superheroes and Snowballs: The Playful World of the UK Genetic Resistance

JIM THOMAS

> Little boy blue come blow your horn!
> The people are pulling up transgenic corn.
> With broomsticks and sickles and gardening hoes,
> They're making the land free of GMOs.
>
> Anon

Prince Charles has not been, so far as anyone suspects, at the vanguard of direct action against genetic engineering. However, when seven fields of genetically engineered rape were uprooted one cloudy June night in 1998 and some "Lincolnshire Loppers" felled GM wheat at the royal agricultural show, it was the Prince who began to receive media questions about whether he supported such "vandalism." He had, only some days earlier, launched a swingeing attack on genetically altered crops himself – not in the contaminated fields of northern England but from the pages of the *Daily Telegraph*, Britain's most conservative newspaper. Genetic engineering, he declared, takes us into realms that belong "to God alone." Pleased by a chink of common sense, most of the British public nodded their heads in vigorous agreement. The newspapers took note.

Since then anti-GM "hysteria," as grumpier industry press officers would have it, has swept Britain. It would be wrong to hold Prince Charles and the newspapers responsible – or the loppers and croppers of the nighttime resistance for that matter. An artful dance of hard work, lucky coincidence, strategy and passion has been behind the gathering momentum of the UK "genetix" movement. The prince and the protests have provided two of many lightning rods by which simmering popular discontent has translated into an astonishing market and cultural turnabout. Voices from within the establishment, audacious actions from the grassroots, and difficult questions at the checkouts have woven an unfolding opera of citizen and consumer power advancing against the monoculture.

A year after the prince's intervention, six hundred paper-suited activists walked beneath fluttering flags and the hot July sun onto the UK's largest GM test site, near Oxford, and destroyed it. The nation's top supermarkets guaranteed to remove genetically engineered ingredients, not only from their processed food but also from the feed of animals providing dairy products, eggs and meat. The UK government and science establishment, bewildered and still retreating, called for a "peace process," and even Monsanto's work canteen was found to be GMO-free! A wise historian never writes history until all the battles are finished and the players are safely dead, but it seems our story has already been told abroad by Reuters, Associated Press and CNN. So, as the gene wars go truly global, take what follows as not a history but a selected handful of tales for kindred movements to learn from.

For me it began with an X. The letter X, borrowed either from a chromosome or the *X Files* televison series (or both), emerged in late 1996 as a handy symbol for those of us seeking to eXpose the hidden eXperiment with our food, environment and democracy. Although there had already been a handful of far sighted anti-genetic engineering activists operating in Manchester and Oxford, the arrival of unsegregated transgenic soya at the Liverpool docks was probably the breach that set the alarm bells ringing. Greenpeace stopped that ship, occupied the dock cranes and then went on to leaflet supermarkets, placing X-stickers on tubs of the UK's favourite margarine, Flora. Greenpeace, who had also marked an entire US soybean field with an X (dubbing them the X-fields), was at that point one among a handful of campaign groups seeking to waken the public to a problem as alien as the novel genes themselves.[1]

For other activists it began with their clothes off. In October 1996, the United Nations World Food Summit in Rome brought together environmental and social justice campaigners from across the globe to host an alternative Hunger Gathering. The UK environmental direct action movement, bruised though triumphant from long-term campaigns against the construction of new roads, was looking for other ways to address the bigger issue of corporate control. Industrial agriculture was a favourite target of discontent and a number of people attended the Hunger Gathering to learn more. They were inspired by campaigners from India, Canada and the US as well as from the strong German and Austrian campaigns against genetic engineering (see Chapter 27 in this volume). When US agriculture secretary Dan Glickman arrived at the Rome summit he was met by UK activists, stripped of their clothes and with anti-biotech slogans scrawled on their bodies – he later recalled this as the point where he "began to realize the level of opposition and distrust in parts of Europe."[2] It was not to be the last time genetix campaigners got undressed to dramatise a point. In late August 1997 five naked activists scaled the roof of Monsanto's advertising company in the city of London demanding an end to "the genetix cover-up." Bare cheeked and raw confrontation, however literal, still remains an important aspect of UK genetix activism – this in a country better known for its reserve.

Where the X, the activists and momentum came together (generally with their clothes on) was in a new forum called the Genetic Engineering Network (GEN) born also in late 1996. GEN was initially a London meeting of groups opposed to genetic engineering. Where it differed from previous groupings was its diversity: ranging from the lobby groups and the membership NGOs to individual activists affiliated with no group as such. GEN still remains principally a non-hierarchical networking hub – distributing information "for action," (rather than for policy, or for its own sake). It doesn't campaign on its own behalf and barely has any structure to speak of, but has concerned itself with sharing strategy and insights between the larger groups and the smaller groups. Its success lies in being a meeting place for genetix campaigners as close friends and individuals not wearing organizational labels – a club of sorts.

There are around fifty independent local anti-genetix groups who campaign around Britain as well as many hundreds of Friends of the Earth, Greenpeace, Women's Institute, Soil Association and Townswomen's groups. GenetiX Update, a GEN newsletter founded in early 1997 tries to serve this grassroots and is complemented by a genetics email list (genetics@gn.apc. org).[3] This became the common electronic backbone that allowed both large NGOs and grassroots groups to track the latest news, share each other's reactions, and speak knowledgeably – often one or two steps ahead of the opposition. Life science companies claim a part in the information economy, but in the UK it was the activists who first grasped the value of information as a currency in the real world and distributed it freely and effectively. The "genetics" email list elicited groans for its daily deluge of news into activist in-boxes, but it provides up-to-date and relevant knowledge nonetheless. Maybe for the first time on such a wide scale "we" became as well informed as "they."

Next came the dressing-up phase – as vegetables, later as mutants, but first of all as superheroes. A country retreat cobbled together in April 1997, known as the Big Gene Gathering, provided an invaluable weekend of training in the issues, strategy and tactics for somewhere over a hundred disparate activists. The whole crowd moved on and most descended the next day on Monsanto's UK headquarters for the first of the Global Days of Action against genetic engineering. They were dressed as "Super Heroes Against Genetics," cheeky postmodern characters taking on the agribusiness villains with capes, masks, tights and underpants. The superheroes took over the boardroom and held a discussion with the management, copied useful documents from filing cabinets, and posted them to their own Superheroes website. A transatlantic telephone call was also held with US activists simultaneously occupying the Monsanto boardroom there! According to Monsanto UK employees, this was the first time that the company realized the problems it was about to face in the UK. Above it all fluttered a freshly painted sky-blue banner: "Gene Wars – The Consumer Strikes Back!"

The superheroes' second big outing was to the office of the "Soya Bean Information Line," where they took over Monsanto's propaganda call centre and manned the telephones themselves. Their third outing was to play that most English game, cricket, against a field of genetically engineered potatoes – hitting the unfortunate spuds for six until they had well and truly been smashed. Thereafter the silly pants and capes disappeared for a while, but they had established a tradition of humour and carnival. This persists, particularly in supermarket actions, customer information raids that often come complete with giant sweet corn, rotund tomatoes or mad scientists. A travelling piece of anti-GE street theatre known as Miss Mopps' Medicine Show adapted elements from the supermarket actions and toured schools, festivals and actions throughout 1998 explaining the issue in catchy songs. The most tireless of the activist entertainers have been the musical group Seize the Day, whose "Monsanto" anthem has been performed on test sites, law courts and even the tear-gassed streets of Seattle. Many of the early activists in the UK genetix movement came from the do-it-yourself culture of protest that erupted in the early 1990s in reaction to draconian anti-civil-rights legislation. This unusual hybrid of resistance and partying was characterized by "courage, creativity and cheek"   a spirit that persists, mutated of course, into genetix activism.

But even dressing up and singing can't beat a bit of honest weeding. "Stop the crop!" has become the rallying call in what became an annual summer race for the fields. The number of genetically engineered test sites "decontaminated" (i.e. hand-weeded) in 1999 was believed to be over seventy – an increase over the forty removed in 1998 and the handful the year previously.[4] These decontaminations have been important not merely for iconic reasons but for physically slowing the introduction of genetic pollution into the British countryside. At the time of writing there is still no genetically engineered crop fully approved for UK commercial growing. Such crops must first undergo open-air testing – both to establish that they work and also to place them on the national seed list. It is these important national seed list trials that have been the focus of most direct action with up to ten national seed list trials in different parts of the country disappearing on the same night.

GMO gardeners have fallen broadly into two types: those who garden openly, often in local communities, and those who garden beneath the stars. Reports of night-time field visits began to filter through to GEN in mid-1997. The procedure became standard that activists could anonymously leave a message on GEN's answerphone and expect their exploits to be publicized or not, as requested. Each time, GEN notched up the tally stick for how many fields had gone. One of the first, The Gaelic Earth Liberation Front (GELF) also sent GEN pictures and an account of their night's work: the removal of a Monsanto sugar beet plot, the only genetically engineered crop in all Ireland. By making Ireland GE-free in one night GELF inspired a rash of later attempts to rid whole counties of the genetic menace in one or two raids.

Possibly the second action to receive any attention was another Celtic attack: the scything of an AgroEvo rape field in Cupar, Scotland, where genetically engineered scarecrows were left as a calling card. The saboteurs later made themselves known, and included a nearby organic farmer and a local author. The leaving of a calling card is a sporadic and risky tradition. The prize for the most original calling card to date goes to croppers who left a "Warning to Humanity" at the site of another decontamination in May 1998, accompanied by a crop circle. Their statement read: "People of Earth let it be known that we have decontaminated the genetically polluted site at Throws Farm in your Earth county called Essex… We trust that we have not gone to this effort in vain and that you will learn from our warning and take heed. Your food evolved the way it did for a reason, meddle with it at your peril! – Extra Terrestrial Prevention Unit, P.S.: Sorry about the crop circles, you don't know how hard it is to park these things."

While the bulk of field actions have been night-time affairs, it is the growing number of deeds done by daylight that have pushed the public campaign forward. Anonymous night-time raids, though important, were easily misrepresented as vandalism and marginalized as extremist in the media. By contrast, the appearance on the news of ordinary people with very good explanations for pulling up crops were harder for the industry to deal with and gained public sympathy for the activists. The first experiments with being open occurred in Scotland, again in Cupar in Fife. Here a local test site campaign had been going for the entire growing season, including meetings with the farmer and a local petition signed by most the community. A few days after the field was scythed, a harvest festival in the local village hall gave way to a procession of around fifty people. They headed to the trial site and removed the remaining plants as local police looked on.

After this, Fife Earth First!, who had helped with the Cupar campaign, became increasingly open in claiming responsibility for other decontaminations throughout Scotland. They gave a distinctly nationalist edge to the campaign, adopting the Genetix X into to the national flag of St Andrew, which they would leave fluttering above vanquished fields or cut into the crop itself. While less discreet than other activists, they nonetheless aimed to evade arrest. A stakeout of the Scottish Crop Research Institute by police waiting in a barn to catch Fife EF! was reported in the local newspaper after the exhaust of their car became hot, setting light to nearby hay and sending the whole barn up in flames. £30,000 worth of damage was caused to the car and the barn. Fife EF! sent a message asking them to park nearer the crop itself in future.

And then there is Genetix Snowball, the most systematic experiment in open crop destruction so far. Initially the project of nine campaigners who wanted to take action openly, accountably and within a code of strict non-violence, Genetix Snowball was launched on July 4, 1998, declared a day of "Freedom from American corporations." Five middle-class women in head-scarves and gardening gloves bent over and weeded Monsanto's genetically engineered oilseed rape in the heart of the home counties, watched by puzzled British bobbies and a bevy of cameras and journalists. The tone could not have been more "English."

Genetix Snowball took its inspiration and unusual name from the Snow-ball campaigns of the 1980s in which thousands of people were arrested for snipping wire at US Air Force bases in protest against nuclear weapons. In both incarnations, the aim was to "get the snowball rolling" so that while each person took on only a small piece of symbolic criminal damage – in the case of Genetix Snowball removing fewer than a hundred plants – they then found two other people to follow their example, inflicting much more serious damage through a multiplier effect of mass public civil disobedience. The snowball campaign was meticulously prepared over several months with its own press officer, support group and even the production of a training video and sixty-page *Genetix Snowball Handbook* "for safely removing genetically modified plants from release sites in Britain."[5]

Like their nuclear predecessors, Genetix Snowball chose an incongruously reasonable demand – "a five-year moratorium on growing and import of GE crops except for government sponsored ecological tests within closed systems" – and went about their campaign in complete openness, writing first with their concerns to all farmers hosting trials, asking for crops to be removed, and offering to do so themselves. The greatest strength of Snowball, however, was the apparent ordinariness of the individuals involved: while they included experienced activists, among those first five women were a music teacher, a woodworker, a journalist and a solicitor. Later actions involved an art teacher, an environmental scientist, a graphic designer and a postman – all willing to talk to the press, keen to explain their actions and eloquent and well in-formed on why they opposed genetic engineering. They were a biotech spin doctor's nightmare. They made crop pulling visible and sensible with their easy "Middle England" media manner. They also showed that they could get away with pulling up genetically engineered crops in front of cameras and police and still get home to a hot organic cocoa in the evening.

Monsanto's response to the first Snowball action was not to press for criminal damage charges – this would have allowed the Snowballers to defend themselves before a jury. Instead they sought a civil injunction. This restrained the five women and their press officer from interfering with engineered plants or encouraging others to do so. Regarding this denial of a jury as a move to "privatize justice," the Snowballers nonetheless attempted to turn the civil proceedings into a full trial on the issue of genetic engineering, turning their media-savvy attack from the crops to the behaviour of the bullying corpora-tion itself. Monsanto tried to expedite proceedings and deny Snowball their day in court, demanding meanwhile the names and addresses of every person who had ever received a snowball handbook. Snowball publicly sent the handbook to the Pope, Prince Charles, Tony Blair, Bob Shapiro, CEO of Monsanto, and the Queen, daring Monsanto to injunct them. The twists and turns of the Snowball court case, which finally ended in the Monsanto in-junction being granted, has been one of a clutch of significant court cases in which direct action has gone hand in hand with skilled use of the law and media to gain important campaign advantages.

Significant amont these is the case of Jacklyn Sheedy and Elizabeth Snook, who helped destroy a GM maize crop in June 1998. The test crop in question had been the focus of an intense local campaign partly because it threatened to cross-pollinate organic sweet corn from a farm only 300 metres away. Six hundred people from the small town of Totnes had marched to the site earlier in the year and the organic farmer, Guy Watson, had taken the government to court for allowing the trial to proceed, claiming that contamination would lose him his prized organic certification. The Watson court case failed to get a judicial removal of the crop but highlighted, especially through the farming press, the real threat of genetic contamination. The unsympathetic judge suggested that Watson take up his fight not with the government but with organic certifiers the Soil Association for having such prohibitive standards.

When Sheedy and Snook were caught removing the crop, they openly admitted their action and were charged with over half a million pounds worth of criminal damage. In pre-trial hearings they claimed, as a defence of lawful excuse, that they were defending the organic farmer and the environment from genetic pollution. They lined up expert evidence ready to put genetic engineering on trial before a jury. Meanwhile over three thousand Totnes residents signed a declaration in support of their action. Only twenty-four hours after their evidence was submitted, the Crown Prosecution Service chose to drop the case against them. Their lawyer, Mike Schwarz, writing in the *Guardian* newspaper, described the decision as "a green light" for activists to take direct action against field trials, adding, "By withdrawing the case from the jury, the crown have decided that there was compelling evidence that the defendants had a lawful excuse to remove the genetically engineered maize. The last thing the crown wanted was to see a jury – a microcosm of society – acquit people who admitted taking direct action against GE crops."[6]

That, however, was what much of the UK public did see, or rather heard, in October 1999 when organic pig farmer Tommy Archer was found by a jury to be "not guilty" of criminal damage despite pulling up a genetically engineered test site. Tommy Archer is a fictional character in the UK's longest running radio soap opera, *The Archers*, originally established after World War II to teach farmers about modern farming techniques. The Archers' GE-trial storyline had followed a decision by a local farmer to plant a GE crop trial and then followed the local campaign against it culminating in its destruction and Tommy's arrest. Fiction bled into truth as ten real field trials had been pulled up only days previously – claimed by a group called "Ambridge Against Genetix" (Ambridge being the fictional village in which *The Archers* is set). The storywriter of *The Archers'* court case was partly advised by Mike Schwarz using much of the evidence that was intended for the Totnes case. Once again, truth became stranger than fiction when Greenpeace cut down one of the UK's largest field trials. Twenty-eight Greenpeace activists pleaded not guilty on the same grounds of lawful excuse, prompting newspaper comparisons between UK Greenpeace boss Peter Melchett and Tommy Archer. In what became a showpiece courtroom event putting genetic engineering itself on trial, Greenpeace chose to be represented by the very same lawyer, Mike Schwarz.

Probably the most visionary of field actions have been "crop squats" – land occupations of former GE field trials, a tactic borrowed from the German movement against genetic engineering. The first crop squat occurred in Norfolk in May 1998 on a former Novartis GE sugar beet site on land belonging to Sir Timothy Coleman, Lord Lieutenant of Norfolk. In a matter of hours a central yurt and marquee had been erected and a number of vegetable beds had been dug. Over a period of two weeks this dusty Norfolk field, more familiar with chemicals and monoculture crops than diversity, was transformed into permaculture gardens with displays on growing food and how to oppose genetic engineering. The site was well visited and popular with local residents and drew plenty of media attention before the notice for eviction finally came. Sir Timothy's Crown Point Estates subsequently broke off contracts with Novartis and decided to plant no more GE field trials.

An urban equivalent to the crop squat has been a series of squatted organic cafes. The most ambitious, known as Toxic Planet, converted an empty office block in central London into a "genetic art experience" – kids got lost in "Maisy's GE maize maze," permaculture and genetics workshops were held inside or even on the street, a room was dedicated to field trial decontaminations, another to a cinema showing radical documentaries. Music, parties and street entertainment brought the neighbourhood and shoppers in, all of whom were fed on organic food and anti-genetic engineering propaganda. These squats have been the radical edge of a wider project to start declaring GMO-free zones – some as small as farms, others whole islands, regions or, in the case of the new Welsh and Scottish assemblies, potentially entire countries.

Perversely, the establishment's response to field trial opposition has been to plant more and make them bigger. While the media and the public were enthralled by activists taking on the crops, what really stung the UK government was criticism from voices it considered "one of ours." In 1998 English Nature, the statutory conservation body, demanded a three-year moratorium on commercial growing, concerned about the ecological impact of changing pesticide regimes associated with herbicide-tolerant crops. The Royal Society for Protection of Birds (RSPB), the UK's largest conservation charity, boasting over a million members, also entered the fray, echoing English Nature's concerns over pesticide regimes.

To pacify them and be seen to be doing something, Environment Minister Michael Meacher announced a four-year programme of "farm scale evaluations" that would proceed before commercialization. Dressed up as a moratorium, the 'farm scale trials" have in fact been the biotech industry's sneakiest move to date. Each field trial is around 10 hectares (25 acres) and there may be up to one hundred per year supposedly to monitor the environmental effect of genetically engineered crops. The trials look at the effects of pesticides, but avoid questions of genetic pollution and gene flow while releasing genetically altered pollen on a larger scale than before. They fail to look for possible effects on soil ecology, bird life or larger mammals and follow the same "contaminate first, ask questions later" rationale as the smaller field trials, claiming all the while that it is for "environmental reasons."

Thankfully neither the anti-genetic engineering movement nor the public has swallowed them. A June 1999 Mori poll found 79 per cent of the public opposed, and almost every "farm scale trial" announced so far has been met with a local campaign. In Swindon a vigorous local opposition forced gentleman farmer Captain Fred Barker to destroy his own crop. In Oxfordshire, a local campaign in Watlington culminated in a rally of over eight hundred people on the field adjoining the genetically engineered rape. The crowd heard speeches from award-winning food writer Lynda Brown, parliamentarian Alan Simpson and well-known environmentalist George Monbiot before donning white decontamination suits and moving en masse into the crop. Around six hundred people, including pensioners, beekeepers, organic farmers and children, spent two hours destroying the crop, white biohazard flags fluttering against the green field as a police helicopter tried to identify faces from above. Some 100 metres away, campaigners had squatted an old farmhouse and turned the garden into a permaculture exhibition, celebrating diverse and organic agriculture. Banners from the farmhouse had been brought to the field. They proclaimed: "Resistance is fertile!"

The following Sunday in sunny central London, Greenpeace and the Soil Association held Britain's first national "organic picnic." Thousands of picnickers joined by musicians and celebrities called on the Prime Minister to scrap the field trials. At 5 a.m. the next morning, as we have seen, the third of these farm scale trials in Norfolk was cut down by Greenpeace activists and a mechanical mower. The activists' arrest and detention, and particularly that of ex-minister and Greenpeace head Lord Melchett, catapulted farm-scale trials and genetix direct action onto the front pages at home and worldwide. A year later a jury acquitted the activists, accepting their defence that they were acting to prevent genetic pollution of neighbouring farms. What is clear is that these so-called "environmental" tests do not wash with even the mainstream environmental movement. Although the farm-scale trials retain the crucial support of English Nature and the RSPB, the rest of the anti-genetix movement is now vigorously opposed to them, as is more and more of the farming community. At the time of writing, roughly a third of them had been dropped or destroyed, local communities have held referenda to force farmers to pull out, and district councils have sought legal injunctions to exclude them from their area. Two of the earliest trials were in a region of eastern England, Lincolnshire, where the local council passed a resolution against GE crops. A third, hosted by an industry spokesman because no farmers could be found, became the subject of yet another vigorous local campaign.

There is, of course, more to the anti-GE party than the activist attendees. In January 1999, Greenpeace and the Guild of Food Writers brought together over a hundred of the UK's top food writers, the tastemakers of Britain, to launch their own campaign for a ban on genetically engineered foods. At a glitzy organic breakfast hosted by the UK's most expensive restaurant, The

Savoy, food writer Darina Allen, whose own cookery school had bordered the Irish GE test site, declared genetic engineering to be the most important issue she would have to face in her lifetime. A month later, during one of the first real media frenzies on the issue, a second grouping, the Five Year Freeze, was launched. This campaign worked to bring as many different NGOs, trade unions, local councils and other bodies under the common demand for a moratorium on the growing, import and patenting of genetically engineered food and crops. At the time of writing, close to two hundred bodies, with a constituency of up to 3 million people, subscribe to the freeze, including the Consumers Association, the Women's Institutes, Islamic Concern, Christian Aid, Action Aid, the Townswomen's Guild, the Local Government Association and many more. The Freeze has been successful in providing a "reasonable" set of demands for civil society and the all-important "Middle England" to put their weight behind, thereby delivering a clear bulwark of disapproval against the vehemently pro-biotech government.

The most powerful and important players, however, have been the ordinary folk, otherwise described by newspapers as "consumers." The astonishing turnaround in the fortunes of Monsanto, AgroEvo and their ilk in the UK has probably been 30 per cent due to the genetix movement, some small amount due to actual or perceived government and industry corruption, and the rest thanks to consumers simply saying no. In 1998 I designed and helped run the Protect Your Food campaign, in which Greenpeace, Friends of the Earth and other groups involved with GEN focused a brand attack on the first labelled genetically engineered product. It was a vegetarian mince product called Beanfeast produced by Unilever, Europe's second largest food producer. We distributed around half a million "disloyalty cards" in supermarkets and received thousands of calls from people who had never campaigned or even complained before this issue came along.

In early 1999, as public awareness on the issue skyrocketed, the head of Unilever UK requested a meeting with Greenpeace to discuss what they should do. Unilever had received thousands of calls on their consumer call lines and sales of Beanfeast had dropped by at least 50 per cent. The next day they went GE-free in the UK. As we always expected, every other major food producer and supermarket followed their example. It became clear that, while the fun had been mostly in the fields, much of the real hard work of the campaign had been done here at the consumer level: information stalls at supermarkets, "gene food tours" for customers around supermarket aisles, public talks and debates in urban centres, and local campaigns to turn school dinners GE-free. It was these less glamorous activities that probably paid the best campaigning dividends.

In the UK there is a long tradition of boycotts and market interventions going back to the "food riots" of the eighteenth century. On these occasions the crowd would seize the marketplace and cast out merchants who were passing on shoddy goods. Metaphorically this was exactly what had happened again as millions of people complained to company call-in lines, returned food or asked difficult questions at checkouts. Some supermarket actions dramatized this: activists filled up their trolleys with genetically engineered products and then all went through different checkouts at the same time, loudly demanding GE-free alternative products while others leafleted the rest of the queue. Astonishingly, almost all these actions received cheers of support rather than annoyance from other waiting shoppers.

Outside a Safeway supermarket in central London two men were arrested for shoplifting after they filled up a basket with GE products, walked out of the supermarket without paying and dumped the goods in a waiting bin, announcing to cheering shoppers what they had just done. Genetix Snowball activists did similarly across the country, replacing the "stolen" goods with organic alternatives. As part of the Protect Your Food campaign, thousands of packets of Beanfeast were labelled "contaminated" by shoppers joining the campaign. Indeed, across the country stickering has been a very popular way of alerting the public to unlabelled GE food. There have been reports of lone labellers making supermarket managers' lives a misery, and on other occasions anti-GE supermarket managers secretly doing the labelling themselves.

Three commercial players in particular could claim applause on the market side of the campaign. The first was Iceland supermarkets, whose chairman Malcom Walker took a personal stand against genetically engineered food – eliminating it from the company's own-brand products a year before any other supermarket. Iceland proved it could be done and their profits shot up by around 20 per cent as a result. The second was the Soil Association, the leading organic certifiers, who decided in 1997 that GMOs had to be entirely removed from agriculture to protect organic standards. The Soil Association, which is also a membership group, launched its first real campaign on the back of the genetic engineering issue and is now regarded as a significant environmental campaigning group in its own right.

A third important player came in the wiry shape and Scottish brogue of Lindsay Keenan, who had been hired by a Glasgow whole-food company to audit its lines for genetically engineered ingredients. He went on to work with the rest of the whole-food trade in an initiative named Genetix Food Alert! to remove engineered ingredients from all their goods, establishing non-GE supply lines as he went. These supply lines were swiftly adopted by other food producers when the public mood became too hostile to sell genetically engineered food.

Meanwhile the many thousands of whole-food shops, wholesalers, organic farmers and whole-food producers became another essential network for distributing anti-genetix information and organizing grassroots campaigning. It was a way of reaching people who already cared about how their food was produced. Whole-food shops similarly offered meeting space for local campaigns, window space for posters, counter space for leaflets and helped provide refreshments and funding for public meetings and debates. Farmers supplying fresh organic vegetables every week directly to subscribers were also keen to carry leaflets, newsletters, stickers and "information for action." Many of the

newer activists and local campaigners involved with the UK genetix movement have come through this route.

Most significantly the whole-food and organic movements have offered
real and available alternatives to the genetically engineered monoculture. This
has allowed the campaign to open up the politics of industrial food production
as a whole. In the UK the organic food market is now experiencing growth
similar to that of computers and telecommunications. The major supermarkets
are competing to outdo each other on the breadth of their organic ranges
with around four to five hundred organic products each. Consumers buying
organics are not merely buying GMO-free but, because of the political position
taken by the Soil Association, are actually making an anti-genetic engineering
statement. Like the organic movement, those involved with the whole-food
trade work to values of community, animal welfare and environmental care
that stand in contradiction to chemical-intensive agribusiness. It is achieving
those values in food production rather than some technical 'GE-free' label
that have become the focus of campaigns. The two largest consumer campaigns against genetic engineering in the UK – Greenpeace's True Food campaign and Friends of the Earth's Real Food campaign – are both calling for
an organic, not merely a non-genetically engineered, solution.

Amongst grassroots activists, who tend to be three steps ahead of the
mainstream groups, the agenda is already moving on beyond this and outwards
to a more fundamental defence of agricultural diversity. It is also challenging
many more aspects of genetic engineering than just engineered crops. A
website, www.primalseeds.org, is dedicated to "untangling ourselves from the
monocult," asking, "How can we create a grassroots movement for local agriculture to protect biodiversity?" Primal Seeds, which links seeds savers,
permaculturists and guerrilla gardeners, arose out of an activist gathering,
Seeds of Resistance, held in Cambridge to parallel the 1999 World Seed
Conference. There activists broke into the official meal of the world seed
trade, asking, "Would you like pesticides with that, sir?" as they
sprayed harmless solutions onto their dinner and offered rare heritage varieties
of vegetables instead.

Almost at the same time the Campaign Against Human Genetic Engineering (CAHGE) was storming the stage at the annual meeting of London's
Galton Institute – also known as the Eugenics Society. CAHGE is the result
of genetix activists and disability activists starting to work together to challenge
the technology of germ-line therapy, which is now becoming a reality. Those
links began to form following a joint campaign in Newcastle against the
building of a new exhibition and research complex dedicated to promoting
the role of the gene in modern society. Disabled Action North East had
joined with local genetix action group GeneNO to oppose the centre, which
had won funding otherwise intended for making Newcastle more accessible
for the disabled. The deep irony of funding a centre for eradicating the
disabled themselves rather than eradicating the problems they face was considered singularly offensive, and they hung banners and closed down opening

ceremonies. Not just the genetic manipulation of humans is being challenged. In the land where Dolly was invented, groups such as Xenotransplantation Concern and Compassion in World Farming have been steadily highlighting the animal genetic engineering underway in British laboratories and campaigning to stop it too.

What I have catalogued here are some of the conscious initiatives of the genetix movement – things that could be copied. What I have left out for the most part are the events, outside of our control, that acted as sparks igniting the good flammable stuff of public concern. It is clear to me that where campaigners worked to put the tinder in place, the lightning still came from elsewhere. Chief amongst these was the Pustzai affair – the case of the Hungarian food scientist from the government's Rowett Institute who reported evidence that genetically engineered food had damaged laboratory rats. Subsequently sacked, vilified by the establishment, and his work overhastily discredited by the Royal Society, Pustzai became a cause célèbre, the newspapers delightedly smelling more than just an infected rat. They found, as they looked closer, that government and the biotech industry were in bed together and ignoring public opinion. The papers did a good job of exposing the extent to which biotech influence had greased the way for genetically engineered food and undermined proper democratic control. The minister for science, Lord Sainsbury, who had his own biotech holdings, came in for scrutiny as did a number of lobby groups linked to Monsanto.

It also became clear that genetically engineered food policy came direct from the prime minister – or the "prime monster" as the *Daily Mirror* renamed him, colouring his face green and putting a Frankenstein bolt through his neck! As Greenpeace dumped 4 tons of GE soya on his front doorstep, their slogan pointed at even more pernicious relationships with counterparts across the Atlantic: "Tony don't swallow Bill's seed!" The decision by four national newspapers to take up their own campaigns against genetic engineering was testament to their confidence that this issue was already of strong enough public concern to justify going out on a political limb.

Subsequently, 1999 became a year of media glare for genetix campaigners, conferring mixed blessings. It was a joy to see every twist, turn, lie and scandal of the biotech industry exposed in headlines, but at the same time the issue itself was taken out of our hands and characterized as merely a food scare, our well-founded concerns portrayed as merely ignorance and fear. Government and industry, too, assumed that media hype was all there was: "It's just you and your mates at the *Daily Mail* whipping up a fuss," was how one government advisor read anti-GE sentiment back to me. However, the wholesale retreat of the food industry from genetically engineered ingredients, and commitments also to remove meat and dairy products from animals fed engineered crops, revealed the true depth of public feeling. The truth is that consumers and activists have circumvented the government and forced the pace of change. And that's who's made the real difference: not princes, newspapers or protesters, just people.

NOTES

1. "Greenpeace Quarantines Genetically-altered Monsanto X-Field in Iowa," Greenpeace International press release, October 10, 1996, available at http://www.greenpeace.org/pressrealeases/geneng/1996oct10.html.

2. See "Remarks as Prepared for Delivery by Secretary of Agriculture Dan Glickman before the National Press Club," US Department of Agriculture, July 13, 1999, available at http://www.usda.gov/news/releases/1999/07/0285; see also Bill Lambrecht, "World Recoils at Monsanto's Brave New Crops," *St. Louis Post-Dispatch*, December 27, 1998.

3. The "genetics" list is archived at http://www.gene.ch/info4action.html.

4. For a current list of destroyed UK test sites, see http://www.primalseeds.org.

5. The *Genetix Snowball Handbook* is available online at http://www.gn.apc.org/pmhp/gs.

6. "Michael Schwarz, who defended two women accused of pulling up GM crops, explains why the decision not to prosecute may boost activists" (Commentary by Michael Schwarz), *Guardian*, April 7, 1999.

26

Seed Satyagraha: A Movement for Farmers' Rights and Freedoms in a World of Intellectual Property Rights, Globalized Agriculture and Biotechnology

VANDANA SHIVA

India has always been agricultural India – and an agricultural society can only survive if its farmers survive. As Gandhi said, "If India's villages perish, India will perish." The history of India's independence has been intimately intertwined with the history of peasant struggles.

Farmers' organizations at the regional and national levels were formed in the late 1970s and early 1980s as a response to the costs generated for farmers by the Green Revolution. However, most farmers' movements were confined to the industrial agricultural paradigm and focused on remunerative prices for agricultural products and subsidies for chemical inputs. This was necessary because the costs of inputs had kept increasing, and thus chemical agriculture, which was being promoted by international agencies and national policies, was making farmers' survival impossible.[1]

The Seed Satyagraha – *satyagraha* is the Gandhian term for nonviolent resistance – has helped launch a post-Green Revolution politics of agriculture. People aware of both trade issues and the promise of sustainable agriculture carried the text of the GATT agreement (which created the World Trade Organization) to farmers' groups across the country to make them aware of things to come. A new partnership emerged between public-interest scientists involved in sustainable agriculture and trade and farmers' organizations that were making demands on government. This partnership moved farmers' mobilization beyond local concerns to national and global concerns. It also moved farmers' demands from narrow demands for more chemicals and more subsidies to demands for protection of farmers' rights and the search for viability of farming livelihoods, not through more chemicals and more subsidies but through internal inputs and biodiversity intensification and rejuvenation.[2]

This campaign has unleashed new creative energies to respond to the global challenges that were becoming life-and-death issues for small Indian farmers. Farmers and activist scientists initiated a multiplicity of levels of

actions, from direct action to political advocacy, from actions to build alternatives to actions to resist globalization. Parliamentarians were urged to act
against industry monopolies that were strengthened by the GATT's Trade
Related Intellectual Properties regime (TRIPs), and farmers were assisted in
the quiet building of community seed banks and sustainable agriculture on
the ground. We helped create everyday options for farmers, avoiding a vacuum
that would ultimately have been filled by Cargill and Monsanto.

THE ORIGINS OF THE SEED SATYAGRAHA

During the late 1980s, two global trends started to emerge. The first was the
rapid takeover of the seed industry and the biotechnology venture firms by
the chemical industry, thus establishing a market monopoly on seed through
the concentration of ownership in the seed sector. The second was the
introduction of Intellectual Property Rights (IPRs) to the Uruguay Round of
GATT negotiations through the TRIPs Agreement, which forces all countries
to adopt Western-style industrial regimes for IPRs to cover seeds and plant
varieties. Learning from history, especially Gandhi's use of the spinning wheel
as both a symbol of resistance and a tool for building creative alternatives, the
Seed Satyagraha would seek to locate constructive action and resistance in the
seed.[3]

As a creative alternative to farmers' dependence on corporate seed, which
eventually leads to displacement of farmers from agriculture, we started
Navdanya, a national movement to protect native seeds and farmers' rights.[4]
Navdanya has helped preserve hundreds of older varieties of rice and other
crops, restored confidence in traditional polycultures in the face of "modern-
ized" monocultural practices, and provided markets for villages that are returning to organic methods of cultivation. We also built a national alliance of
organic producers called ARISE (Agricultural Renewal in India for a Sustainable Environment).

Navdanya and ARISE do the work of putting a nature-centred, farmer-
centred alternative in place to resist the anti-nature, anti-farmer policies of
globalized agriculture. By placing the farmer at the centre of conservation,
these efforts relocate control over the political, ecological and economic aspects
of agriculture in the hands of farmers. By conserving native seeds, we are
helping conserve indigenous knowledge and indigenous culture.[5]

During 1991–92, we held training workshops and study camps for the
National Farmers Union, the Bharatiya Kisan Union (BKU), for the Karnataka
Rajya Ryota Sangha (KRRS) and for the Chattisgarh Mukti Morcha (the
peasant–worker alliance in the Chattisgarh region which is in the Vavilov
Centre for rice diversity). We also started to produce two newsletters, *Bija*
(The Seed) and *Krishi Samachar* (News for Farmers), to bring timely information on international and national policy to farmers' groups. The awareness
generated among farmers about global issues created the possibility of large-
scale mobilization of farmers against IPRs and globalization.

On October 2, 1991, the farmers held a massive rally in Bangalore and declared that they would step up their agitation against both the structural adjustment policies of the government and the GATT and TRIPs agreements that would entrench these policies, as well as legitimize Northern companies' patenting of Indian farmers' biological resources. Patents on seed were seen as a hijacking of farmers' intellectual property, which is the result of centuries of innovation.

In order to understand the effects of these proposals better, the Research Foundation on Science, Technology and Ecology organized a National Seminar on "Proposals of GATT" in Bangalore in February of 1992. The seminar was attended by representatives of farmers' organizations from around the country, environmentalists and others. Following deliberations, the meeting unanimously decided to reject the GATT proposals on agriculture and Trade Related Intellectual Property Rights (TRIPs).

During the same period, under provisions of India's structural adjustment programme, the government of India decided to import wheat, even though there was enough wheat available in the country. The cost of the imported wheat was far higher than even the high domestic prices, which originally were the excuse the government had given for the imports. Cargill Inc., based in the USA, was one of the main beneficiaries of the Indian purchase. Enraged farmers held a rally at the New Delhi Boat Club to protest against the government's willingness to pay increased prices to transnational corporations (TNCs) rather than to the Indian farmers.

In August 1992 the Indian government made a further attempt to reduce fertilizer subsidies. The prices of fertilizers doubled overnight and, again, farmers protested. Thousands of farmers were arrested in Uttar Pradesh. Four farmers were shot by the police during a protest at Ramkola in Deoria district in September. The continued resistance of the farmers forced the government to provide subsidies to enable farmers to meet this enhanced cost.

Meanwhile, as the liberalization policies of the government were attracting the large seed houses of the world into the country, the public was increasingly becoming aware of the effects that the GATT proposals and TRIPs would have on agriculture. On October 2, 1992, Gandhi's birth anniversary, the farmers resolved to use Gandhi's weapons against this form of recolonization. At a massive rally at Hospet in Karnataka, over 500,000 farmers pledged to fight for Gandhi's concept of *Swaraj* (self-rule) and launched the Seed Satyagraha to resist policies aimed at handing over food and seed production to multinational corporations.

The farmers' demand that they be consulted during the formulation of a national agricultural policy went unheeded by the government. Their repeated protests against the entry of the multinationals in the seed and food production sectors were ignored. The farmers were left with little choice but to resort to direct action. In December of 1992 over five hundred farmers led by farmers' leaders Babagouda and Hanumangouda stormed the Cargill Seeds

India Pvt. Ltd office premises in Bangalore to force the government to take notice of their demands. In an atmosphere reminiscent of the Gandhian era, they served "Quit India" notices on other transnational corporations involved in the agriculture sector.

Private seed company officials went on record saying that patents were good for the farmers. Cargill in fact went so far as to say that the Indian farmers had always been and would always be dependent on imported seed for food production. This statement is absurd. India has been the source of origin of many of its major food crops, while the rest have come from Africa, Central and South America over centuries of free exchange between farmers.

Cargill Seeds India faced another farmers' rally in March of 1993. The government did its best to sabotage the voice of the farmers of India by arresting ten thousand who were on their way to Delhi to attend the rally, and by granting permission to hold it only at the last minute after changing the venue. However, over 200,000 farmers and their representatives attended, and made the rally a huge success.

In July 1993, the farmers again went into action and demolished part of Cargill's seed unit at Bellary. Not having learned their lesson, Cargill instigated the uprooting at night of native crops from over 200 hectares of land around their contracted fields in a bid to prevent cross-pollination. On seeing his farm devastated next morning, one farmer died of shock. Several others were ruined. The leaders of farmers' organizations all over the country set out to launch a "Kick Out Cargill" agitation.

In August the farmers of Karnataka presented district collectors with Collective Intellectual Property Rights claims to biodiversity and natural resources to mark the anniversary of India's independence. This presented a challenge to the commonly held belief that patent protection can only be afforded to private parties. It further challenged the granting of a legal personality to corporations, but not to communities that identified resources, developed knowledge about their utilization, innovated over centuries, and nurtured both the resource and the knowledge about it.

Fed up with the government's meaningless assurances on support prices, subsidies for fertilizers and power and, most importantly, the GATT draft on agriculture, nearly 100,000 farmers staged an indefinite sit-in (*dharna*) at the historic Red Fort grounds in Delhi starting on September 17, 1993. Chowdhary Mahendra Singh Tikait, leader of the Bharatiya Kisan Union, warned the government that accepting the GATT draft would be tantamount to accepting recolonization.

All these activities culminated in a mass rally in Bangalore in October, the first anniversary of the Seed Satyagraha. Representatives of farmers' organizations from all over the world also met to pledge themselves once again to continue the struggle for freedom for agriculture and agriculturists from corporate domination. When Cargill and Monsanto announced a common venture to promote genetic engineering of agriculture, we redoubled our efforts to create a movement against Monsanto as well as Cargill.

MONSANTO QUIT INDIA!

By the summer of 1998, widespread field trials of genetically modified Bt cotton were under way at forty sites through nine states of India, unbeknownst to nearly everyone in India. In May, Monsanto had acquired 26 per cent of Mahyco, the largest Indian seed company. An alliance of environmental and sustainable agriculture organizations launched a "Monsanto, Quit India" Campaign on August 9, the anniversary of Gandhi's historic call for the British to "Quit India."

In November 1998, a news report broke the story of the Bt cotton trials to the general public, revealing that the trials were being carried out without permission from the state governments or with the consent of local communities. The government of the state of Karnataka, in the southwest of India, announced the establishment of a committee to investigate the trials. Meanwhile, farmers in Karnataka took matters into their own hands. Action was initiated on November 28, when crowds burned the transgenic cotton crop in the fields of farmers, who had only come to know days before that they had been growing genetically modified cotton. Further crop burnings were carried out in Karnataka and Andhra Pradesh, as farmers came to know that the trials they were accommodating in their fields were of the genetically modified "Bollgard" variety.

In December, the Andhra Pradesh government asked Mahyco–Monsanto Biotech Ltd, the joint venture developing Bt cotton in India, to stop all private field trials in the state, adding that if the company insisted on continuing trials it would be permitted to do so only at public research institutions under the close supervision of state scientists. The truly criminal nature of the Bt trials was revealed when farmers in four states told researchers from the RFSTE that the Bt cotton had been planted several months before permission to conduct the field trials was granted. Further, the permission was granted to the company by a committee of the Department of Biotechnology, under India's Ministry of Science and Technology, without the consent of officials of the Department of Environment, Forests and Wildlife.

Discussions with the farmers made it clear that none of those involved in the trials was briefed on the potential impacts of GMOs on neighbouring fields or the surrounding ecosystem, or on how the Bollgard cotton was supposed to control bollworms. The farmers did not know they were participating in field trials of genetically modified seeds. There were no post-harvest management or disposal precautions in place, hence farmers could have sold the cotton produced by unapproved genetically modified plants along with their normal cotton in the open market. As it turned out, the yield from the Bt varieties was poor to negligible in most areas.

In one district of Andhra Pradesh, near the city of Hyderabad, nearly five hundred cotton farmers had committed suicide the previous year due to debts linked with the high costs of hybrid seed and pesticides. In 1998, the cotton seed sold by private corporations failed in over two hundred villages, provoking another rash of suicides. Farmers in this region understand the social costs of non-sustainable pest-control strategies because farmers have been paying with their lives. There is an increasing level of awareness of how genetic engineering, Bt cotton and Monsanto exacerbate these problems. After a number of fields of genetically modified cotton were uprooted by farmers' organizations, the farmers had a unanimous resolution passed in the regional parliament against genetic engineering and forced the state government to ban all future trials. By the end of 1998 Bt crops in six villages had been uprooted by the state government itself.

More than twenty farmers groups, thirty NGOs and eminent scientists formed a coordination committee to combine legal action, public hearings and scientific research in the campaign against Monsanto. We held training programmes on the uselessness of pesticides and genetically engineered crops in controlling pests, and also demonstrated ecological pest control methods such as the use of neem and pongamia extracts, as well as mixed and rotational cropping. While some of the high-profile actions in Karnataka used undemocratic methods to destroy some farmers' fields without their consent, we have tried in Andhra Pradesh to nurture the movement through months of background work and months of follow up. A sustainable movement can only be built democratically through dialogue, awareness and creative, constructive actions, including initiatives on sustainable alternatives for farmers.

OPPOSING TRIPs IN THE INDIAN PARLIAMENT

We have also had to engage in a democratic struggle in parliament and in society against the imposition of IPR regimes, to ensure that farmers' rights are protected in our national laws and seed monopolies are not established. The first goal was to prevent the changes in India's patent laws mandated by the WTO without full democratic discussion.[6]

There are two major problems with the WTO requirements on intellectual property. First, the TRIPs agreement allows for patentability of all life forms, including animals and plants. Second, it forces a monopoly control on seeds and plants, only offering the country a choice as to whether monopoly control will take the form of patents or breeders' rights. The only system recognized at the international level is the system of plant breeders' rights as codified in the International Convention for Protection of New Varieties of Plants (UPOV). Plant breeders' rights as recognized in UPOV give monopoly markets to breeders of new varieties. Farmers now have to pay royalties for saving seed on their own farms even under breeders' rights regimes. Thus, both systems threaten farmers' rights.

In 1994 the Agriculture Ministry prepared a draft Plant Variety Legislation to implement TRIPs. I was called in as an expert, and in that role I invited all national and regional farmers' organizations for consultations with the Agriculture Ministry. The 1994 draft had no reference to farmers as breeders and was a direct interpretation of UPOV. We insisted that the option for India could not be UPOV, which protects only the rights of industry, not the rights of farmers.

The next draft prepared by the Agriculture Ministry still did not consider farmers' rights as breeders' rights that stand prior to the rights of the seed industry. It also did not have farmers' rights translated into corporate responsibility and corporate liability, an issue which has gained urgency with the epidemic of farmers' suicides and the negotiation of the international biosafety protocol. Hundreds of farmers have committed suicide in India as a result of privatization and globalization of the seed supply system, leading to increased debts for seeds and agrichemical inputs and increased vulnerability to crop failures linked to monocultures and new seeds.[7]

The farmers' suicides raise the critical issue of liability. IPR regimes have created a property system in seeds in which the "owner" has absolute rights and absolute freedom from responsibility. In industrial agriculture, farmers bear all risks and corporations derive all profits. Sustainability of agriculture requires that risk sharing and profit sharing should be more equitably distributed between farmers and the industry. Without a balance of rights and responsibility, farmers will increasingly be pushed to debt, displacement and suicide and agriculture will increasingly be pushed into non-sustainable monocultures and monopolies over seed and the entire food chain.

The widening debate over genetic engineering in India has brought some improvements in the bill. A more recent draft of the Protection of Plant Varieties and Farmers' Rights legislation "provides that registration of a variety will not be allowed in cases where prevention of commercial exploitation of such variety is necessary to protect public order or public morality or human, animal or plant life and health or to avoid serious prejudice to the environment. The Central Government can exclude any genus or species from the purview of protection in the public interest."[8]

THE FIGHT AGAINST BIOPIRACY: THE BASMATI PROTESTS

The attempted appropriation of basmati rice varieties by a US company also became a test case of the recognition and defence of farmers' rights. The basmati patent, granted in 1997 to the Texas-based trading company RiceTec, is in fact a denial of farmers' breeding that has gone into the embodiment of basmati characteristics in farmers' varieties. The patent drew outrage from the Indian farmers' organizations, especially in the states of Uttar Pradesh, Haryana and Punjab where basmati is grown.

On April 3, 1998, farmers and other organizations held a protest outside the US Embassy and submitted a letter of outrage to the US ambassador, which stated, in part:

> The moral and economic poverty of the US is being exhibited by the promotion and protection of piracy from the Third World. The WTO ruling against India[9] and the persistent US pressure on the Indian government to change its patent laws is an assault on Indian democracy and an encouragement to the biopirates. It is the US

Patent Law which is weak and needs change. US patent laws should recognize the traditional knowledge system of the biodiversity rich Third World countries in order to deny "novelty" and "non obviousness" criterion to the patent claims based on such knowledge. We want to say firmly to change your patent laws first and stop exerting pressure on the government of India for changing its patent law. India is a sovereign country and the people of India will never tolerate such pressure tactics. They reject the ruling given by the WTO against India. It is the US and not India which is criminal and protects the criminals....

We will never compromise on this great civilization which has been based on the culture of sharing the abundance of the world and will continue to maintain this trend of sharing our biodiversity and knowledge. But we will never allow your culture of impoverishment and greed to undermine our culture of abundance and sharing....

We have had enough of your lies, lies and damn lies. The people's protest today is a demonstration of the fact that the American and the WTO–TRIPs regime is a biased and an unequal patent regime, and is not suited to India's socio-economy, its national and public interests.

THE CHALLENGE FOR THE FUTURE

In the coming years, the movement for sustainable agriculture and farmers' survival in India faces a number of challenges:

1. Farmers must be aided in making a rapid conversion from practices that are pushing them to suicide to sustainable organic agriculture.
2. Farmers must be assured legal protection in National Plant Variety Legislation. If democratic processes fail to protect farmers' rights in seed legislation, the Seed Satyagraha will need to become a non-cooperation movement which encourages farmers to engage in civil disobedience against IPR laws.
3. Liability needs to be built into IPR laws so that corporate rights are matched by corporate responsibilities.
4. Producer–consumer partnerships and fair trade systems are needed, locally, nationally and globally, to respond to IPRs, genetic engineering and the globalization of agriculture.

In 1999, on the first anniversary of Monsanto Quit India, more than a thousand groups – including local village communities, women's organizations, environmental groups, and campaigns against corporate globalization – launched Freedom Week in India, from August 9, Quit India Day, to August 15, Independence Day. Freedom Week was aimed at redefining and reclaiming fundamental freedoms, including the right to food, the right to save seed, the right to protect biodiversity and indigenous knowledge, the right to livelihood and culturally appropriate and ecologically sustainable food systems. We issued calls to Monsanto and Cargill to quit India and to the WTO to quit India's food and biodiversity. The declaration we issued for that week stated:

> The freedoms we strive for are:
> Seed freedom – We are committed to keep our seeds free. We are committed to

conserving our diverse seed varieties. We reject genetically engineered seeds. We reject patents on seeds and patents on life.

Food freedom – We will defend our food rights and food freedom. We will defend our diversity of foods. We will defend our right to nutritious and safe food.

Biodiversity and Water freedom – We believe that biodiversity and water are community resources of the people and cannot be owned as private property.

Knowledge freedom – We will defend and rejuvenate our indigenous knowledge. We reject the patenting and piracy of our intellectual heritage. We demand the revocation of all biopiracy patents. We demand an amendment of western style patents.

Nature's freedom – We reject the imposition of plastic wastes from food packaging that leads to the destruction of our environment. We will fight to defend our local small-scale agro-processing systems so that both nature's freedom and people's livelihood are protected.

Future's freedom – We want to leave for our children a world of culture as biologically diverse as the one we inherited.

People's freedom – On this 15th August, our Independence Day, we declare our right to be free from hunger, disease, malnutrition, poverty, inequality and environmental hazards.

This campaign marked the beginning of a full-year campaign of non-cooperation with the forces that work against these freedoms. We continue to call for Cargill and Monsanto to Quit India, and are going beyond this to create a systemic response to stop the commercialization of genetically engineered crops in India with a broad mobilization of women, consumers, scientists and farmers' groups throughout the country.

NOTES

1. Vandana Shiva, *The Violence of the Green Revolution: Third World Agriculture, Ecology and Politics*, London: Zed Books, 1991.

2. National Workshop on "Globalization of Agriculture and the Survival of Small and Marginal Peasants," organized by the Research Foundation for Science, Technology and Ecology (RFSTE), New Delhi, on May 30, 1998.

3. Vandana Shiva, "The Seed and the Spinning Wheel: The Political Ecology of Technological Change," in *The Violence of the Green Revolution*; Vandana Shiva, "Diversity and Freedom," speech delivered on the presentation of the Right Livelihood Award, December 9, 1993.

4. "Cultivating Diversity: Biodiversity Conservation and the Politics of the Seed," New Delhi: RFSTE, 1993; "Sustaining Diversity: Renewing Diversity and Balance through Conservation," New Delhi: RFSTE, 1994; *The Seed Keepers*, New Delhi: RFSTE, 1995.

5. *The Seed Keepers*, p. 11.

6. Vandana Shiva, "WTO Rules against Democracy and Justice in the US–India TRIPs Dispute," Briefing Paper no. 1 on Trade and Environment Series prepared for the 2nd Ministerial Conference of WTO, Geneva, May 18–20, 1998.

7. Vandana Shiva and Afsar H. Jafri, "Seeds of Suicide: The Ecological and Human Costs of Globalisation of Agriculture," New Delhi: RFSTE, 1998; National Workshop on "Globalisation of Agriculture and the Survival of Small and Marginal Peasants," organized by RFSTE in New Delhi on May 30, 1998.

8. The quotation is from Article 29(1) of the draft being debated as of February 2000.

9. In response to a petition by the United States, declaring that Indian law did not offer sufficient protection for intellectual property, the WTO gave India until 2005 to change its patent laws to conform with TRIPs requirements. New mechanisms to begin receiving patent applications for seeds and other life forms were ordered to be set up immediately.

27

Europe: Hostile Lands for GMOs

Why do Europeans reject genetic engineering
so much more fiercely than Americans?
A European's personal view

THOMAS G. SCHWEIGER

June 25, 1999 was a historic day in Europe. The environment ministers of the fifteen European Union (EU) member countries met in Luxembourg to discuss how best to address the opposition of European citizens to the introduction of genetically modified organisms (GMOs) in their food and agriculture. When Europeans woke the next morning they found the outcome of the night-long debates of this "Environment Council" splashed over the front pages of almost every newspaper in Europe: "GMOs: Moratorium!"

Across the Atlantic, in the homeland of most of these apparently unwelcome GMOs, the declaration was noted, too: with alarm by the biotech industry and with concern by the Clinton/Gore administration, which had put all its support behind the biotech industry, and saw relations between the United States and Europe move a step closer to a trade war. Anti-genetech campaigners in the US once again expressed their envy at the situation in Europe and the successes of environmentalists there.

It was indeed a day of victory for all the environmental, animal welfare and consumer groups in Europe that had fought for years to stop the use of GMOs in Europe. Impressive campaigns had been waged in countries throughout the continent since 1996, when genetically engineered crops first made headlines. It all started in Austria – a country hailed as one of the most environmentally friendly in Europe – when an environmentalist discovered that genetically engineered potatoes had been "released" by the government without the necessary safety checks and approvals. Subsequently, 25 per cent of the electorate signed a public petition asking for a ban on GMOs in Austria and a prohibition on patenting life. In the years that followed, high-profile public debates made headlines in most European countries for months at a time.

The situation in Europe could not be worse for the biotech companies. Opinion polls repeatedly confirm that some 80 per cent of Europeans do not want GM food and crops; farmers don't use GM seeds; supermarkets and retailers try not to sell products containing GMOs because consumers don't want to buy them; more and more brand-name products use "GMO-free" as a quality label; several products are banned in various countries either by government degree or court rulings. European Union legislation on GMOs is being toughened rather than loosened: a de facto GMO moratorium has been in place since 1998, and a political moratorium, expected to last at least until 2002, is now in place as well. In August of 1999, the Deutsche Bank issued a report advising investors to stay away from ag-biotech companies, as the risk of loss is far too high. There are simply no markets for GMOs in Europe, and the industry is set to go down the same path as the nuclear industry. Indeed, there can hardly be a more hostile environment for GMOs.

This is a markedly different scenario to that which exists in the United States and Canada. Industry representatives keep saying that they cannot understand all the "fuss" in Europe, as consumers and farmers in the US have embraced the new technology and "novel foods" with enthusiasm, they claim. Furthermore, the fact that US consumers have been eating GMOs for years without any problems should convince the Europeans that the stuff is safe, shouldn't it? On the other hand, anti-GM campaigners in the US are more than impressed with what their colleagues in Europe seem to have achieved. In the US, they say, they have still a long way to go.

But why is it that Europeans are so much more critical and uniformly opposed to the introduction of genetic engineering in their food production than people in the US? Is it just technophobia, as some suggest? Is it true that the American citizen "happily embraced" the new technology, as the industry claims, or is it rather a lack of knowledge on the part of the consumers, as campaigners say? Is it just a European obsession, this fear of "unnatural" food? Are Americans just more advanced? Are NGOs in Europe just so much more effective, or is it more about cultural differences?

There is no simple answer to these questions. Genetic engineering touches all aspects of life, and people are concerned for widely varying reasons. Before examining some of the underlying sentiments in Europe, let us first review the events of the late 1990s in Europe and see what actually happened.

HOW IT ALL STARTED

It all started with Monsanto. This US company had been at the core of all the controversy in Europe since the beginning. Well known for being one of the most aggressive and uncompromising companies, it was Monsanto's handling of their introduction of Roundup-tolerant soybeans in 1996 that first alerted Europeans that something was going on. At the time, opposition to the GM soybeans was not widespread at all: only a few insiders knew what it was all about and even powerful environmental groups like Greenpeace began to work on this issue rather reluctantly and only on a very small scale.

When the news broke that this new and rather unknown sort of product was to be introduced to Europe, the consensus was that (quite apart from the questions of safety) what was needed was a clear labelling of all GM soy and products thereof. The consumer organizations, which were not outrightly opposed to GMOs, asked for this in order to let the consumer know what they were buying. Even politicians, eager for biotech products to be brought to Europe, warned Monsanto that labelling would be a sensible thing to do. Seeing that no mandatory labelling was in the offing, Monsanto threw all warnings to the wind, rejected pleas for labelling, and instead deliberately mixed its GM soy with traditional soy before exporting it to Europe.

It was this policy, regarded by many as an arrogant – and typically American – approach to consumers' wishes for transparency, that caused the first outrage. Europeans felt they were to be force-fed by a US multinational which deliberately disregarded European sentiments. The fact that the company actually mixed GM soy into non-GM soy was seen as a vicious move in order to destroy the possibility of choice for Europeans.

The media now picked up on the issue. What was it that Monsanto was trying to hide? If they refused to label their product, and would hide it among traditional soy, then something must be wrong. This was especially alarming to consumers, when they realized that soy was a component of between 60 and 80 per cent of the processed food they consumed daily; it was virtually everywhere.

First, demands for a boycott were raised. Oil mills, alarmed by the possibility that they might not be able to provide GM-free soy to their clients, started looking for sources of GM-free soy – which at the time proved difficult to find – or alternatives like rapeseed (canola) oil. Environmental NGOs even stepped in and helped broker some "GM-free" deals. The big "soy debate" was everywhere and the hitherto unknown name "Monsanto" became a household name.

It so happened that at exactly that time the European Union passed legislation on "novel foods," defining the criteria for risk assessment, European authorization and labelling. The recent row around Monsanto's soy suddenly created public interest in this regulation, which so far had gone through a lengthy process without much attention. Less than a handful of environmental experts were lobbying in Brussels on this issue. Now Europeans were to find out that their politicians were adopting a law which not only allowed GM foods but did not even foresee mandatory labelling and segregation of genetically engineered products like Monsanto's controversial GM soy. These were the first signs of public mistrust of the regulatory system for genetically engineered products.

Next came an even bigger political row. European citizens soon learned that a GM maize (corn) produced by the Swiss pharmaceutical and chemical giant Ciba–Geigy (also known for various environmental disasters, and now merged with Sandoz to form Novartis) had received no clearance from government authorities for commercial growing in Europe, despite doubts about its environmental and health safety. The matter was to be resolved at a meeting of European environment ministers in June of 1996. By this time,

environmentalists in most countries had started campaigning against the introduction of the GM maize and informed their governments about its various
hazards. When the meeting was held, fourteen of the fifteen countries objected to the authorization of the maize, and only France, the original appli-
cant and sponsoring country, still favoured its release.

But if Europeans had thought that this would mean the end of the story,
they were mistaken: under European Union law the mighty European Commission – often termed the "unelected European Government" – overruled
the ministerial Council and authorized the GM maize. Once again outrage
followed, which led to a stiffening of the opposition. Joining the opposition
were the governments of Austria and Luxembourg (and initially Italy), which
imposed national bans on the GM maize despite the Commission's ruling.
Such a measure is highly unusual in the "common market" of the European
Union and is only allowed as an emergency measure when public health and
safety or the environment is at risk. Obviously, this is what these governments
believed to be the case. It was an enormous boost for the critics and added
weight to their arguments.

By now most environmental NGOs had picked up on the topic and had
started to organize campaigns. In the forefront were Greenpeace, Friends of
the Earth and several national NGOs. Over the next three years the topic of
genetically modified crops and GM foods made headlines in most countries;
even in countries like France and the UK, whose governments were originally adamant supporters of GM food, public opinion completely turned
around. At more or less regular intervals new scientific findings made their
way into the public debate, showing how unsafe or simply unknown these
products were. Clumsy and continuing arrogant reactions by the industry and
some politicians led to further decreases in public trust. When asked who they
trusted most to tell them the truth about genetic engineering, polled citizens
listed environmental and consumer organizations foremost; politicians, scientists and industry enjoyed hardly any trust at all.

Consumers started to boycott GM products, retailers took them off their
shelves, and imports from the US and Canada of soy, rape and maize to
Europe fell dramatically because they were not guaranteed GM-free. Eventually the mounting political pressure led to the adoption of the political de
facto moratorium in June 1999.

WHAT ARE EUROPEANS ACTUALLY OPPOSED TO?

Monsanto has had to take the bulk of the criticism even within industry
circles. Their competitors accuse them of having destroyed the European
market for genetically engineered products for years to come through their
insensitive behaviour, and their handling of the soy crisis is now being termed
the biggest business blunder of the decade in biotech industry circles.

Everyone would agree that Monsanto's reaction and approach to the criticisms made things very much easier for anti-GM campaigners. They could

not have wished for a better "enemy." But is it fair to say that the situation in Europe evolved in the way it has just because of Monsanto? Would a different, a "better," handling of the introduction of GM food, maybe even by a European company, have really made a difference? Or are there other, deeper issues? What does the opposition to genetic engineering in Europe really rest upon?

It seems that the answer to this question lies much deeper than merely the issue of genetic modification. It is foremost a cultural matter. The use of this technology contrasts with so many core values of so many different people that it is necessary to examine these one by one, in order to assemble a picture of what it is that Europeans are so concerned about. One might compare this picture with a puzzle, comprising a number of different pieces that only together form an image.

Hypothesis 1: Food

Europeans don't like their food to be tampered with

Most European countries boast a highly developed traditional food culture and people take great pride in their national cuisine. The idea that someone wants to mess with their food in some totally unknown fashion – and with no benefit to the consumer – is simply unacceptable to most Europeans. If this "someone" is seen to come from the land that invented the unsophisticated "fast food" then it must be something "anti-cultural." It is no surprise that environmental groups managed to join up with leading chefs in Europe in their opposition to GM foods.

If the first GM products that reached the Europeans had not been foods, and indeed staple foods like maize, soy and rapeseed, with no added value for the consumer, opposition to genetic engineering might not have been as fierce as it turned out to be. But food is very close to everyone, and it is something that we all deal with on a daily basis. If every time we go to the supermarket or we eat a cookie we are reminded of the possibility that this product might contain unsafe material, but we cannot tell because it is not labeled properly, then this raises awareness dramatically and on a daily basis.

Hypothesis 2: Food and public health

Europeans have learned that too much "technology" and "industry" in your food is a serious threat to your health

As it happened, Europeans have had to deal with a number of food in recent years. The most prominent was the BSE ("mad cow") crisis in the UK, but hormone-treated meat and dioxin-tainted chicken had also alarmed consumers. In many cases, the handling of the crisis by the authorities was even more of a scandal than the original problem.

What Europeans have learned most from these scandals is that they cannot trust the authorities, nor scientists and politicians, when they declare something to be "safe." An ever more industrialized agriculture and food production is inherently unsafe – that is what most people on the street would now agree upon. Thus the very strong trend towards organic produce in Europe.

When Europeans learned about genetically engineered food they would have likely been very sceptical from the outset. But when the first signs of problems were highlighted in the media, and when industry scientists and politicians as a reaction just brushed concerns away and declared that everything was perfectly "safe," consumers knew what to expect. The fact that most GM crops contained antibiotic-resistance genes just proved the case. There was a clear link to concerns about public health and the problem of antibiotic resistance in medicine. It was incomprehensible that the EU started to ban the use of antibiotics in animal feed and at the same time antibiotic resistance genes were to be introduced into our food. Environmental campaigners found support from medical doctors for their arguments seeking precaution.

Hypothesis 3: Food and agriculture

Europeans know that their eating habits can have a positive impact on conservation

Europe is much more densely populated than the United States, and space is much less available than in North America. As a result, agricultural and natural areas are closely interlinked. While the US and Canada could afford to set aside huge areas of land as unspoiled nature reserves and national parks, in Europe it is impossible to find an area that has not had some agricultural influence. Hardly any national park in Europe consists of "virgin" nature. Thus, agricultural activities have a very direct effect on the natural ecosystems in Europe and people are very much aware of this. Many species have been lost in Europe because of the intensification of agricultural systems and the lack of refuge areas. Hardly any conservation group in Europe can afford not to deal with agricultural practices. People have learned to pay much more attention to what is happening in the agricultural sector. Pollution of the environment through agricultural practices – like the use of pesticides or too much fertilizer – have been commonplace.

And now the new fear of "gene-pollution"! "Jumping genes" made headlines when it was proved that genetically engineered traits could be passed on to neighbouring fields and even to wild relatives. It seemed clear that to introduce GM crops into the environment was an unsafe thing to do. Obviously this worried people enormously, especially as it was clear that such "genetic pollution" could never be cleaned up again. And, again, industry scientists had to admit that they had got it wrong: for years they had argued that genetic outcrossing was impossible. Now that it did happen, their only response was: well, yes, but it won't do any harm. Who would believe them?

Anti-genetic engineering campaigners soon found the support of conservationists, even semi-official ones like English Nature, who are worried about the possible negative effects of GM crops on the ecosystems around them.

Hypothesis 4: Organics are better

Europeans champion the trend towards organic agriculture. They cannot, therefore, also endorse the opposite concept: genetically engineered crops

Knowledge about the close relationship between agriculture and natural areas is also partly responsible for the boom in organic agriculture. People realized that organic agriculture contributes not only to personal health through safer, more "natural" food, but also to a better environment and a healthier "nature." Throughout the 1980s and 1990s, organic agriculture in Europe has made a staggering comeback and it is by far the fastest growing sector in the agricultural industry. Organic produce has long left the odour of "alternative" behind and has won over all the main supermarket chains. Demand in many countries has been higher than the supply, and this in turn has boosted the trend within farm communities to switch over. Organic is not only healthy, it is trendy, modern and makes a whole lot of economic sense.

There is no need to explain to anyone in Europe that organics and genetic engineering just don't go together. They are different concepts, different philosophies – "natural" versus "technical." When the biotech industry and some politicians tried to brand GM as compatible with organics, it just proved once more how little they really understood about the sentiments of people and consumers. They didn't stand a chance and only ridiculed themselves in the eyes of the public.

So when GM crops were to be introduced, they were immediately and automatically seen as a threat to organic agriculture. "Genetic pollution" of organic farms made headlines across Europe when a farmer in the UK lost his "organic" label because his harvest had been polluted by the pollen of neighbouring fields where GM crops were test-grown. The biotech industry once again showed its talent for PR blunders – and its true face – by insisting that the guidelines for organic agriculture would simply have to face reality and adapt to GM crops.

Organic farmers' organizations were in the forefront of the anti-GM campaigners. But even traditional farmers' organizations were worried. Swedish farmers, for example, made it clear from the beginning that they would not accept the environmental and economic risks of using GM crops when there was no conceivable benefit from them.

Hypothesis 5: Segregation and labelling

European consumers cherish the labels on their food. It is a long-standing right to be able to make an informed choice. The biotech industry was seen as trying to take away this right from them

One of the results of the organic trend was that strict segregation of products and clear labelling have become commonplace in European supermarkets. Consumers want to know what they buy and they know the label is there to tell them. Whether it's a free-range egg, beef from "happy cows," or bread that was backed with organically grown cereals, the labels, voluntary or state-mandated, would tell them every day what they wanted to know.

So for the European consumer the biotech industry's opposition to clear labels indicating genetic modification, was a clear breach of trust and confidence. If the producers don't want to adhere to total transparency, then they are defying a long-standing consumer right to know and make an informed choice. On top of all that, the industry is obviously not prepared to accept any liability for its products, which is seen also as a breach of relations between producers and consumers. The mighty European consumer organizations, although not outright opposed to GMOs, joined forces with anti-GM campaigners in calling for clear segregation and mandatory labelling of GMOs.

Hypothesis 6: Anti-Americanism

GMOs were perceived as an American invention and American companies were seen as trying to meddle with European values and cultures. That could not be allowed to happen

Europeans have a somewhat schizophrenic relationship to North America. On the one hand we happily welcome all the commercial culture that sweeps across the Big Lake, be it Hollywood or Coca-Cola. On the other, Europeans see themselves as much more sophisticated, better educated and certainly much more culturally aware and "politically correct" than the average US citizen. US politics are often viewed very critically around the world and seen as no less than bullying.

It somehow fitted into the view of the Europeans that those most aggressively trying to force GMOs onto the European consumers were US companies. The whole technology is perceived as being an American invention, designed to gain even more of a market monopoly around the world. This sentiment is also constantly fed by calls from European politicians that Europe's biotech industry needs to catch up with the United States. Monsanto's position in the centre of most of the controversies seemed to confirm this. European companies kept a much lower profile – especially after Ciba–Geigy's dramatic disaster with its maize had made headlines.

Within the GM debates there is also a strong sentiment of self-determination in the face of the US. It is felt that the United States is trying to force a US technology sold by US companies down the throats of the Europeans and that our right to say "No" is being compromised. The attitudes of both US officials and company representatives nourish this feeling: the European markets must be opened to US products, whether Europeans want them or not. US agriculture secretary Dan Glickman was quoted in the European media complaining that consumer interests get in the way of US trade interests. But in Europe people believe that consumer demand in the first place is the basis of all trade.

The hormone-beef case before the WTO woke many a European up to the harsh realities of world trade: free-trade philosophy overrides all other concerns, be they health or environmental matters. But Europeans think it should not be possible to force a product on to European markets if it is considered unsafe or unwanted by Europeans. The fact that a WTO ruling gave this right to the United States only proved that the free-trade rules need an overhaul. GMOs were seen as the next test case.

Not unnoticed in Europe were the calls from the South, also protesting against this newest form of American imperialism. Most Europeans have a feeling of guilt somewhere in the back of their minds when it comes to Europe's former colonies; that is not true of the US. So the fights in India and Brazil against Monsanto generated sympathy in Europe and strengthened our own determination. Just as Europeans feel that organics and GMOs don't go together, so they are convinced that "fair trade" (another darling in Europe) and genetic engineering also contradict each other.

Hypothesis 7: The myth of "sound science and technology"

Europeans have a critical mind and do not easily trust "sound science" and technological advancement. They prefer to be on the safe side. Sound science and technology are not "gods" in Europe

Therein lies the ground for another major cultural difference between the US citizen and the European. While the American often seems to consider techno-logical advance as the sole guarantee for future happiness, Europeans take a more balanced look at developments, realizing that technology can also do a lot of harm. Not everything that is scientifically feasible is necessarily a sensible thing to do. Our experience with nuclear power proved that sentiment only too right.

Europeans also lack the generous trust in their authorities that the US citizen is reported to enjoy. If the Food and Drug Administration in the US has said that something is alright then it must be so, and it appears that hardly anyone dares to doubt it – with only a few notable exceptions. Europeans are much more critical towards whatever their authorities say. A green light from a government agency does not mean that everyone trusts their opinion. And Europeans have seen their doubts being confirmed over and over again these past years: the UK government's handling of the BSE crisis, by brushing aside for years any warning or doubt, is just one example of many.

Europeans have long learned that what is considered "sound science" today can easily become a hazard tomorrow. The idea of the "precautionary principle" is much more established in the philosophy of the European mind. On this very basis, the French Conseille d'État, the highest administrative court in France, ruled in 1998 to suspend a government approval for growing GE maize in France.

Hypothesis 8: It is just wrong

Europeans instinctively feel that GM food is dangerous and provides no benefits. So it's just wrong

Whilst Europeans don't pretend they can judge the scientific debates raging around the safety of these products, they do see that GM provides no benefit to anyone except the industry, but it puts the risks on the shoulders of the farmers, consumers and the environment. So the whole thing just "feels wrong" – and if something feels wrong, then Europeans are prepared to fight it.

THE ROLE OF NGOs

Another difference that is often highlighted in these debates is that NGOs in Europe have picked up the issue, but their counterparts in North America have only begun to do so. The reason may lie in the very differences in attitudes and cultural perspectives discussed above. NGOs are always also a mirror for the society they are based in. As mentioned above, every conservation organization in Europe has dealt with agricultural practices. Consequently they were quickly aware of what was going on in that sector. In the US the main groups might not have had that sensitivity.

Environmental campaigners in Europe just felt that the time was right in the mid-1990s to start working on the issues of genetic engineering – and this intuition proved right, as the enormous public response shows. I remember very well the situation one day in early 1995, when we were repainting the office of the Austrian environmental group Global 2000 in Vienna. Global 2000 had no record then of being involved in genetic engineering and there had been no public debate in Austria at the time; but when the people in the room chatted while painting the walls, we realized that we all had a hunch that a genetic engineering campaign would be appropriate just then – the time felt ripe. When we got up and running a few months later, the public response was overwhelming and it fast became what would prove to be the largest campaign for Global 2000 for years to come.

Back in the early 1980s, when genetic engineering was just a purely academic subject and happened only in laboratories, the first public debates were organized by a few determined people and specialized NGOs formed. Germany was in the forefront at that time; the then newly founded Gen-Ethisches Network is still active today. As the name suggests, the emphasis was always on the ethical problems associated with genetic engineering, but then most related aspects could be subsumed under that title. In the UK, Genetic Concern has a long history, but in the early years it led a mostly academic life.

When the first field tests took place in Germany in the late 1980s actions were staged to uproot the plants. But that was still a very early stage in the debate, and commercial introduction – and thus consumers' direct contact with GMOs – was far away. The topic in Germany more or less faded away for over a decade, as other issues gripped the headlines. Only now are Germans waking up to what has happened in the meantime. Countries like Austria, France, Luxembourg, Italy, the UK, Denmark and even Greece have woken up during the past four or five years and the public debate has led the respective governments to take protective action in banning certain GMOs on their territory.

The NGO scene in Europe is very diverse and differs widely from country to country. Cross-border contacts have often been limited. The same has been true of close contacts between the different NGO sectors: environmentalists, conservationists, development and animal welfare NGOs have often worked alongside each other, but with each jealously guarding its own agenda, with hardly any contact, much less cooperation, between them.

Genetic engineering has changed that. The different movements in Europe have found a unifying topic like no other. It is probably safe to say that never before have so many NGOs from different fields cooperated so closely. This, of course, is due to the fact that genetic engineering touches virtually all areas of life. Environmentalists, conservation groups, animal welfare campaigners, development and Third World activists, organic and mainstream farmers and consumer organizations formed the core of the anti-GM campaigns throughout Europe. Each approached the topic from its own point of view, but all shared the common aim of stopping the technology from infiltrating the food and agricultural sectors.

Coordination and cooperation have worked extremely well since the mid-1990s and now serve as a model for other pan-NGO cooperations. Close contact and cooperation now reach beyond that core grouping and also involve medical and patient organizations, public health lobbies, parts of the scientific community, churches and ethical institutions and many more. A prime example was the setting up of a common lobby office in Brussels to combat the European "Life Patents Directive" during 1998. On a daily basis a Europe-wide network termed "GENET" exchanges information and strategies, mostly via electronic communications amongst a number of NGOs and individuals, and keeps everybody up-to-date on all developments.

Another fairly new development is that national NGOs have started to focus increasingly on what is happening on the European level. Campaigning and lobbying have been effectively coordinated to reach the European Commission in Brussels, the parliament in Strasbourg, as well as the national governments. Campaigning has used all necessary and possible means: direct actions as well as highly sophisticated lobbying; media and PR campaigns as well as legal actions at the high courts; direct informative contacts with organizations' members and individual consumers, as well as high-profile scientific debates. Environmental organizations have found themselves representing consumer interests; consumer groups have started to talk about public health issues; doctors have complained about agricultural problems; farmers show increased concern for the environment and animal welfare, and so on.

It is certainly true that the issue of genetic engineering has been one of the hottest environmental issues in Europe for the past several years. NGOs and public pressure have succeeded in pushing the issue ever higher up the political agenda. Even heads of state have to deal with the issue now, as was shown to the world during the G8 summit in Cologne in early 1999. There, some of the most powerful political leaders across the globe identified GMOs as one of the most urgent threats in the new millennium.

At the same time, environmentalists have activated consumer power in a way that has rarely been seen before. Consumers can see the fruits of their actions in the supermarkets, and it gives them a sense that they can be more powerful than the mighty multinational companies, if they only use their purchasing capacity in the right way.

HOPE FOR NORTH AMERICA?

There are increasing signs that the very positive climate for biotech is changing in North America as well. More activists have started to focus on this issue, and it seems that there is growing interest in a controversy that is raging across the Atlantic in the Old World, a controversy that could steer the US and EU towards another trade war. As the media and the people take a closer look at the problem, they come to see more and more of the controversial issues, leading them to take a second look at their own attitudes and practices.

Of course there is the commercial aspect also. After a few years of planting GM crops, US farmers realise that the promised benefits (increased yields, etc.) are not really materializing, that they are becoming engaged in patent disputes with the seed sellers, and on top of it all they can't export their harvest to Europe. The isolation of the US in international negotiations towards the Biosafety Protocol, intended to regulate trade in GMOs, made negative press even in the US. The efforts in India, Brazil, Japan and Africa to remain "GE-free" are noted with concern both in commercial and in political circles.

An industry representative has recently admitted to a journalist in private that they sense the climate is changing in America, too. In a few years, he said, we'll have the same troubles over there as we have here in Europe. So, there is hope for the Americans, that for once a trend will move from the Old to the New World, from East to West.

28

No Patents on Life:
The Incredible Ten-year Campaign
against the European Patent Directive

STEVE EMMOTT

> Because biotechnology involves reproduction as well as production, the issue of
> control for purposes of profit is central.
>
> Brewster Kneen, *Drawing the Line: The Ethics of Biotechnology*[1]

Until 1980 it was generally assumed that living materials, whether plants,
animals or microorganisms, were, by their very nature, not capable of being
invented and therefore not patentable. Indeed, there exists a wholly separate
body of intellectual property law to provide a different form of protection to
plant breeders. However, in that year the US Supreme Court decided in the
Chakrabarty case that a genetically engineered oil-eating bacterium could be
patented (see Chapter 23 in this volume). The biotechnology patent race was
under way.

Because the rules up to that point had been based on inanimate inventions,
the language did not really exist to translate patent law into this new area,
where the transformation of genetic material involves the creation of new or
modified life forms. For Europe, the obvious route would have been to
amend the 1973 European Patent Convention (EPC), which was, after all,
the only existing common body of patent legislation.[2] However, it can take
ten to fifteen years to renegotiate international treaties and the biotech industry
decided to go for an alternative that it thought would be a quick and un-
controversial fix.

Industry persuaded the European Commission, the executive arm of the
European Union, with the sole right to initiate legislation, to introduce a
directive on the "legal protection of biotechnological inventions."[3] Without
bothering to explain how this would fit in with the EPC rules, the Commission proposed to make living material explicitly patentable. Their main
justification for this was the need to "catch up" with the US and provide the
same degree of intellectual property protection for European biotech companies

as their American competitors enjoy. "If we don't get this Directive, we will relocate to the States" was a frequently heard threat.

DOES ANYBODY OBJECT?

It did not seem to have occurred to the legislators at that stage that any of this would be controversial. This was, they said, merely a tidying-up exercise designed to smooth out any anomalies in a dry and dusty corner of the law, and to help bring about common standards of interpretation in cases which were unclear. I remember trying to voice fundamental objections to granting "patents on life" at a meeting in London with the UK Patent Office, only to be told that these were interesting points but had nothing to do with them or with patent law.

In June 1988, seventy people representing fifty NGOs and development agencies from twelve European countries met in Humlebaek, Denmark, for the first meeting of the Seeds Action Network in Europe. A workshop was held on patenting life and the participants resolved to launch coordinated efforts to defeat the European directive.

The EU patenting directive was officially published on October, 12 1988 by DG-III, the European Commission's services for industry. Genetic Resources Action International (GRAIN, at that time called the ICDA Seeds Campaign) and the elected representatives of European Green parties in the European Parliament (EP) agreed to co-organize a public hearing. Some two hundred people came and debated all the points of view – something the Commission had not bothered to do when drawing up its proposal. Speakers included NGOs, biotech industry representatives, seed companies, animal welfare activists, legal experts, Third World representatives, farmers' organizations, and... the drafters of the directive themselves.

At the close of the meeting, NGOs met for a strategy session and formalized their agreement to mount strong public action against the directive. They pledged to "start a broad campaign against the patenting of life at the local, national and international levels."[4] A subsequent critique by GRAIN, the Green Group in the European Parliament, the UK Genetics Forum, Compassion in World Farming and others gave birth to a coalition of NGOs and committed individuals forming the "No Patents on Life" campaign in Europe, with links to other NGOs across the globe.

The draft patent directive met a storm of protest from an amazing array of organizations – not all of them by any means committed opponents of gene technology. Farmers could foresee that their seed supplies would become patented; patient support groups could imagine that new medical research would be subject to limits imposed by commercial viability; church leaders and ethicists objected to the concept that life-forms could be privately owned; development charities feared for the impact of the piracy of the Third World's genetic resources; animal welfare groups could count the increase in animal experiments, and so on.

The next several years were exhausting, but NGOs stuck firm to their commitment. Efforts to publicize the issue of life-patents, and the EC directive in particular, mushroomed. This included countless conferences and seminars, publications, and lots of educational materials. By the turn of the 1990s, big environmental groups like Greenpeace and Friends of the Earth started getting involved, and the scope of people trying to stop the directive broadened.

THE EUROPEAN PARLIAMENT MOVES

What was happening in the Parliament all this time? Both a lot of scuttle and a lot of dead air. Thanks mainly to the incessant activities of the Green Group in the Parliament, the directive was subjected to hot committee debates, three readings in the plenary and numerous amendments. Throughout the process, the Parliament and Commission constantly clashed. From the very beginning, the Parliament had hesitations about the directive. It presented over forty amendments and took a strong stand on a number of issues. These included the question of a farmer's exemption (the right of farmers to reuse freely and exchange patented seed, which the Parliament wanted to see guaranteed) and efforts to stop the patenting of human materials and of medical treatments. After the signing of the UN Convention on Biological Diversity in 1992, the Parliament sent the directive back to the Commission and requested an assessment of whether and to what extent this directive would support and not run counter to the intellectual property provisions of the Biodiversity Convention. A legal fudge resulted in a ruling that there was no conflict.

The institutional ballgame changed, however, when the Maastricht Treaty on further integration of the European Union was signed late in 1993. Among other things, this gave the European Parliament more legislative power, requiring the Parliament and Commission to enter into direct negotiations to settle legislative disputes. This conciliation procedure went on for several months before closing in January 1995 with the adoption of an "agreed" text. However, rather than a balanced compromise, the hardliners won and the patent pushers got virtually everything they wanted. It allowed for patenting of human genes and germline therapy, and there was no exemption for farmers reproducing patented livestock, despite the Parliament's insistence.

THE FINAL COUNTDOWN?

On February 16, 1995, NGOs were alerted that a final vote on the compromise text would take place in Brussels on March 1. This was it, after seven years. We knew that all the odds were stacked against us and in favour of the directive's adoption. A compromise had formally been reached, the Parliament seemed ready to let the whole thing go, and the Council of Ministers was planning to do no more than rubber-stamp the directive into existence a week later. What happened in the final two weeks is now history.

NGOs reared into action to inform the deputies about the profound social opposition to the Directive. Members of the European Parliament (MEPs) were bombarded with letters and phone calls from all angles. GRAIN circulated an open letter that was signed, in one week, by over a hundred agencies representing well over 3 million people worldwide. The letter was delivered to all MEPs the day before the vote.

In the meantime, several developments outside Parliament were to have a decisive impact. On February 21, 1995, the European Patent Office announced a landmark decision in favour of Greenpeace International regarding a patent on genetically engineered herbicide-resistant rape (canola).[5] The original patent, granted in 1990 to Plant Genetic Systems (Belgium) and Biogen (USA), was cut down by the deletion of six claims. The EPO's Technical Board of Appeal ruled that although the patent could cover genetically engineered plant cells, it could not extend to a whole plant, its seeds or any future generations of plants grown from the cells. The EPO finally seemed to prohibit patents on plants and animals per se.

The Austrian Parliament also moved to tie down their government's position against the Directive. The fact that yet another European Union government was going to slate a "No" to the Directive in the Council of Ministers had an important impact on many Europarliamentarians. In Italy the mood was turning against this legislation, and members of the Italian Parliament signed a resolution instructing their government to "promote urgent initiatives toward the European Union and other international organizations in order that the principle of patenting life forms – self-reproducing animals and plants – be refused."[6] The resolution was signed by representatives of every political group in the Italian Parliament except the Fascists – a daunting feat in the Italian political culture. This aligned the Italian parliamentarians with countries such as Austria, Denmark, Luxembourg and Spain, which already had expressed formal reservations about the European Directive.

NGO lobby efforts built up as March 1 came closer. The day before the vote, Greenpeace hung a huge banner – and several people – from a passarelle linking two Parliament buildings over a major thoroughfare in Brussels proclaiming "European Parliament: Reject Patents on Life!" The audacious action stirred up the Parliament, the media and everyone who gathered to watch. The morning of the vote, the *Financial Times* ran a news item "MEPs Urged to Vote against 'Patents on Life'," reporting on the letter GRAIN had circulated and the events at the EPO and Italian Parliament. Every MEP had a copy by 8:00 that morning. Then the real countdown started. Nervous NGO representatives wearing T-shirts that read "Vote No! to Patents on Life" mingled with MEPs as the hours passed and final speeches about the directive filled the air. Then came the vote itself. Shortly after the noon bell, the president read out the verdict, and at first no one in the room believed her. "Four hundred and fifty-one votes expressed: 188 in favour, 240 against, 23 abstentions." After a moment of stupor, the Greens jumped up and cheered, lots of people were clapping from other political groups and NGOs exploded with disbelief and joy. It was over. The directive was pronounced dead.

What happened? Everything was stacked in favour of the directive. An analysis of the voting shows that the majority of the Socialist Group had abandoned the Directive's Rapporteur (the MEP designated to report on the legislation), who was himself a Socialist, and voted massively against the directive. But there were also "Nos" from Conservatives and Christian Democrats. Many parliamentarians realized at the last moment that this directive was too biased towards industrial interests and did not take into account the growing social and ethical concern over this piece of legislation. It was, proclaimed GRAIN, "a vote of Conscience over Capital."

In the days and weeks following the vote, there were numerous efforts to put the EP's ruling in a bad light. Industry sources and several newspapers and weeklies blamed the Parliament for its "emotional" and "irresponsible" decision. None of their arguments made much sense. The typical threat that the biotech industry would massively move to the US and Japan did not stand up for obvious reasons. The rules of the directive would clearly have applied to any company, anywhere in the world, wanting to obtain a patent in Europe. Shifting one's physical location doesn't solve that problem. If adopted, the directive would have increased legal uncertainty, because it tried to undermine existing EPC restrictions by squeezing in human materials, medical treatments, and plant and animals. Even some in the pharmaceutical industries called the rejection of the directive "the lesser of two evils," worried that the final compromise text provided too many ambiguities and too little clarity.

This was the first time the European Parliament had used its newly acquired post-Maastricht power to block a bad piece of legislation. Within the European Commission and some member state governments, voices were heard to say that this power should be taken away, returning the Parliament to the largely decorative, consultative and powerless function it held before. That would have been moving back in history and fortunately nothing came of it.

We won a battle, but not necessarily the war. Patenting efforts at the national level and at the European Patent Office continued. The importance of the decision of the European Parliament should have been in its message: patenting in the field of life sciences bears major social, political and ethical implications, and both legislators and the courts should proceed, if at all, with great caution. The decision of the European Parliament also sent a strong signal beyond European borders, especially in discussions on patenting in the Third World. When a bloc of "advanced" countries tells the rest of the world "be careful with patents on life and go slow," this impacts directly on decision-makers in countries like India and Brazil.

THE DIRECTIVE RISES AGAIN

Nine months after the vote, the Commission proposed a "new" life-patents directive. Like the earlier one, it was presented simply as a harmonization measure – a technical tidying up to bring patent law in all member states of the EU into line.

George Poste, then chair of research and development at the pharmaceutical giant SmithKline Beecham tried to reassure MEPs in a letter that "this Directive does not extend patent law … it simply translates the European Patent Convention and current legal practice into European Union law. All that industry is seeking is to continue what has been possible since the first gene patents were granted 20 years ago."[7] His position is typical of industry representatives, who hoped MEPs would not notice the major changes the directive really implied. In reality, the directive sought yet again to change the terms of reference for patenting, and significantly broaden the scope of products and processes that are deemed patentable in Europe.

This time around, the biotech industry had learned that it could not take success for granted and mounted a major lobbying offensive. It was led by the pharmaceutical companies, as the industry believed that the potential benefits of medical research were the most compelling arguments for gaining public acceptance of biotechnology. The new directive, like the old one, would give industry patent control of the whole "supply chain" from the basic genetic material, through the processes that make use of genes and gene sequences, to the products that result, to successive generations that carry the genes and into other products that incorporate or express the same genetic information. No wonder Poste of SmithKline believes that "Genes are the currency of the future."

The new directive resembled the old one in many ways – different language, same intentions. Although there was some tightening up, such as in the arena of germline therapies, the new text sought to expand what it deemed patentable. In a number of instances it sought to reverse EPO case law, particularly the EPO's ruling that plants and animals are not patentable. It was clear that the fight to reject the new directive would be much tougher than the previous one, and the Greens and the NGOs agreed that they would seek to amend it by incorporating wholesale exclusions:

Human genes and germline therapy

Probably the most compelling issue that persuaded parliamentarians to vote against the first directive was the question of patentability of human genetic material and germline gene therapies. MEPs were very clear that human genes and gene sequences should be placed outside the scope of patentability. They had three basic grounds for such objections. First, on philosophical, moral or religious grounds, humans and their genetic makeup should not become commercial commodities. Second, human genetic material already exists and is there to be discovered; it cannot therefore be invented. This is a long-established principle of patent law. Third, granting a monopoly patent on a gene to one company or institution effectively stops others from using that knowledge freely to develop other beneficial products such as medicines or treatments.

Although the first draft directive ruled that parts of the human body should be unpatentable *in situ*, the new text was much more explicit, stating specifically that useful materials isolated from the human body would be patentable, even if identical to naturally occurring materials. With these words, the directive deliberately blurred the boundaries between discovery and invention. The technical processes of isolating a gene and replicating it in the laboratory may well be patentable inventions, but to proclaim that this confers a patent right over the gene itself is a breathtaking leap of illogic.

In 1994 the French changed their patent law to prevent this very situation from arising. Its new law said that on grounds of public policy and morality "the human body, its elements and products and knowledge relating to the overall structure of a human gene or element thereof may not, as such, form the subject matter of patents."[8] The Commission cited this as an example of the legal uncertainty which arises from the lack of a directive; in practice it sought to override the French law. The EPO began to accept such patent applications in an ad hoc way, including patents for human stem cells derived from umbilical cord blood,[9] and for the hormone Relaxin, used to assist in childbirth.[10] The announcement from the Roslin Institute in Scotland of the successful asexual cloning of a sheep – and the probability that human cloning could be achieved using the same techniques – soon spurred a re-examination of the whole question of patents involving human genetic material.

Germline therapy was another key area which parliamentarians felt had not been dealt with satisfactorily the last time. In this instance, the Commission had made the ban more comprehensive, and the new text stated clearly that "methods of human treatment involving germ line gene therapy" were to be unpatentable.

Plants and animals

Although the intellectual property section of the World Trade Organization (WTO-TRIPs) does not oblige member nations to extend their patent regime to plants and animals, the Commission continued to insist that the EU should do so. Article 4.2 read "Biological material, including plants and animals, as well as elements of plants and animals obtained by means of a process not essentially biological, except plant and animal varieties as such, shall be patentable." The term "not essentially biological," in patent law, refers to microbiological processes which are *prima facie* patentable under the European Patent Convention (EPC). Here again the Commission extended the demarcation lines, stating that products as well as microbiological processes shall be patentable.

Farmer's privilege

The new draft directive also reflected the acceptance by the commission that there should be an exemption for farmers who breed to replenish their livestock, as well as a plant propagation privilege, which was outlawed in the earlier text. It still prohibited off-farm exchange or the cooperative sharing of seed or livestock without attracting royalty payments. This exemption would encourage the growth of strict contractual arrangements imposed by the seller

on the farmer along the lines of those used by Monsanto in the US for its
Roundup Ready soybeans. The Monsanto contract imposes a technology fee,
and requires an agreement from the farmer not to use farm-saved seed, to use
only Monsanto's own branded herbicide, and to pay predetermined legal
damages for any breach of the contract terms. Such contracts completely
bypass the attempt to write a farmer's privilege into patent law.

Animal welfare

One of the concessions obtained by the EP in the first round was a ban on
patenting processes for modifying animals in ways which would cause them
suffering or handicap. It was, however, subject to an overriding "cost–benefit"
test which opened the door to such patents if there was "substantial benefit
to man or animal" and/or the suffering was disproportionate to the objectives
being pursued. The commission retained this text more or less unaltered.

Biopiracy

One of the most serious, but least understood or debated, issues of concern
related to the directive is the question of biopiracy: the appropriation of
peoples' knowledge and genetic resources by corporations and the consequent
privatization of both through the application of patents. Most of the world's
existing wisdom in the fields of agriculture and medicine has come about
through the free exchange of knowledge and resources between communities,
individuals and countries. The biotech industry would have us believe that
strong intellectual property rights regimes are essential for innovation to thrive,
yet reality belies the theory.

Apart from its small concession of the farmer's privilege, the new draft
directive did not address the implications that patenting plants and animals has
for farmers and communities around the world. It originally contained no
reference to the principles set down in the Convention on Biological Diversity which provide for the equitable sharing of benefits arising from the
knowledge of local communities, their innovations and practices. It did not
address the issue of prior informed consent for the use of genetic resources
collected from their territories. Nor did it consider the implications for farming systems and agricultural innovation, which will be stifled by the privatization of agricultural resources.

Patenting their own resources in order to protect their livelihoods, even if
they were interested in doing so, is far beyond the means of most farmers and
innovators in developing countries. At present, many people give away their
seeds, their plants and their knowledge without realizing that these may become the proprietary rights of some industrial company. Even if they are
aware of this situation and could afford to get caught up in the patenting
game, for many indigenous communities privatizing genetic resources is simply
not an acceptable path to follow. It is often against their moral, spiritual and
ethical beliefs.

THE VOTE

The campaign came to a head on July 16, 1997 when the European Parliament passed the first reading of the Commission's latest proposals. Despite a flurry of last-minute lobbying by NGOs the vote resulted in a two-thirds majority in favour of the directive, including 66 amendments out of over 250 put forward.

Willi De Clercq of Belgium, the Liberal chair of the Legal Affairs Committee, highlighted the heated controversy caused by the proposal, saying it was "the largest lobby campaign in the history of the EU."[11] Industry had clearly won votes on two counts. First, they struck at the hearts of Members of the European Parliament (MEPs) with the slogan "No patent, No cure," borne on the yellow T-shirts of an army of disabled people on wheelchairs, bussed in for the occasion. Second, the routine economic rhetoric of jobs, growth and competitiveness featured prominently during the pre-vote debate. Those opposed to the directive were disappointed that most MEPs were taken in by these hollow arguments, failing to grasp the seriousness of granting monopoly control of the world's genetic heritage to powerful transnational interests. The extent of the losses incurred by those opposing the directive can best be exemplified by a modest amendment offered by the Green Group to prohibit development and patenting of genetic weapons. Even this was defeated by a substantial margin.

The Commission now needed to get the approval of a majority of the fifteen EU member states for the revised text. Despite heavy counter-lobbying, it succeeded in doing so in November 1997. Only the Netherlands voted against, with Belgium and Italy abstaining. The Belgian minister was set to vote against the directive, but had his arm severely twisted after an intervention from Belgium's only significant biotech company, Plant Genetic Systems.

The directive then came back to the EP for a final vote. For the NGOs, any hope of defeating the legislation depended on getting further amendments passed – this time with a much stiffer voting threshold. Only then would there be another conciliation process and a chance for history to repeat itself by last-minute rejection. The Commission and the biotech industry lobby were desperately anxious to avoid such a scenario and Willi Rothley, the German Socialist who was again appointed as the Parliamentary Rapporteur, travelled through Europe persuading his Socialist colleagues to back him and to reject all further amendments. Industry lobbying was also aimed at the other major power bloc in the Parliament, the Christian Democrats.

The final vote was in Strasbourg on May 12, 1998. A procedural motion by the Greens to reject the whole document was refused by 432 votes to 78 and Green MEPs then held up proceedings for about twenty minutes by unfurling a large banner demanding "No to Biopiracy," and waving flags showing a skull-and-crossbones motif. Every amendment, including the one to reinstate the previously adopted anti-biopiracy provisions, was defeated by more than two to one and the directive was finally approved.

DID WE ACHIEVE ANYTHING?

There are several positive provisions in the Articles of the Patent Directive as adopted which did not appear in the original version published by the Commission in January 1988. This might not look very much and should not be taken as approval of the outcome, but it is the end product of nearly ten years' work by a lot of people.

1. The directive is without prejudice to member states' obligations under the Convention on Biological Diversity.
2. The human body is not patentable.
3. The industrial application of a gene sequence must be disclosed in a patent application.
4. Inventions contrary to public order or morality are unpatentable, including expressly: processes for human cloning; processes for modifying the germline of humans; uses of human embryos for industrial or commercial purposes; and procedures likely to cause suffering to animals without substantial medical benefit.
5. An ethical group will evaluate all ethical aspects of biotechnology.
6. There is a limited farmer's privilege to reuse the product of his or her harvest on the farm, as well as to breed patented livestock for agricultural activities not including commercial breeding.
7. Implementation of the directive was delayed almost ten years from the date of December 1990 originally proposed by the Commission.

THE FINAL CHAPTER?

This is not quite the end of the story. Motivated by continuing anger at the political and economic manipulations that had finally pushed the directive through, and by a sense of having been cheated because our best arguments had been brushed aside, activists explored legal challenges to the validity of the legislation. Although there is no practical way for NGOs to object to EU legislation, any member state can do so. The Netherlands was the obvious candidate, having consistently opposed the directive because its national laws prohibit patents on plants and animals.

The Dutch animal welfare organization, Dierenbescherming, obtained a legal opinion from a renowned German constitutional law professor and finally persuaded the Dutch government, after much heart-searching, to file a case against the European Parliament and the Council of Ministers. The papers were lodged at the European Court of Justice on October 19, 1998, one day before the deadline for opposition expired; the case threw into doubt, once again, the validity of the directive. The Dutch complaint is still awaiting a ruling, and Italy and Norway have since filed objections also.

But the European Patent Office jumped the gun. On September 1, 1999, the EPO's Administrative Council issued an amendment to the agency's implementing regulations, effectively incorporating the legal protection of biotechnological inventions word for word from the directive. The amendment was voted in during mid-June when nobody was looking and got a 75 per cent majority vote from all the signatory states present. This left few remaining options for challenge by dissatisfied member states. Germany has legally changed its vote to a 'no' and others could still do so. However, the fact that EPO rules are now aligned with the directive removes a key procedural argument – that of conflict between the directive and the original treaty establishing the EPC.

Meanwhile, the arguments long put forward by the NGOs resurfaced in some unlikely areas. US president Bill Clinton and UK prime minister Tony Blair reportedly sought to intervene to make human genes unpatentable,[12] and the Parliamentary Assembly of the Council of Europe (a non-EU body representing forty European countries, with a mandate to monitor bio-ethics) passed a resolution in September 1999 saying "The Assembly therefore believes that neither plant-, animal- nor human-derived genes, cells, tissues or organs can be considered as inventions, nor be subject to monopolies granted by patents."[13]

AND THE FUTURE? SOME PREDICTIONS

The issues surrounding life-patents will simply not go away just because the biotech industry and the law-makers wish them to. The intellectual property fight will continue in fora such as the UN Convention on Biological Diversity and the WTO TRIPs Council. Acts of biopiracy will continue to anger the traditional custodians of indigenous knowledge and biological resources. Local peoples will develop their own community rights which will bring them into conflict with the Western view of "property." Case law will uncover more of the absurdities of trying to own the essence of something that can reproduce itself. Medical researchers will prove that other people's monopoly patents are stifling their search for new cures and medicines. The coming explosion in human genetic manipulation will put the ethical issues of patenting and control firmly back in the headlines.

Will common sense win in the end? As the croupiers say in roulette, "Mesdames, Messieurs, faites vos jeux."[14]

NOTES

1. Brewster Kneen, Ivan Illich and the EECCS Bioethics Working Group, *Drawing the Line: The Ethics of Biotechnology*, Occasional Paper no. 5, Brussels: Ecumenical Association for Church and Society, 1997.

2. The European Patent Convention (EPC) is administered by the European Patent Office (EPO) in Munich, Germany. It is a stand-alone international treaty which predates the development of the European Union's single market and is not legally part of the EU institutional framework. Patents are available within European territory for all comers on equal terms, whether their home country is England, the US, Japan or elsewhere.

3. The three main European Union institutions are the Commission, the Council and the Parliament. The Commission is the executive arm of the EU but it also has the sole power to initiate legislation. The Council of Ministers represents the interests of the fifteen Member States and is the primary legislative authority. The Parliament is the junior partner but in a growing number of areas, including patent law, it has legislative co-decisionmaking powers alongside the Council, requiring two "readings," and the possibility of a third after negotiations to resolve any outstanding disputes.

4. Reported in "The Directive is Dead," *Seedling*, vol. 12, no. 1, Barcelona: GRAIN, March 1995.

5. EPO Case T356/93.

6. Reported in *Seedling*, March 1995.

7. Quoted from lobbying material sent to MEPs.

8. French Law 94–653 of July 1994.

9. European Patent 0 343 217 granted to the US Biocyte Corporation. A Green MEP and various NGOs filed a legal opposition and the patent has now been revoked, subject to appeal.

10. European Patent 0 112 149. The Greens have appealed against this case; ref. EPO T272/95.

11. De Clerq, speaking in parliament after the vote.

12. BBC News Report "Plan to Block Patenting of Human Genes," September 20, 1999

13. Paragraph 12 of text adopted by the Assembly on September 23, 1999.

14. "Ladies and gentlemen, place your bets."

29

No to Bovine Growth Hormone:
Ten Years of Resistance in Canada

LUCY SHARRATT

In early 1994 recombinant Bovine Growth Hormone (rBGH) was poised to become the first genetically engineered agricultural product on the market in Canada. But by the end of the year grassroots opposition and a parliamentary inquiry forced a moratorium on the use or sale of the drug. Five years later, the Canadian regulatory authority, Health Canada, denied Monsanto's application for approval of rBGH. Monsanto was surprised and shocked by the announcement that its product was refused.[1] The company immediately vowed to appeal the decision but so far as the National Farmers Union is concerned, "Monsanto may protest, but the facts are clear and the issue is settled."[2]

The Canadian government's decision to reject rBGH is the outcome of more than a decade of persistent grassroots resistance, strengthened by diligent inquiry by some scientists within Health Canada, key news media exposure, and a national political debate that included hearings before two parliamentary committees. Protest by individuals and groups in communities across Canada built into a powerful national movement for a ban on rBGH.

FARMERS AND CONSUMERS REVOLT

As early as 1986 officials in Health Canada had concluded that there was "no demonstrable human health risk" from consuming milk from cows injected with rBGH. That decision – unaccompanied by a scientific rationale – persisted through to the early 1990s.[3] In 1988 recombinant Bovine Growth Hormone was tested on dairy herds in four provinces and the milk was quietly added to the general commercial supply. Dairy processors were first to expose the sale and unwitting public consumption of test milk, a story that was picked up by the media in November 1988. This was the Canadian public's introduction to rBGH, and negative reaction was immediate and widespread. When consumers in British Columbia found out about the research trials, one processing company received more than six hundred phone calls in one morning, jamming seventeen trunk lines.[4] This was the beginning of a massive, spontaneous public opposition that swelled and was sustained through ten years, forcing the continued delay of approval and, finally, Health Canada's rejection of rBGH.

In the province of Ontario processors in county milk committees wrote letters to the Ontario Milk Marketing Board (OMMB) persuading the Board to ask researchers to stop putting milk from their experiments onto the market. When Ontario dairy farmer Lorraine Lapointe learned that rBGH test milk was being sold she fought for an inquiry into what she saw as negligence on the part of the OMMB for failing to inform farmers, processors and consumers. Lapointe lost her first bid for an inquiry but, after months of effort that took her away from farm and family, she successfully appealed and represented herself in a two-day hearing of the Farm Products Marketing Commission.[5] The National Farmers Union supported Lapointe but some of her neighbouring farmers were fearful of the controversy. Lapointe later told a Senate hearing: "When I started digging, I did receive a lot of threats but it did not matter because this is very important, not just to our health but to the health of the dairy industry."[6] The OMMB published the inquiry's final recommendations in their newsletter only after pressure from Lapointe. Dairy farmers in her region later elected Lapointe to the Ontario Milk Marketing Board, making her the Board's first female member. Her efforts were typical of the hard work undertaken by individuals in communities across the country, engaging fellow citizens and demanding accountability from authorities.

Lorraine Lapointe, like many Canadians, was compelled by the contradictory image of milk that was promoted as "nature's most perfect food" being produced by injecting dairy cows with a genetically engineered hormone thought to be both harmful to the animals and potentially hazardous to human health. "Milk. 'Pure and Natural' that's what we've been told for years. I believe it, and want to continue to believe it," argued Richard Lloyd of the National Farmers Union.[7]

The National Farmers Union was strongly opposed to rBGH because of concerns for animal health and farmer livelihood. Concerns for small farmers were central to Canadian evaluations of the risks and benefits of rBGH; confronted with these arguments Monsanto's presentation of rBGH as an efficient production 'tool' began to unravel.

The conflict between Canada's supply management system and the logic of rBGH highlighted the fundamental question of economic need for this non-therapeutic drug. The particularity of Canada's system partially accounts for the success of Canadian resistance. In Canada all milk is pooled in a single system where demands for labelling would require segregation and the creation of dual marketing, the expense of which would likely be passed to the consumer. Moreover, the dairy industry operates on the principle of supply management where farmers must buy "quota" which gives them the right to produce and sell a set volume of milk. If a farmer wants to expand production, he or she must purchase additional quota from the pool. Victor Daniel, an Ontario cattle breeder and firm opponent of rBGH, says he was greatly inspired by the generosity of public health agencies, non-governmental organizations, and private citizens in support of Canada's supply-managed dairy industry.[8] The controversy over rBGH developed in the wake of a national debate over free trade that politicized many Canadians and created concern for the future of national industries.

With arguments against rBGH shaping public opinion, author and activist Brewster Kneen sought out Lorraine Lapointe and the two initiated a "Pure Milk Campaign," the first organized effort to disseminate information and encourage opposition. The pamphlet of the campaign portrayed rBGH transforming dairy cows into milk machines. Individuals and groups collected thousands of signatures on the first petitions calling for a ban on rBGH.[9] Brewster and Cathleen Kneen publish *The Ram's Horn,* a newsletter on food and agriculture issues that played an important role in informing people across the country of the emerging controversy and connected rBGH to the greater issues of genetic engineering and increasing corporate control in agriculture.

Brewster Kneen argues that there were three important strategies at play in the emerging movement in Canada.[10] First was the importance of using the name Bovine Growth Hormone (rBGH) to prevent sanitizing the drug with the more neutral-sounding name "recombinant bovine somatotropin" or rBST. Calling the product Bovine Growth Hormone was important for immediate portrayal of the issues and posed a significant challenge to industry communications strategies. Second, the articulation of farmers' concerns needed to remain central, ensuring that rBGH was not narrowly defined as a food safety issue. The final factor was the insistence on labelling rBGH milk if the drug was approved; these were intuitive reactions that became significant aspects of a formidable resistance.

PARLIAMENT ENTERS THE DEBATE

In November 1993 the United States Food and Drug Administration approved rBGH for sale in the United States. By January 1994 rBGH was weeks away from approval by Health Canada,[11] but in March the House of Commons Standing Committee on Agriculture and Agri-food began hearings into rBGH. The hearings subjected rBGH to a systematic examination by Members of Parliament (MPs) and gave the public a new opportunity to find information and focus resistance. Newly elected MP Wayne Easter, a retired dairy farmer and past president of the National Farmers Union, says that the decision to conduct hearings into rBGH "was a direct result of the efforts of the lobbyists representing the pharmaceutical companies meeting with MPs urging support for approval of rBST."[12]

The House of Commons committee heard testimony from Health Canada officials, farmers, processors, Canadian consumer groups, critics from the United States, and representatives of Monsanto and Eli Lilly (Lilly had also submitted an rBGH product to Health Canada for review but later withdrew it pending Monsanto's approval). The first of many turning points in the hearings was a reference by the National Dairy Council of Canada to research on the possible link of rBGH with cancer.[13] Committee members were shocked that no long-term studies into human health risks had been conducted. MPs were also surprised to learn that Health Canada bases its product reviews on data submitted, and owned, by the drug manufacturers, a situation that the then director-general of Health Canada, S.W. Gunner, admitted he found troubling.[14] MPs were ready to challenge the ability of the health department to review corporate data in the public interest because of a disturbing track record in Health Canada that included the collapse of the Canadian blood supply system, the health risks of silicone breast implants, and the lingering memory of thalidomide. The past and present failures of the Canadian government to protect public health haunted the rBGH review throughout the 1990s and opened opportunities for greater scrutiny of the regulatory process.

Mistrust of the manufacturers and of Health Canada pervaded the parliamentary committee hearings and was proven justified in a number of instances, the most striking of which was the testimony of Dr Len Ritter, the former director of Health Canada's Bureau of Veterinary Drugs (BVD) – the bureau responsible for rBGH review. Ritter, while on unpaid leave from the BVD, appeared before the committee as a representative of the Canadian Animal Health Institute (CAHI), an industry lobby group of manufacturers and distributors of veterinary drugs. Jean Szkotnicki, executive director of the CAHI, introduced Dr Ritter as the former director of the BVD, who "may be able to enlighten the committee on the review conducted by the bureau."[15] Ritter replied that he was responsible for the programme within Health Canada for some years and declared, "Do we have a safe product? The answer is clearly and emphatically yes." Ritter was accused in the House of Commons of acting as a lobbyist for rBGH manufacturers. He was later cleared of conflict of interest charges by the deputy minister of health but some observers maintain this exoneration was falsified and believe the issue is unresolved.[16]

The House of Commons hearings raised the political profile of rBGH. Throughout the committee proceedings Canadians wrote letters to their MPs and called or visited their MPs' offices, alerting members outside the Agriculture Committee to the controversy.[17]

The swell of public concern and vocal opposition also persuaded dairy processors that a decision to approve or reject rBGH could not simply be left in the hands of Health Canada.[18] Processors represented by the National Dairy Council of Canada were forced to intervene in the review because of their real fear that milk sales would drop with the approval of rBGH. The dairy council demanded that the government be liable for lost income to processors if milk sales dropped after approval. More importantly, the council called for a two-year moratorium on rBGH. The Dairy Farmers of Canada, though extremely reluctant to do so, were compelled by public pressure, including letters requesting them to oppose rBGH, to recommend a moratorium on rBGH use until consumer confidence could be assured.[19]

In its final report, the House of Commons Agriculture and Agri-food Committee recommended that the government implement a two-year moratorium to allow for further examination of the issues, including possible health risks. With public attention continuing to follow rBGH, the government was persuaded that this recommendation could not be rejected outright.

Across the country people discussed rBGH in their homes and communities and became active. Officials in both Agriculture and Health Canada were flooded with letters and position papers opposing rBGH that came from a surprisingly diverse representation of private individuals, farmers, and citizen and community groups – groups like the Ukrainian Women's League, who started their own letter-writing campaign. The Toronto Food Policy Council, an agency of the City of Toronto Board of Health, had been introduced to rBGH by Brewster Kneen and had taken an early interest in the issues, successfully linking the food safety concerns of urban consumers with the concerns of farmers.[20] In 1994 the Toronto Board of Health recommended to the government that rBGH not be approved. In a letter to the minister of health the Board also demanded that the Canadian covernment label all genetically engineered foods entering the market and conduct socio-economic impact studies before any new food production technologies were introduced.[21]

Citizens across the country worked in their communities to raise awareness and invite opposition. Among them was Ontario cattle breeder Victor Daniel. Daniel, with the support of his local breeding association, argued that the use of the drug would compromise the ability of breeders to keep accurate performance records and would therefore destabilize the cattle breeding industry. This was an unexpected and persuasive argument that Daniel presented to the Toronto Food Policy Council. He soon became a council member and met with MPs on the group's behalf. As an individual breeder, Daniel also met with senior bureaucrats at Agriculture Canada, who he says acknowledged the seriousness of some of his main concerns.

For seven years Richard Lloyd of the National Farmers Union researched the issues around rBGH. Lloyd's work was central in maintaining continuity in resistance from year to year. Lloyd supplied critical and timely information to public interest groups, community leaders, and politicians and kept the farmer perspective up front. In spite of difficulties organizing farmers who have on-farm responsibilities, Lloyd and others in the National Farmers Union organized a farmer lobby aimed at members of parliament.

In August 1994 the minister of agriculture responded to the Agriculture Committee's recommendation for a moratorium with a delay on the use and sale of rBGH that, at the suggestion of the drug companies, was negotiated with manufacturers. The government announced a "voluntary moratorium" until July 1, 1995. Though this decision did not meet the demands of opponents and left room for approval of rBGH, it provided an important new opportunity for the public to fight for, and achieve, a final ban.

IF AGAIN YOU DON'T SUCCEED, DELAY, DELAY AGAIN

A survey conducted by Industry Canada at the end of 1994 showed that 83 per cent of Canadians knew what rBGH was; up to 34 per cent said that they were unlikely to buy milk if rBGH was approved. [22] In November 1994 the national television programme *The Fifth Estate* aired a documentary on rBGH that included allegations of an attempted bribe by Monsanto of Health Canada. And in December the European Union announced a moratorium until the year 2000.

Public pressure for an extension of the voluntary moratorium increased as the July expiration approached. It was at this point that many of the individuals and groups opposing rBGH came together in a coordinated effort. The NFU was joined in its efforts by the Council of Canadians, a 35,000-member citizen advocacy group that had formed in 1985 to fight Free Trade and promote democracy. The Council became a support centre for the vital and diverse resistance that was still growing across the country and was central in coordinating strategy and organizing groups. Council campaigner Alex Boston offered information and advice to local activists but also coordinated a national strategy with the National Farmers Union as well as the Canadian Farm Women's League, the Canadian Environmental Network, and the Canadian Institute for Environmental Law and Policy.

The Council of Canadians and the National Farmers Union designed a postcard campaign and more than 100,000 individuals sent postcards to the minister of health that began, "I want to continue to trust Canadian dairy products." Within three months of their campaign the Council of Canadians had created a database of 2,500 Canadians who were actively opposing rBGH. At this time people were meeting with local politicians, circulating information, and asking their municipal councils and school boards to take positions. Across the country people placed petitions in health food and farm supply stores. One health food magazine alone collected 500,000 names on petitions. [23]

With a common message, the diverse interests together lobbied MPs. The shared strategy brought together 350 organizations including farmers, breeders, processors, health professionals, environmentalists, lawyers, school districts, public health agencies and consumers, demanding an extended moratorium. [24] In a Council of Canadians press conference, members of parliament from all parties demanded an extension of the moratorium and presented Council petitions signed by 200,000 Canadians. "If this government does not extend the moratorium, this will truly be a fundamental miscarriage of the democratic process," argued Council campaigner Alex Boston.

The newly energized and coordinated resistance propelled rBGH into the media. Almost daily through the spring of 1995 members of parliament made statements and asked questions about rBGH in the House of Commons. [25] Richard Lloyd of the National Farmers Union believes that by the end of June, rBGH was on the top of the parliamentary agenda and the debate had reached the prime minister's office. [26] The Council of Canadians approached the Standing Committee on Health and, half a month away from the end of the voluntary moratorium, the committee recommended that the minister of health should "do all within her power to seek a prolongation of the voluntary moratorium on rBST, for a minimum of two years to allow Members of Parliament to further examine the human health implications."[27]

The prospect of a ministerial request for an extended moratorium was met with threats from Monsanto and Eli Lilly that they would pull their investments out of Canada. But such corporate pressure tactics only fed suspicion and further turned public opinion against commercialization. "What they are saying is blackmail and nothing less," responded MP Wayne Easter.[28]

No extension was legislated or negotiated by the Canadian government, but a few months after the end of the moratorium the minister of agriculture reported that regulators in Health Canada had asked Monsanto for more data on animal health, delaying approval once again. The minister declared that this delay – which would amount to a year or longer – was proof that the regulatory system was working.[29] The details that later emerged from inside Health Canada tell an entirely different story.

INSIDE HEALTH CANADA:
INTIMIDATION, BRIBES AND THEFT

While the controversy over rBGH raged in public, in the news media and in parliament scientists inside the Bureau of Veterinary Drugs (BVD) in the Health Protection Branch of Health Canada were fighting their own battle against pressure to issue a Notice of Compliance that would permit the sale of rBGH.

Dr. Margaret Haydon is a veterinary scientist who has worked in the BVD for fifteen years and began reviewing rBGH products in 1984. Dr Haydon testified at a 1998 labour-board hearing that she was pressured by the then director, Dr Len Ritter, to issue a "Conditional Notice of Compliance" for Eli Lilly's rBGH product.[30] Dr Haydon had never heard of this kind of approval and refused to comply. Years later in a letter to the Senate Agriculture Committee, the assistant deputy minister of the Health Protection Branch, Dr Joseph Losos, explained that this was merely a misunderstanding on the part of Dr Ritter, who had recently come from a bureau where conditional notices were legal. Documents received through Access to Information rules show, however, that as late as 1997 managers in Health Canada planned "to investigate the scenarios under which Human Drugs [Directorate] may be approving drugs to be used prior to full approval."[31]

The same year that Dr Haydon was pressured to sign a conditional notice of compliance, some of her files – amounting to most of her work on rBGH over the previous ten years – were stolen from locked cabinets in her office in what she says must have been an "inside job."[32] Days after she noticed files were missing, her office was broken into a second time and files were stuffed back into her cabinet. She reported the incidents to the police but an investigation produced no suspects. Dr Haydon was subsequently removed from the rBGH review with no explanation.

In 1996 Health Canada introduced a programme of "cost recovery" that requires drug companies to pay a portion – between 70 and 90 per cent – of the costs of each product review. The Bureau of Veterinary Drugs reduced its budget in anticipation of this income.[33] For Dr Haydon and a number of other scientists in the department, cost recovery increased the direct influence of chemical and pharmaceutical corporations over their work. At least one drug company wrote to Health Canada demanding that, as a paying client, the review of its product be expedited. Drug companies have access, through participation in a Joint Program Management Advisory Committee, to the names of the scientists reviewing their data, and some companies have written to department managers requesting that particular scientists be removed from reviews if they were seen as working too slowly. Despite such direct attacks, the scientists persevered.

Signs of pressures inside Health Canada were beginning to be reported in the print news in late 1997. Dr Haydon and her colleague Dr Shiv Chopra were interviewed on national television and asked why they thought they were under pressure to approve drugs quickly. Dr Chopra replied, "Well, what do you think? Money. For multinational companies that produce those things."[34] His frankness and that of Dr Haydon were rewarded with threats from managers in Health Canada. Dr Chopra was invited to speak about the Canadian regulatory system at a community panel on genetic engineering in Ottawa but on the day of the event he received a registered letter, email message and fax from his superior instructing him not to attend on threat of losing his job. In September 2000 a federal judge ruled that Drs Chopra and Haydon were justified in voicing their concerns, declaring that it was "un–reasonable" for the department to reprimand the scientists for going public.[35]

In 1998, Dr Haydon, Dr Chopra and four colleagues in the Bureau of Veterinary Drugs filed a grievance with their union, the Professional Institute of the Public Service. They complained that they had been reassigned away from controversial reviews to new duties under the pretext of relieving a backlog. The six alleged that they were transferred away from important files in retaliatory measures for their efforts to ensure rigorous product review. When mediation with management failed, the scientists took their complaint to a labour-board hearing where they charged that they were being harassed to approve rBGH and other veterinary drugs despite their safety concerns. Providing an unprecedented window into the review process, the grievance hearings also presented new space for resistance. The Council of Canadians and the Sierra Club of Canada held a press conference in support of the scientists. The media covered the two-day hearings and ten members of the public attended both days. The public and media were given a second opportunity to hear the scientists' story when their testimony was requested at Senate hearings only one month later.

THE SENATE OF CANADA HEARS FROM THE SCIENTISTS

In May 1998 the Senate of Canada unanimously passed a motion urging the government to defer licensing of rBGH until long-term risks to human health were determined. A month later the Senate Committee on Agriculture and Forestry began hearings into rBGH. Two Senators had heard of a report within Health Canada that assessed the status of the rBGH review, but Health Canada officials provided the Committee with only a heavily censored version of the "Gaps Analysis Report" – until the full report found its way onto the website of the National Farmers Union. The details of the report focused Senators' questions on the human health risks of rBGH as well as the conduct of Health Canada managers.

At the core of the Gaps Report was the analysis of Monsanto's short-term toxicology study, a ninety-day test on just thirty rats. The study showed that orally administered rBGH was absorbed into the bloodstream with adverse effects.[36] The results of Monsanto's test had never been published, nor made publicly available.

Dr. Chopra and the three other authors of the Gaps Analysis Report concluded that "the nature of the product (being a hormone) and its chemistry should have prompted more exhaustive and longer toxicological studies in laboratory animals,"[37] and that "[b]oth procedural and data gaps were found which fail to properly address the human safety requirements of this drug under the Food and Drugs Act and Regulations."[38] With the Gaps Analysis in hand and no satisfactory response from Monsanto, the Senate Committee recommended that the company conduct long-term human health studies.

The new information and damning critique of the Gaps Analysis had international significance and became the basis of a challenge to rBGH approval in the United States. The US Center for Food Safety and more than two dozen US consumer groups demanded that the Food and Drug Administration reverse its decision because of the new details they say were not considered in the US review.[39]

As well as bringing to light previously unknown problems with Monsanto's data, the Gaps Analysis revealed that the degree of industry influence inside Health Canada was greater than critics had ever suspected.[40] The report also showed that research conducted by Victor Daniel and the Toronto Food Policy Council had become a part of Health Canada's review process; Council letters to Health Canada with pointed questions challenging the science used in the review appeared in the appendices of the report.[41]

The Senate Committee requested the testimony of the scientists who worked on the rBGH file and those who authored the Gaps Analysis Report. Among them were Drs Margaret Haydon, Shiv Chopra and Gerard Lambert – three of the six Health Canada scientists who had filed grievances against their managers. Before appearing at the hearings, the three sought, and were granted, assurance from the minister of health that their testimony would not put their jobs in jeopardy. In opening testimony the three took the unprecedented step of reading aloud the oath under which they were to answer the senators' questions. The scientists then recounted the pressure tactics of their managers, in close relationship with Monsanto representatives, and explained their concerns with rBGH. The drama of their testimony was reported in the national newspaper the *Toronto Globe and Mail*, and on national television news. The media highlighted the story, first reported in 1994, of a meeting attended by Dr Margaret Haydon where she and Dr Bill Drennen, then director-general of the Health Protection Branch, say a Monsanto official offered a bribe of $1–2 million in return for approval of rBGH without further delay.[42]

CLOSING THE GAPS

A panel of the Canadian Veterinary Medicine Association confirmed that there were a number of legitimate animal welfare concerns associated with the use of rBGH, including increased risks of lameness and clinical mastitis (a painful udder infection), as well as shortened lifespan.[43] The panel's conclusions vindicated farmers and confirmed some of the concerns articulated by the Health Canada scientists, in particular Dr Haydon. When Health Canada managers finally rejected rBGH they used these conclusions on animal health to explain their decision.

On January 14, 1999, the *Toronto Globe and Mail* reported that Health Canada would be rejecting Monsanto's application for rBGH. At a tightly controlled news briefing on January 15, before the panel on human health had concluded, Health Canada officials announced that rBGH would not be approved "at this time."[44] "With all of this scientific information available, we saw no reason to delay the decision any longer," said Health Canada spokesman Joel Wiener.[45] Following the high-profile Senate hearings, this rejection was national news.

Victor Daniel of the Toronto Food Policy Council describes Health Canada's decision as "the right answer for the wrong reasons."[46] Rather than responding directly to the concerns of Canadians, after ten years of delays forced by public pressure made approval too politically difficult, Health Canada took the last of many opportunities to refuse rBGH.

CONCLUSION

The resistance to rBGH in Canada was a grassroots movement that persisted and grew over ten years. The diversity of the opposition was its greatest strength; farmers spoke out against animal ill health and threats to the dairy industry, consumers demanded safe milk, and government scientists exposed industry pressure and inadequate science. Each voice in opposition was a strong and legitimate voice for a constituency of people who were actively opposed to rBGH. The Council of Canadians structured common opposition at strategic moments and helped sustain the diversity of the movement, giving local resistance a more prominent face and pushing rBGH into the media.

With a truly grassroots and national movement against rBGH, Monsanto was unable to target individuals or groups to discredit. Canadians organized to defeat rBGH without a national organization concerned with food issues or a visible consumers' movement. The scrutiny of rBGH by both MPs and senators restored hope in Canada that the mechanisms of the parliamentary system can function for the public interest.

NOTES

1. CBC-TV, *The National Magazine*, January 14, 1999.
2. Peter Dowling, quoted in National Farmers Union news release, "Ban on Bovine Growth Hormone a Victory for Canadian Cows, Farmers, and Consumers," January 15, 1999.
3. "rBST (Nutrilac) 'Gaps Analysis' Report," rBST Internal Review Team, Health Protection Branch, Health Canada, April 21, 1998.
4. Kempton Matte, National Diary Council of Canada, testimony to House of Commons Committee on Agriculture and Agri-Food, Issue 4:5, March 8, 1994.
5. Lorraine Lapointe, telephone interview, March 15, 1999.
6. Lorraine Lapointe, testimony to Standing Senate Committee on Agriculture and Forestry, 29 October 1998.
7. Richard Lloyd, telephone interview, September 12, 1995.
8. Victor Daniel, telephone interview, March 20, 1999.
9. Brewster Kneen, *Farmageddon: Food and the Culture of Biotechnology*, Gabriola Island, BC: New Society Publishers, 1999.
10. Ibid.
11. Mike O'Neil, telephone interview, March 9, 1999.
12. Wayne Easter, telephone interview, September 19, 1995.
13. Mike O'Neil, telephone interview, March 9, 1999.
14. S.W. Gunner, testimony to House of Commons Standing Committee on Agriculture and Agri-Food, March 7, 1994.
15. Testimony to House of Commons Standing Committee on Agriculture and Agri-Food, Issue 5:32, March 8, 1994.
16. Richard Lloyd, telephone interview, March 1, 1999.
17. Mike O'Neil, telephone interview, March 9, 1999.
18. Rod MacRae, telephone interview, March 25, 1999.
19. Peter Oosterhof, House of Commons Committee on Agriculture and Agri-Food, March 7, 1994.
20. Richard Lloyd, telephone interview, March 1, 1999.
21. City of Toronto, City Clerk's Department, Letter to Minister of Health, Diane Marleau, June 1, 1994.
22. Optima Consulting, "Understanding the Consumer Interest in the New Biotechnology Industry," November 1994.
23. Richard Lloyd and Helen Forsey, "Cows on Steroids? No Thanks!" *Natural Life*, September/October 1998.
24. Council of Canadians, *Canadian Perspectives*, "Local Action BGH Update: Time to Turn Up the Heat," Autumn 1995: 14.
25. Editorial, "BST and Corn," *Ontario Corn Producer*, August–September 1995.
26. Richard Lloyd, telephone interview, September 12, 1995.
27. House of Commons Committee on Health, June 15, 1995.
28. Doug Saunders, "Drug Firms Upset about Expanded Ban," *Toronto Globe and Mail*, June 27, 1995, p. 2.
29. "Approval of BST Use Delayed for One Year" *The Western Producer*, October 5, 1995.

30. Notes from the Public Service Staff Relations Board grievance hearings, Lucy Sharratt, September 15–16, 1998, Ottawa.

31. George Paterson, memo, "re: teleconference with AAFC re: rBST, May 15, 1997," May 2, 1997, via Access to Information.

32. Notes from the Public Service Staff Relations Board grievance hearings, Lucy Sharratt, September 15–16, 1998, Ottawa.

33. Ibid.

34. CTV-TV, *Canada Am*, June 11, 1998.

35. Caroline Alphonso, "Supporters Hail Ruling for Public Whistle Blowers," *Toronto Globe and Mail*, September 18, 2000.

36. "rBST (Nutrilac) "Gaps Analysis" Report.

37. Ibid., p. 30.

38. Ibid., p. 4.

39. "Legal Challenge Filed with FDA to Remove Monsanto's BGH from the Market," December 15, 1998, Release, Environmental Media Services, Washington DC.

40. Rod MacRae, telephone interview, March 25, 1998.

41. Ibid.

42. CBC-TV, *The Fifth Estate*, November 29, 1994.

43. Canadian Veterinary Medical Association, "Report of the Canadian Veterinary Medical Association Expert Panel on rBST – Executive Summary," November 1998.

44. Canadian Health Coalition, letter to Minister of Health Allan Rock, January 26, 1999.

45. Health Canada, News Release, "Health Canada Rejects Bovine Growth Hormone in Canada," January 14, 1999.

46. Victor Daniel, telephone interview, March 20, 1998.

30

Cooperatives:
A Source of Community Strength

ROBIN SEYDEL

In living systems, both large and small, the whole is generally acknowledged as greater than the sum of its parts. This truism is easily recognizable in ecosystems, where tiny microbes help feed the roots of giant trees, which then shade and support a varied community of plants and animals.

Human existence mirrors nature's cooperation. Our hearts pump blood, lungs expand and contract, and food is digested to provide nutrients for our cells to create the energy we need. These and other systems, functioning together, allow that special spark we call our human spirit to define our essence. Just as the human spirit confers upon us a greatness that is more than the sum of our physical parts, our communities traditionally have been more than just aggregations of individual, isolated consumers. Unique local flavours, colours, customs and cultures, developed over eons of participatory activity, are the source of the great beauty of diversity and the soul of community.

Today, cooperative action can provide the inspiration, strength and community cohesiveness needed to overcome the plans of profit-hungry corporations and their view that we – indeed all living things – are merely a set of patentable, genetic raw materials, alterable at will, without respect for species, spirit soul, or the greatness of the whole.

COOPERATIVE CONTINUITY

Cooperative efforts have long been the basis for the development of community, facilitating the maintenance of balance and helping overcome challenges to our survival. In the earliest human communities it was group hunting, gathering and mutual protection. Later, agricultural lands and efforts were shared, feeding people and growing more settled communities. Organized cooperative activities were commonplace in early US history: barn raisings and volunteer firefighters come immediately to mind.

During the mid-nineteenth century in Rochdale, England, a group of textile workers, tired of the abuses of the industrial, corporate system, codified cooperative operating principles. Based on cooperative concepts developed and used by traditional societies since earliest times, these updated principles of community ownership – including dividend returns based on patronage rather than investment and equal access to and control of resources through democratic governance – continue to inspire cooperatives today, and have nurtured models of economic justice in the midst of increasing corporate control.[1]

Historically, cooperatives have experienced resurgence during periods of heightened social awareness, helping to combat society's injustices and providing needed goods or services at affordable rates. Surviving the ravages of time and economic pressure, a number of food and housing coops formed in the United States during the Depression era and World War II are still serving their members' needs. Today, around the world, cooperatives provide food, put out fires, supply electricity, support farmers, create housing and jobs, and provide access to stronger positions in the marketplace.

During the late 1960s and 1970s, a wave of cooperative activity in the United States grew out of people's desire for access to whole, unprocessed, unadulterated, organic foods. Many of these newer coops struggled and died during the 1980s and 1990s due to changing cultural values and corporate competition as the natural foods industry became "big business." Some surviving coops, stressing the bottom line and conventional business thinking, are reluctant to engage in community organizing activities. Others have utilized and built upon their unique position by encouraging action on important food safety, health and environmental issues. Rooted in local communities, they provide a rich medium for empowerment and have a distinctive role to play in the struggle to overcome the aggressive activities of the biotechnology industry. Acting on the cooperative principle of education, cooperatives can create an atmosphere of healthy dialogue and free exchange of information unencumbered by corporate "spin." They provide a forum for opinions that run counter to the vested interests expressed through conventional information outlets. Even more important, they provide a channel through which people can act on those opinions by producing, marketing or purchasing products that help sustain their beliefs.

BUILDING COMMUNITY:
WHAT'S GOOD FOR THE PEOPLE IS GOOD FOR THE COOP

An example of the success that cooperatives can have when they engage in community organizing on issues as compelling as biotechnology can be seen at La Montañita, a natural foods cooperative that has survived since the mid-1970s, in Albuquerque, New Mexico. The member loyalty that has kept this coop financially strong, despite formidable corporate competition, is due as much to its educational and action-oriented community development activities as it is to the coop's dedication to its primary mission of providing members and shoppers with the finest natural foods at the best possible prices.

La Montañita offers numerous member benefits, such as a credit union, group health insurance and free shopping and delivery services for people in need (elders, shut-ins, those with debilitating illnesses, etc.), and activities like community dances, fiestas, a free university, Earth Day celebrations and more. Utilizing its resources to the best of its ability, the coop rotates daily donations of all good but no longer saleable, items between local homeless and domestic violence shelters, saves compost for area farmers, does other recycling, and supports the activities of area organizations and public schools with volunteers and donations. The coop works to support the local, sustainable, farming community, encourage organic food production and maintain the integrity of organic standards. These activities help solidify the local community and create much needed support for the coop.

Education is one of the cooperative principles, and the coop newsletter, the *Coop Connection*, was created to meet this responsibility. During more than a dozen years of publication, it has become a widely read and distributed community forum, with over 12,000 readers, which provides news on the business of the cooperative, as well as information on the production, distribution and consumption of food; recipes and nutritional tips; and consumer advocacy on a variety of health, safety and environmental issues including chemical usage, food irradiation and biotechnology.

BOVINE GROWTH HORMONE: FIRST RESPONSES

When genetically engineered recombinant Bovine Growth Hormone (rBGH) first appeared on the market, the *Connection* gave readers their first information on the subject, information that was more complete and in-depth than any other accessible source. The coop took a strong stand on rBGH due to health and safety concerns. Every dairy supplier was contacted. Local and national dairy producers were polled as to their positions on rBGH, and their signed and sometimes notarized letters were kept on file. Responses were printed in the newsletter and issue updates followed on a monthly basis for nearly a year and continue still.

Products from Borden, Häagen Dazs, and Creamland dairies – companies whose letters were in clear support of the use of rBGH – were taken off the shelves and replaced with organic and rBGH-free products. In many cases lower than normal mark-ups on these items were used so they would be more affordable to our shoppers (New Mexico is one of the poorest states in the US). Carefully worded information signs were posted on dairy, cheese and ice cream cases. Purchasing staff held their breath, fearing that sales would take a dive and that the coop might not be able to withstand the loss in revenue. During the first month dairy sales were in transition but holding steady.

Using member volunteers, the coop teamed up with local animal-rights organizations and other concerned groups, both local and national, forming a coalition that organized several demonstrations in front of the local offices of the Food and Drug Administration (FDA). Sending out press releases and distributing printed leaflets, the coop worked hard to have the most extensive library of information available on rBGH. By the second month, as the news filtered out, waves of concerned consumers came to the store looking for rBGH-free products.

As we were often quoted in local mainstream news stories, the US Department of Agriculture (USDA) soon came knocking. They quickly recognized that we had done our homework, and on some aspects of the issue had even more information than they did. Returning to their offices with a wad of literature they had not previously seen, the inspectors knew we were sincerely concerned, knowledgeable, and sought only to protect New Mexican farmers and consumer health and safety. Corporate competitors jumped on the bandwagon and joined the local coalition, but did not pull products known or suspected to have rBGH off the shelves.

Dairy sales boomed and coop purchasers had a hard time keeping milk and other dairy products in stock. We made up in volume what we lost in lowered mark-up on individual products. We had done what we believed was the right thing, "putting our money where our mouth was," as one shopper said, and it paid off. Sales were strong. We had kept the faith with the people in our community and it built our reputation as more than just another health food store. The coop had become a trusted community information resource and a bastion of integrity and community in a world where those things often seem to be in short supply.

COMMUNITY DIALOGUE DEVELOPS POLICY

In order to continue to provide the most up-to-date information, we tapped a variety of sources both locally and nationally. Regular writings on biotechnology from a number of knowledgeable people graced the pages of our newsletter. We continued to clarify our position on genetically engineered food and other food safety issues. When a number of members requested that we reinstate Häagen Dazs ice cream, we did so after printing the wide-ranging pro and con letters in our newsletter; freezer doors were well posted with rBGH info. Consequently, we don't sell all that much of it, and because it is the slowest mover in its category it is consigned to one shelf at the bottom of the freezer. We continue to carry it as a nod to the diversity of the community we serve.

This community dialogue helped the coop define what has become our standard mode of dealing with this kind of issue. Staffers and members agree that it is our job to provide the best consumer information available on any given issue or product, and that it is our responsibility to clearly post that information in our store. It is not, however, our job to make people's choices for them. Unlike the biotech industry (and their handmaidens in various government agencies) that would force us to use their gene-altered products through a lack of labelling, freedom of choice is an individual right that we will not trample.

This policy would hold until the "Boca Burger" debacle several years later. After *Consumer Reports* announced that its DNA testing showed that Boca Burgers, a popular brand of soy-based "veggie" burgers sold in our frozen foods department, tested positive for genetically altered soy,[2] the company went into high gear to combat this bad publicity. Offering deep sale prices and plenty of dollars to stores that would do in-store sampling, our purchasers took the bait and agreed to a sale and a sampling. Concerned community members, including staff, made a fuss and forced the Boca Burger demo in one of our two stores to end almost as soon as it began. A few days later at a purchasers' meeting a further elaboration of our policy was developed. The staff recognize that our store is owned by the people who shop there and if enough members want a specific product, and it continues to sell at a reasonable rate, we will do our best to provide it for them. However, we will not promote products known to contain genetically modified organisms (GMOs) in any of our literature, at in-store samplings or in any other way, no matter how much money is offered by the manufacturer.

COMMUNITY EDUCATION:
BEYOND THE STORE SHELVES

After the initial debates over rBGH products, we rather quickly recognized the enormity of the myriad aspects of biotechnology. Beyond milk, Roundup Ready soybeans, altered cotton, canola oil or tomatoes, the effects of this technology could have even further-reaching consequences than those of that first ever radioactive mushroom cloud, which rose over our state more than half a century ago.[3] When Dolly, the first cloned sheep, was announced, we decided that we should not miss the opportunity to discuss the issue in its larger context. Networking with national and international groups we joined the international "Global Days of Action" against genetic engineering. For a number of years we had sponsored an Earth Day celebration, always the Sunday closest to the traditional April 22 date. International organizers had scheduled the Global Days of Action to begin then as well. For us it was a perfect match.

Our celebration had grown over the years to a festival attended by between four and five thousand people. With more than fifty environmental organizations from around the state, solar energy displays, an entertainment tent that hosted a diversity of musical tastes – from blues to Celtic, New Mexican folk to world-beat, flamenco dancers, traditional Native American musicians and dancers – along with barnyard animals, displays from animal rescue and rights organizations including live wolves, raptors and other species, local artisans and craftspersons and many local farmers and food producers, it had become a favourite community event. That year, we set up a second "Teach-In" tent and invited members of our community as well as members of the national biotechnology activist community to speak. We did three workshops over the course of the day, including a short history of biotechnology, an overview and issue update, and a discussion of its effects on local sustainable agriculture with a panel including a representative of the USDA, an organic inspector from the state-run Organic Commodities Commission that certifies local organic farms, and farmers from two Indian pueblos and a traditional Hispanic community.

The last workshop addressed the ethical dilemmas presented by this emerging technology. Religious leaders representing several faiths, including a Native American elder, a rabbi, a priest and a minister, were joined by philosophers and educators. The director of the division of the Human Genome project that is located nearby at Los Alamos National Labs, as well as members of the Sandia National Laboratory's ethics committee were invited. Although both national labs confirmed speakers, one canceled just before the event and the other did not show up that afternoon.

We left plenty of time for dialogue during each session to ensure that community members had time to share their own thinking and concerns. Members of the press were invited to attend and the tent was filled throughout the day. A short piece documented the event in the local morning newspaper the following day. Due to the organizing efforts of national activists, the Associated Press wire service picked up the story and articles mentioning our teach-in were carried in newspapers of several distant communities, brought to our attention by coops in those areas.

Our Earth Day celebration has continued to spotlight biotech issue as well as other threats to organic sustainable agriculture. Speakers over the years have included Brian Tokar, Global Days of Action organizer Ronnie Cummins, and Hope Shand, whose research and inspiring address on the Terminator technology stirred increased concern among the New Mexican farming community.

The coop, through its newsletter, and in-store activities, has continued to address all aspects of GMO commercialization. The events at the November 1999 WTO meeting in Seattle raised interest and activities to a higher level. A GMO Task Force, comprising coop members, volunteers and other concerned consumers began meeting several times a month. Committees of the Task Force include community education and outreach, a study group (to keep us updated on the latest information), a media committee, and volunteers willing to help our purchasers document the policies of manufacturers – especially those whose products contain soy, corn and canola – along with the availability of third-party DNA testing and other pertinent topics.

Actions include regular tabling both at coop locations and at numerous other events, support for legislative action, numerous lectures and community dialogues in schools and churches, and demonstrations. Radio coverage has generated much community discussion, most notably a two-hour call-in show on the local public radio station that featured a member of our Task Force, a representative of Native Seed/SEARCH (a regional non-profit dedicated to the collection, preservation and utilization of our region's traditional seeds) and a local organic farmer who is a board member of our state's organic certification commission.

FRIEND OF FARMERS

New Mexico is a mostly rural state blessed with many small and medium-sized family farms and a vibrant organic farming community. It is clear that, due to the use of Bt and issues around pollen drift, GMOs pose a serious threat to organic agriculture. Corporate intimidation of farmers (through licensing agreements, lawsuits, etc.) around the cultivation of patented seed stocks threatens the independence of family farms. Food security through local production, and a healthy rural economy that both supports and is supported by local farmers, go hand in hand with and are integral to community sustainability. In keeping with the cooperative principle of sustainable development, it is also necessary to combat corporate agribusiness and GMO food production. Here again it is evident that what is good for the community is good for the coop. Providing local farmers with better access to the marketplace is one part of the job that cooperatives can do well and a piece of the whole local picture that La Montañita takes very seriously.

Many farmers tell us that they sell more of their crop at the coop than all their other sales combined. We regularly pay local farmers better than market rates for their harvests. We also offer reduced mark-ups on locally produced items, and thus can often sell local, organic food at prices that match or better conventionally produced agribusiness products at mainstream supermarkets. We support farmers markets, local CSAs[4] and farming events with free advertising in our newsletter. For several years we have been a major sponsor of the New Mexico Organic Farming Expo and for five years have organized a benefit harvest festival in conjunction with the non-profit Rio Grande Community Farms to help send local farmers to the national Eco-Farm Conference. Together, these actions have earned us the reputation as a friend of local farmers. Supporting farmers' markets and assuring consumers of a stable supply of local products has placed us well to participate in and support the actions of farmers working to combat biotech threats to their survival. It has also strengthened the local economy; an important strategy in overcoming the "life science" industry's attempts at global domination of the food supply.

SUSTAINING COMMUNITY: A COOPERATIVE GOAL

The natural foods industry, with its continuing buyouts, mergers and consolidations, is a microcosm of the larger economic landscape. As our world becomes increasingly controlled by transnational corporate interests, the economic justice inherent in the cooperative model becomes ever more important. By embracing the diversity of our community, encouraging democratic practices in the control of and access to economic resources, and reflecting our values in all our activities, we recognize and act upon the unique qualities that embody the cooperative spirit. It is this very nature that holds several of the keys for our future survival.

A strong local economy is the basis of a sustainable community and a principle of cooperative philosophy. The economic independence that cooperatives encourage is a powerful shield for communities from the ravages of corporate globalization. A balance is struck as the coop supports and facilitates the fulfilment of its community's needs. In return members of that community support the cooperative that they own. This is the heart and soul of a cooperative's existence and distinctively places coops as centres of resistance and community strength, as we seek to stop future releases of "modified" organisms and protect all of creation from genetic contamination.

NOTES

1. The Rochdale Principles, as updated and adopted by the International Cooperative Alliance Congress in Manchester, September 23, 1995, are: (1) voluntary and open membership (including non-discrimination); (2) democratic member control (usually one member, one vote); (3) member economic participation (equitable ownership and democratic control of common property); (4) autonomy and independence (agreements with other organizations ensure democratic control and cooperative autonomy); (5) education, training and information (for employees and the general public, particularly young people); (6) cooperation among cooperatives (creating local, national, regional and international cooperative structures); and (7) concern for community (sustainable development through policies accepted by coop members).

See Center for the Study of Cooperatives, *Making Membership Meaningful*, Saskatoon, SK: University of Saskatchewan, 1995; W.P. Watkins, *Cooperative Principles Today and Tomorrow*, Manchester: Holyoake Books, 1986; *National Cooperative Directory* (annual), Randolph, VT: Coop News Network.

2. "Seeds of Change," *Consumer Reports*, September 1999, p. 45.

3. The Trinity Test Site in Almogordo, New Mexico, was the site of the first ever test of an atomic bomb on July 16, 1945, just prior to its use on Hiroshima and Nagasaki.

4. Community Supported Agriculture (CSA) is a system in which community members support farmers by purchasing membership shares in the spring, giving farmers the necessary funds to run their farm as well as ensuring a market for their harvest. Farmers deliver weekly or bi-weekly deliveries of vegetables, fruit, eggs and other farm products to members.

31

McDonald's, MTV and Monsanto: Resisting Biotechnology in the Age of Informational Capital

CHAIA HELLER

BIOTECHNOLOGY AS A MODE OF PRODUCTION

A thing is a history of a thing, and more. Indeed, history is a tangled web with frayed edges, each woven into what came before. And so it is with biotechnology. To understand it, we must understand its history, the wider universe of people, places and things that brought it into being. Biotechnology is bigger than the instruments, organisms and scientists who move strands of DNA from one cell to another. It is a mode of production, a way of thinking about and producing nature and society that both constitutes and is constituted by society itself.

The story of genetic biotechnology begins not with a great man or a laboratory, but with the wider machinations of capitalism at the end of the twentieth century. But rather than begin with the usual catalogue of ecological ravages left in the wake of industrial capital, this story begins after the fact: after the pollution and the decimation of species, after the big question mark of global warming. This story begins in the morning, when capital awakens, staggering to the window with a hangover, after the party has ended, to look out over a world it has created in its own image. Out in the distance, it can see the rusted momentoes of a better time: once-glistening Cadillacs lying on their sides, their hubcaps stolen and sold by someone struggling to make it to the end of one more day in post-industrial USA.

Biotechnology is more than a scientific practice. It is a network of actors, organisms, tools and discourses that circulate through the corporate, state and international trade apparatuses that emerged after the dust settled over post-World War II capital. While some claim that it is "nothing really new," that its transgenic creations represent a continuity with such previous biotechnologies as plant and animal breeding, they deny the underlying issue: transgenic biotechnology emerges out of a different *world* than plant breeding or beer making. It emerges out of a different set of economic, political and social demands and commitments. Biotechnology is a new form of production that emerged as capital hit the limits of industrial production and began to enter what may be called its *organic* phase: a phase in which capital targets the reproductive dimensions of cultural and biological life as loci for intensified production and commodification.[1]

In this phase, a service economy marshals what I will call the organic reproductive processes of everyday life including food, health and recreation, transforming them into franchised formations ranging from fast food and HMOs to MTV and Disneyland. Biotechnology emerges as part of this trend, reducing cultural and biological life processes into an ever renewable and flexible raw material for production. As Fredric Jameson points out, post-industrial capitalism is characterized by a global reach as well as by the penetration of capital into nature on a transformational scale never before thought possible.[2]

Recombinant DNA emerged as a possibility during a time when private industry was turning to new technologies to increase its returns on investments.[3] Changes in US patent law in the early 1980s that recognized genetically modified organisms as "inventions" provided private industry and public universities with a commercial incentive to develop and eventually market their transgenic products.[4] With rights to intellectual property assured, the biotechnology industry began to emerge as a particularly pliable and lucrative postmodern industry in the 1980s.

But public response to this new form of production has not been what corporations had anticipated. In the United States, after the initial public concern of the 1970s over genetic engineering had died down, corporations did not expect to encounter further difficulties in marketing transgenic medicines and foods to consumers in an increasingly global market. The Reagan and Bush administrations believed that they could guarantee the success of the biotechnology industry by giving a green light to US corporations to research, develop and market their products without being hindered by regulations.

By the late 1980s, however, the first signs of resistance began to appear on the horizon. In 1987, a group of Earth First! and eco-anarchist activists in the San Francisco Bay Area of California destroyed an entire field of experimental genetically modified strawberries, causing thousands of dollars in damages.[5] In 1992–93, the resistance went international when thousands of peasants in India protested against Cargill, whose costly patented seeds threatened their agricultural viability and autonomy.[6] In the fall of 1996, when the first shipments of genetically engineered crops arrived by boat in Europe, Greenpeace International organized a campaign in which activists besieged ships containing transgenic crops as they arrived in European ports. Demonstrating at warehouses and processing plants in Germany and Switzerland, activists created a media spectacle that catalysed a Europe-wide movement that is still gaining strength today. More recently, at the great demonstrations against the World Trade Organization in Seattle in November 1999, activists from around the world protested against agricultural biotechnology as part of a wider critique of globalization.

As the Seattle demonstrations illustrated, resistance to technologies such as agricultural biotechnology is often broader in scope than resistance to the production of genetically modified organisms. It is also a refusal of a new way of ordering social and biological life. If technology is the network of actors, institutions, instruments, organisms and discourses that constitute technological practice, if this network both constitutes and is constituted by a wider society, then it is clear that activists are not simply resisting biotechnology. They are, in fact, resisting a world.

Let us look closely at this world. Let us see biotechnology not solely as a science or technological practice, but as part of a wider historical process: as a new way of producing nature and culture marked by degrees of *flexibility*, *organicity and recombinance*. To examine this new mode of production, we must step back, before the emergence of biotechnology, to discuss a profound sea-change that has been occurring within capitalist production, a sea change that began thirty years ago.

FLEXIBILITY, ORGANICITY AND RECOMBINANCE: FEATURES OF A NEW INFORMATIONAL MODE OF PRODUCTION

First, to examine the modality of flexibility, we must explore a dramatic shift that has been taking place in the structure of the capitalist system. The Fordist system of production (a system of assembly-line mass production that introduced the Model-T Ford), which held prominence up through the heyday of US economic growth between 1945 and the Vietnam War, has been undergoing a period of thorough restructuring. Political and economic practices of sedentary and rigid industrial manufacture have been gradually displaced by a more "flexible" approach to capitalist production and accumulation.[7]

The decline of US corporate productivity and profitability after 1966 co-incided with the recovery of Western European and Japanese economies. For US industry, this meant increased competition at a time of declining profits and increased inflation, weakening the US dollar as a stable international reserve currency. For the first time in decades, the US witnessed the demise of its power to regulate the international financial system. To address this crisis, capitalists began to soften up the industrial system to meet the demands of an increasingly competitive global market. The Fordist system, which had reigned since the turn of the twentieth century, had become increasingly anachronistic due to inflexible systems of manufacture, including fixed capital investments, mass production systems, and inflexible industrial design. Fixed labour markets, labour allocation, and labour contracts with a unionized workforce rendered the system too rigid to be able to compete with foreign labour structures.[8]

To compete with emerging industrial powers around the globe, US corporations literally became "foreign corporations" by going transnational, appropriating the labour, resources and markets of countries throughout Asia, South America and Africa. The desire of First World capital for flexibility coincided with the need of newly independent Third World nations to achieve economic stability and autonomy. As these governments attempted to rebuild the infrastructure of a post-colonial economy, they allowed First World industries to set up shop within their borders. This transfer of US industry to the Third World attests to the fact that there is nothing so "post" about "post-industrialism": First World countries are still "industrial"; they just moved their industrial infrastructure to the Third World.

SERVICE INDUSTRY AND THE
ORGANIC PHASE OF CAPITALISM

Emptying out the industrial core within the US left behind an enormous productive void that would be filled by new forms of service production. With the emergence of a capital-intensive service industry, corporations increasingly began to commodify dimensions of everyday life, reducing reproductive dimensions of social life to mass-produced informational product. Indeed, today, capitalism is penetrating *organic* domains of "life, culture and imagination": qualitative dimensions of social life that include the biological reproduction of the species and the cultural reproduction of everyday life, in all of its symbolic meaning.

In the pre-war industrial era, there still existed a more distinguishable line between the commercial marketplace and the realms of home and neighbourhood. Stepping out of the factory, the worker could still return each night to a neighbourhood community and private life that were marked by degrees of organicity, solidarity and self-management.[9] If a service economy existed alongside industrial capitalism, it existed as a secondary source of job provision and capital accumulation. Over time, the tension between forms of commodified and non-commodified cultural practice began to wane, as reproductive cultural practice was increasingly coopted by service industry. The commodities manufactured by an informational capitalist system increasingly took the form of services: restaurants, childcare, medical care, poverty service (including a panoply of social welfare agencies and prisons) and financial services.

Present-day service capital represents the dispersed concentration of capital into standardized franchised units. Whereas Fordist production created one central site of factory production, informational capitalism disperses small, service-oriented "factoryettes" in the form of franchised chain stores, chain restaurants, and chain service stations. Replacing the idiosyncratic flavour of local service provision that was the hallmark of the industrial era, we see the emergence of "enchainment": the strangling of a local economy by a ring of chain stores and businesses. Whether clustered together in shopping malls or along downtown strips, the appearance of franchised chain stores, from fast-food and film processing to auto-repair stations and daycare centres, represents the move of capital into its more serviceable mood.

Such cultural enchainment is hinged on the idea of intellectual property. As a semiotic factory, a chain store owes its economic and cultural potency to its ability to reproduce a set of patented symbols, images, texts, building design and production protocols dispersed in the form of franchised service factoryettes. For reproductive cultural practice to become capital intensive, both the service product and the process must be transformed into intellectual property. The corporation of the informational age, then, must sell more than mere service product. McDonald's success lies in its ability to transform hamburgers into a patented semiotic field of informational signs, symbols, images and texts. In the post-war period, we see mega-corporations mass-produce service commodities whose value is linked not necessarily to their general *content*, but to their *form* or informational value. In the end, it's the golden arch, not the beef, that makes McDonald's a world power.

While service production itself is not a historical novelty, it undergoes a dramatic transformation when it reaches the post-industrial period. What is new is the intensifying concentration of capital into the service industry, the national and global expansion of service production through standardized, franchised chain-store formations, and the transformation of service commodities into patentable information.

By translating both its processes and its products into "informational" or intellectual property, the service industry is a form of informational capital relying on informational technologies. Here, it is critical to broaden the definition of informational technologies beyond its current association with technologies in microelectronics, computing, telecommunications, optoelectronics and genetic engineering. In this definition, we may include a domain of information-based service production that is not classifiable as primarily telecommunications or biotechnology: a domain of franchised cultural production ranging from McDonald's and WalMart to Disneyland.

As an informational technology, the franchised service industry not only commodifies domestic *products* within the home, but moves domestic *processes* outside of the home as well, relocating what were local cultural practices into often translocal franchised service centres such as fast-food restaurants, daycare and elder-care centres, shopping malls and franchised playgrounds.

As the service industry has proven, the production of practice is often more lucrative than the production of product. By enclosing upon the domain of service, informational capital triumphs over the limits of sedentary Fordist production. For example, once a consumer has purchased a sedentary product such as a colour television, the transaction becomes rigid and complete.[10] The television manufacturer must wait five to ten years before the consumer will be obliged to remake the purchase. In contrast, when a consumer rents an ephemeral video, the transaction is not complete, but is fluid and open-ended. The video-rental agency need only await the following evening for the consumer to return the video, ready to rent another.

In turn, informational capital, as a form of flexible production, appropriates the attentiveness and situational orientation of the domestic sphere, an orientation that was dismissed by both classical liberal theory and the ration-alizing logic of industrial capitalism. By rearranging information to create specialized products, an information-based service economy mimics both the sensibility and the mechanisms of domestic cultural reproduction.[11] Just as an attentive parent must cope with the particularized needs of their loved ones, corporations like Burger King are ready to meet consumer needs as well, presenting the world with the first flexible burger. The 1970s' Burger King jingle is truly the anthem of the service era it ushered in: "Hold the pickles, hold the lettuce/ special orders don't upset us/ all we ask is that you let us/ have it your way…"

THE RISE OF RECOMBINANT USA

If rationalization and homogenization are the hallmarks of industrial capital-ism, then *recombinance* is the emblem of informational capitalism. Recombi-nance is a productive modality characterized by the continual remelding of architecture, graphic design, radio, television, and film that have come to constitute the spectacular stage.[12] Within recombinant production, the assembly line is reversed to create the "anti-assembly line." Whereas the Fordist assembly line moved in a linear direction from "standardized parts" to create "unified wholes," the post-Fordist anti-assembly line rearranges unified wholes to create a pastiche of informational parts. For example, as the Fordist assembly line transformed standardized car parts into whole cars, the post-Fordist anti-assembly line rearranges unified wholes, such as songs, to create a pastiche of song parts (creating endless remixes of songs, as in much of sampled hip-hop).

Recombinance provides the flexibility required by a standardized service industry, furnishing the informational "movable joints" for the production of otherwise rigid and homogenous service commodities. The patented recipe for one standardized service product such as a Burger King "Whopper" becomes an endless series of variations on the Whopper theme. The modality of recombinance articulates itself today through postmodern architecture, art, design, music, fashion, and even social theories, which extract signs and symbols from various cultural moments to create wholes comprising compo-nents that often share no common history or development. Within re-combinant production, cultural artefacts are reduced to information bits to be cut and spliced together to create novel commodities.

Recombinance characterizes a range of telecommunicative practices ranging from virtual reality to the hypertexts of the CD-ROM, practices in which "users" of interactive media engage in recombinant practice, selecting, manipulating and synthesizing bits of text or image to create textual novelties. In the music industry, we see a shift from artist-driven music production to a music produced primarily by sound engineers who sample and mix pre-produced music tracks, which represent the hallmark of recombinant culture. And recombinance is prolific: one song can provide the seed for a limitless crop of arrangements to be sold to dance clubs, dentists, retailers, and shopping malls. More and more, we do not create integrally, conveying the gradual and coherent unfolding of one idea or image to another. Those who grew up on a diet of MTV and video games are often more comfortable selecting items from pre-prescribed digital menus or just moving things around.

Yet we cannot base a critique of recombinant culture on a static notion of cultural integrity or purity. It is vital to distinguish commodified recombinance from the forms of spontaneous collective synthesis that are integral to all cultural practice. As the field of anthropology has acknowledged in recent years, culture-making has always been a hybrid process, a continually developing synthesis of information, language and identity that emerges within and between peoples of different populations and cultures. Rather, it is the *primacy* of profit-driven recombinance, as a principal form of capital-intensive production, *over* non-commodified forms of local, fluid and hybrid cultural production that is problematic. Of great concern is the declining tension between holistic and local forms of cultural generativity and moments of translocal commodified recombinant production and consumption. The shift of capital towards an elastic and limitless production of recombinant informational service products flags a sharp curve in the capitalist road.

BIOTECHNOLOGY AS INFORMATIONAL CAPITAL

So far we have explored the wider context of the post-industrial era that gave birth to an informational capitalism that reaches into the "organic" reproductive dimensions of everyday life, culture and imagination, creating new flexible and recombinant forms of production. Now we may ask: what do McDonald's and MTV have in common with biotechnology? To answer this question, we must first recognize McDonald's, for instance, as constituting more than just a centre for retail and other services. We must understand it as a new recombinant and flexible mode of production, as an informational technology for producing new understandings and practices of culture through the production of service. Once we understand what McDonald's is and what McDonald's does, the connection between recombinant DNA and "recombinant USA" becomes very clear.

Like other informational technologies, biotechnology is particularly flexible, allowing capital to meet the ever-changing demands of an increasingly competitive global market. With a virtually endless supply of "biological information" in the form of genes, bacteria, viruses and other organisms for recombinant production, biotechnology allows capital to transcend its reliance on an exhaustible and expensive resource base dependent upon fossil fuels and primary materials. In turn, its scale and dexterity permits industry to transcend the inflexible factory design of the Fordist era. Exchanging the factory shop floor, with its heavy machinery and army of (organized) labourers, for the sterile laboratory, with its delicate instruments and handful of expert (often unorganized but well paid) research directors, postdoctoral fellows and technicians, the biotechnology lab is truly a postmodern factory.

The movement of major corporations into the domain of such reproductive cultural practices as seed development represents a move towards an
increasingly informational mode of capitalist production. Corporations such as
Monsanto, an agrochemical multinational invested heavily in agricultural biotechnology, manufacture tailor-made, specialized agricultural products to suit
particular niche markets. Shifting from industrial chemicals to information-
based service production that targets "organic" or biological reproductive
cultural practices such as agriculture, Monsanto recast itself as a "life sciences"
company, investing enormous capital on the development of agricultural biotechnology. Monsanto insists that it is the Microsoft of the biotechnology
world,[13] providing its patented "software" of genetically programmed seeds
that are only compatible with its patented "hardware," or chemical inputs.
Monsanto's Roundup Ready soybeans are genetically engineered to be compatible with its popular herbicide glyphosate (marketed as Roundup),
compelling consumers to buy both products at the same time. Such product
coupling allows corporations to create products so specialized that they require consumers to buy accessory "kits" only they can provide.

In this way, agricultural biotechnology is a form of information-based
service production not unlike domestic, leisure and tourist services. Corporations such as Monsanto use the same franchise logic as McDonald's to extend
their information-based service empire, transforming farmers into franchise
middle-managers who will buy their cloned and patented informational
product. Like McDonald's middle managers, farmers "lease," rather than own,
the materials obtained from the "parent" corporation. Farmers who buy
Monsanto seeds are obliged to sign one-time use agreements promising not to
replant seeds for the next year's season.

Under the reign of agricultural biotechnology, farmers become increasingly
deskilled and dependent upon the activities of service providers, such as
agrochemical companies who begin to provide seeds, and products that farmers
formerly provided for themselves. And just as corporations such as fast-food
franchises provide food service that capitalizes on workers' needs for quick,
accessible and transportable food by workers in a flexible economy, Monsanto
promises to produce specialized seeds and chemicals that capitalize on farmers'
needs to adapt flexibly to an ever-changing agricultural economy.

However, capital must confront one problem in its attempt to appropriate
biological reproductive processes: unlike service commodities such as hamburgers, a seed has the potential to reproduce itself biologically. At stake is the
attempt to exploit the limitless dimension of biological reproductivity, even
while stripping biological organisms of their autonomous generativity.[14] Indeed,
informational capital must be able to co-opt the farmers' cyclical biological
need for seeds while also destroying seeds' generative properties.

Technologies such as the "Technology Protection System" (renamed the
"Terminator Technology" by anti-GMO activists) signal the concentrated
efforts of corporations to enhance and produce new forms of organic obsolescence. Having sought the patent for a technology that produces genetically

engineered seeds whose offspring will be sterile, Monsanto hopes to be able one day biologically to obstruct farmers from saving seeds for the next year's planting season. Although the corporation publicly announced plans to abandon the technology due to "sensitivity" to public opinion, Monsanto will likely re-embark on the project once public reaction has died down.[15] Once agri-biotechnology multinationals have developed a commercially viable seed sterilization technology, they will eliminate the exorbitant financial costs (as well as costs related to bad PR) of hiring private police and lawyers to identify and sue farmers accused of breaking "one-time use" seed agreements.

BACKED-UP NATURE: BIOPROSPECTING AND THE ADVANCED CAPITALIZATION OF NATURE

Biotechnology is the systematic conversion of biological nature into informational capital. As an expression of what may be called "the advanced capitalization of nature," biotechnology represents the attempt of informational capital to profit from and transcend the limits of a biological nature that has been greatly compromised by industrial capitalist production. In the twenty-first century, capitalists deploy informational technologies to "back up" nature by identifying, patenting, and profiting from whatever is left of the earth's diverse life forms to use for future industry. While the Human Genome Project (HGP) (the internationally funded programme to map the ten thousand genes in the human genome) attempts to map out future colonial territories within the cells of human beings, attempts are being made to map out future colonial territories within the biological nature of plants, animals and other organisms. Such prospectors include botanical gardens in the US and the UK, the National Cancer Institute, independent biologists serving national and international institutes, and private companies.

While bioprospectors race to save the best of the last of "nature," a host of "anthro-prospectors" race to save the best of the last of human cultures that are at risk of being driven off native lands and driven into assimilation within centres of urban poverty. However, rather than fight against such injustice, anthro-prospectors are collecting genetic samples from individuals living within cultures identified as being "at risk." Beginning in 1991, the Human Genome Diversity Project (HGDP) emerged as an anthropological correlate to the HGP. However, whereas the HGP focuses on mapping one complete genome, the HGDP focuses on exploring the subtle genetic variation between human populations (see Chapter 18 in this volume).

The question of biodiversity, both human and non-human, is steeped in conservationist rhetoric. Corporations and governments posit the *effects* of capitalist plunder as a legitimizing cause of the need for further corporate control of what is left of the earth's biological life.[16] Genes of plants, animals and humans considered potentially valuable are stored in "national biodiversity inventories" while awaiting potential commercial applications. It is here, within the endless rows of frozen test tubes, filled with the "best of the last" specimens of biological life, that the relationship between science and informational capital comes to light.

POLITICIZING BIOTECHNOLOGY:
"THIS IS WHAT DEMOCRACY LOOKS LIKE"

Once we understand biotechnology as a new way of producing and ordering society and nature, we may begin to comprehend better the relationship between Monsanto, McDonald's and MTV. We can see these structures as producers of a recombinant, flexible, yet tightly controlled and standardized world of informational capital.

What are the features of the "biotechnological world" we seek to resist? As we have explored, it is marked by the increasing harnessing of reproductive processes of everyday life and by a transition to a new flexible, organic, and recombinant form of service production heavily reliant on informational technologies. But even more broadly, this world is one in which citizens are disempowered, obliged to accept or merely protest the whimsical decisions of leaders who make public policy in representative democracies such as the US. This shift from industrial to informational capital, this transition to what could rightly be called a "mall world" of recombinant and standardized service, is not the result of citizen action. It is being chosen, regulated and controlled by a Mafia of capitalist and governmental leaders around the world who represent less than 1 per cent of the people on the planet. *This* is not democracy. It is a world that is being managed by an elite few, not only the leaders of nation-states but increasingly, the managers of capital attempting to assume a new governmental form.

Thus, in addition to the economic, cultural and technological facets of the problem of informational capital, there is also a *political* dimension that is crucial to explore as well. Political life, both in its statist form and in its real original form (to be addressed later) is being reconstituted and further degraded by the rise of informational capitalism.

First, political structures such as the state are targets of informational capital. The emergence of such transnational institutions as the World Trade Organization (WTO), dominated by a US corporate lobby, ushers in an era in which the political sovereignty of the nation-state is being overridden by new capital-driven institutions that are neither corporations nor states: they are *meta-states*, expressions of informational capital as it emerges into an autonomous administrative and juridical governmental power that will greatly shape both state and corporate practice. As a truly postmodern entity, the WTO is pure service: it is a bureaucratic service for the extension of capital across national boundaries to ensure the most flexible systems of production, importation and marketing possible. One could also say that the WTO is *meta-capital*. It is capital-plus. Its endless documents of trade and intellectual property "agreements" represent the textual infrastructure for flexible capitalist accumulation.

While movements against biotechnology often engage a political critique, contesting, for instance, the WTO's attempt to wrest juridical power from the state, such critiques often defend state sovereignty, asserting the need to reform or abolish the power of the WTO. In so doing, they fail to question the legitimacy of the state as a political institution, missing the vital opportunity to transcend the state's hierarchical and centralized logic and structure. In turn, movements against biotechnology often express an *anti-corporate* rather than an anti-capitalist stance. Citing corporations, instead of the capitalist system itself, as the main source of the problem, activists attempt to turn the "capitalist clock" back to a kinder and gentler form of capitalism. Unfortunately, this critique also fails to recognize the need to move beyond a logic based on hierarchy and centralization, and thus cannot move beyond a capitalist system that was born out of a logic of unlimited growth, accumulation, profit and domination.

High noon will always eventually turn to midnight. There is a logic to a clock: its gears, springs or silicon chips modulate its movements in particular ways. Like a clock, capitalism and the state are constituted to move in a particular direction: towards ever greater levels of centralization, hierarchy and, ultimately, non-democracy. Indeed, when we look historically at the modern nation-state, we see that it rose in tandem with capitalism, and out of the same logic of domination. Rather than simply attempt to turn the clock of domination back, we must develop a new sense of time and history, built not out of the dustbin of capital and state-driven events, but out of the potential within the human spirit and the revolutionary impulse itself. We can think beyond what is immediately before us, drawing from the logic of a different "clock" that has been beating in the heart of humanity since the beginning of time.

Politicizing biotechnology entails moving towards a logic based on human freedom. It requires reclaiming the original meaning of politics developed by the Greeks many centuries ago: the power to assemble as *citizens* to govern our own communities.[17] According to social ecologist Murray Bookchin, the political life of free citizens cannot be reduced to "statecraft," or to the managerial and authoritarian practices of the state that are so often confused with real politics.[18] For Bookchin, real political power is the power of citizens to make decisions *in general* about their lives. It is the power to gather as members of communities to discuss, decide and determine the public policies that will shape how we work, produce and live together. Until we have this power, we will be left only to stand on the sidelines of society, fighting for rights, choices, alternatives and improvements within a system we know to be on a collision course with most of humanity and with the rest of the natural world.

If our concerns about biotechnology are concerns about *the social production of society*, then we must begin to ask who indeed should produce society? Should McDonald's, MTV and Monsanto produce the food, art and very stuff of our lives? What would it take for people, in the towns in which they live, to be able to decide for themselves the kinds of lives they would like to lead? The question I am posing is not solely "economic." It is not enough to fight against informational service capital, waging individual campaigns against WalMart or Monsanto, trying to keep chain stores out of our communities or to ban genetically engineered food from our supermarkets. While such campaigns are necessary in the short term, they must be filled out with a broader political analysis and reconstructive vision. In turn, it is not enough to create asylums of cultural and economic autonomy such as local food coops or organic growers' associations. While they provide (often privileged) groups of individuals with shelters of sanity and health, they cannot counter a system that reaches deep into the lives of poor people across the world who are forced to participate in both service and industrial production to survive.[19]

The real antidote to capitalism is to refuse its tendency to translate the world into its own terms. Capital transforms us into workers, producers, commodities and consumers so effectively, so seamlessly, that we see ourselves primarily as economic agents, as resisters to, or producers of, economic practice. The dissolution of the idea of citizen into the idea of consumer, with the new notion of consumer/citizen, signals the final collapse of humanity into *homo oeconomicus*. But we are also, as Aristotle said, *zoon politikon*, a political animal. We are beings with the potential to think, discuss, decide and determine all aspects of our lives, including matters of economics. The fact is, we cannot fight economics with economics. We can only topple an economic system by pushing back with *political* power. The enormous dislocation of peoples, capital and goods throughout the world can be countered with a movement for *a new kind of political locality* based on principles of confederation, cooperation and direct democracy.

If we are to retrieve the notion of citizenship from the category of consumer and from the category of the state as well, we have to ask ourselves *what kind of citizens*, what kind of political life, do we want to retrieve? Can we only resuscitate ourselves as citizens bound by national borders, passively represented by politicians, citizens who are dominated by the nation-state? Or may we re-establish ourselves as new kinds of citizens empowered to participate directly in the management of our everyday lives? It is time that we begin to build a *direct democracy*: one in which citizens meet directly, face to face, to determine democratically their own lives.

In the 1999 anti-WTO demonstrations in Seattle, the call for direct democracy was in the air. Direct confrontation with state military forces in the form of police and the National Guard led to a five-day period of radicalization among young activists, for many of whom this was their first encounter with military repression. The real, yet more abstract, fight against the WTO concretized itself into a struggle against the non-democracy of the state and capital, as activists found themselves beaten, injured by chemical weapons, jailed, tortured and deprived of their civil rights in a "progressive" First World city – merely for engaging in peaceful protest and for taking to the streets as citizens to express their freedom of speech.

There were countless marches that week as courageous activists risked their safety to take to the streets, refusing the curfew and no-entry zones dictated by the city of Seattle in conjunction with the federal government. During one march, a chant arose, poetically and spontaneously, that captured the imagination and passion of the other activists who were undergoing a life-changing transformation. After days of collective democratic decision-making and peaceful intelligent protest, after days of seizing the right to think, decide and take public action, they began to chant over and over, *This is What Democracy Looks Like*. On an intuitive and rational level, activists knew that democracy is a direct act; it is the movement of real people as they take action to participate in a face-to-face embodied way, in determining their own lives.

This chant, *This is What Democracy Looks Like*, entails a new way of thinking about political reconstruction. It means not only that we take to the streets, but that we take to our communities, demanding and rebuilding a real and passionate political life. This chant inspires us to develop a new understanding of citizenship that is not defined in relation to a state or nation, but is instead defined *in opposition to* nations and states. It is time to redefine citizenship in relation to local communities and to regional, continental and even global confederations.

People challenging globalization often use terms such as "local" and "global" when discussing how to transcend the current system. Yet the local/global dyad fails us as we attempt to map out the units of political organization that will constitute the new society. While the idea of "thinking globally and acting locally" rightly asserts the need to rebuild local communities within a humanist and internationalist context, the idea must be elaborated in distinctly political terms. While the term "local" could be translated into the city, town or village as a political body, the *global* does not translate into a clear and concrete political structure.

What, then, would be the political structure that would embody the concept of the "global"? The spirit of humanism and internationality that is contained in the idea of the "global" could be translated into the political structure of the *confederation*. Indeed, the confederation is the next valid level of political organization as we move beyond the local level. A more meaningful way to counter "globalization" is to counter what is "global" with what *ought* to be local and confederal.[20]

This approach to the question of political reconstruction is called libertarian municipalism.[21] Developed by theorist Murray Bookchin, libertarian municipalism is a way of thinking about political transformation that proposes a way to counter globalization by establishing self-governing local towns, cities and villages, linking them together to form confederations. Within libertarian municipalism, members of communities reclaim existing local political forums such as city and neighbourhood councils, gradually transforming them into citizens' assemblies. Using local electoral campaigns as one way of educating the public about direct democracy, libertarian municipalism proposes that citizens begin to seize publicly their potency as political actors, wresting decision-making power from states, corporations and meta-states such as the WTO. As members of municipalities form local groups engaged in the process of political transformation, they may confederate with other free municipalities to create a true *rapport de force*, a coordinated and united counter-power to the state and to capital as well.

To resist biotechnology is to dismantle the technical, social and political networks that both constitute and are constituted by this technology. We, too, must wake up one fine morning at the beginning of a new century and take a good look at what we see. Fixing our vision beyond the transgenic fruits ripening on our window sills, we must look even further still past the army of service workers running back and forth from job to job before picking the kids up from daycare at the end of one more day in Recombinant USA. If we look very hard, we will begin to see it all: a society based on systems of domination and social hierarchy, a network of state and capitalist institutions, and a world of people who are resisting the system to maintain their courage, imagination and intelligence through it all.

A movement that challenges biotechnology is a movement that challenges a world. It provides a critique not only of a particular science practice, but of a *society* that constitutes and is constituted by that practice. But more than critique, it proposes a better world. In such a movement, people inspire and inform themselves and others, helping to make the world legible and re-makeable. A truly humane movement against biotechnology gives people hope for the future, as well as the knowledge and confidence to build a future worth fighting for. It is a movement for real political power, not just over technology, but over everyday life in all of its fullness. *This* is what democracy looks like.

NOTES

1. The notion of an "organic" phase of capitalism must be distinguished from Marx's concept of the organic composition of capital. Marx counterpoised the organic composition of capital (the labour variable) to the variable of capital intensity. In contrast, I am pointing to the shift from a primarily industrial capitalist modality to one in which capital targets biological and reproductive dimensions of cultural practice as primary loci for creating surplus value.

2. Fredric Jameson, *Postmodernism; or, The Cultural Logic of Late Capitalism*. Durham, NC: Duke University Press, 1991; quoted in *Making PCR: A Story of Biotechnology*, Chicago: University of Chicago Press, 1996, p. 21.

3. Susan Wright, "The Social Warp of Science: Writing the History of Genetic Engineering Policy," *Science, Technology, and Human Values*, vol. 18, no. 1, Winter 1993.

4. Andrew Kimbrell, *The Human Body Shop*. San Francisco: Harper, pp. 188–203.

5. Homo Fragaria (pseud.), "The Strawberry Liberation Front," *Earth First!*, vol. 7, no. 6, June 1987, p. 1; Brian Tokar, "Engineering the Future of Life?" *Z Magazine*, July/August 1989, p. 110–16.

6. See Chapter 26 in this volume, and Vandana Shiva, "Quit India! Indian Farmers Burn Cargill Plant and Send Message to Multinationals," *Third World Resurgence* 36, August 1993, pp. 40–41.

7. David Harvey's *The Condition of Postmodernity* (Oxford and Cambridge, MA: Blackwell, 1990) offers a comprehensive and wonderful discussion of capitalist "flexible accumulation." This work offers a creative and integrative look at the shift from Fordist to post-Fordist capitalism.

8. Ibid., pp. 125–41.

9. Murray Bookchin, Lecture at the Institute for Social Ecology, Plainfield, VT, July

7, 1999.

10. Harvey, *The Condition of Postmodernity* , p. 156.

11. As feminist philosopher Joan Tronto points out, in the domestic sphere women demonstrate an ability to think and act in ways that are fluid. The care that women give to children and their families is particularistic and specialized, standing in sharp contrast to the universalistic stance that classical liberal philosophers such as Locke or Smith relegated to the public sphere. Whereas classical liberalism promotes the idea of "men" caring *about* universal and unchanging values such as "justice, liberty and freedom," women are largely relegated to the work of caring *for* the particularistic and ever-changing needs of people within their households. See Tronto's "Women and Caring: What Can Feminists Learn about Morality from Caring?" in Alison Jaggar and Susan Bordo, eds, *Gender, Body, Knowledge: Feminist Reconstructions on Being and Knowing*, New Brunswick, NJ: Rutgers University Press, 1989, pp. 172–88.

12. For a discussion of "recombinant culture," see Critical Art Ensemble, *The Electronic Disturbance*, New York: Autonomedia, 1994.

13. Revelling in high expectations for its own growth, the Monsanto agents came up with "Monsanto's Law," which states:

> The exponential growth in the computing power of silicon chips, described by Moore's Law, led to the development of the information technology industry, creating aggregate global value in trillions of dollars. At Monsanto, we believe that a similar non-linear trend in biotechnology capabilities is creating comparable growth potential in the life sciences. We believe that these genomic technologies will continue to double in capability every 12 to 24 months – a statement we're calling "Monsanto's Law." (*Monsanto Company* Annual Report, 1998, inside front cover)

14. For an in-depth discussion of generativity in relationship to biotechnology, see Vandana Shiva, "The Seed and the Earth: Biotechnology and the Colonisation of Regeneration," in *Close to Home: Women Reconnect Ecology, Health and Development Worldwide*, Philadelphia: New Society Publishers, 1994.

15. In reality, Monsanto never really controlled the patent for "Terminator" seeds. They tried for nearly two years to purchase Delta and Pine Land, the company that developed the Terminator along with USDA scientists, and this purchase fell through in December of 1999. Delta's executives deny they ever slowed development of Terminator, and some thirty related patents are currently held by the largest transnational biotechnology companies. See RAFI Communiqué, "Suicide Seeds on the Fast Track: Terminator 2 Years Later," Rural Advancement Foundation International, February/March 2000, at http://www.rafi.org.

16. Arturo Escobar offers a compelling and critical analysis of biodiversity discourse. See "Cultural Politics and Biological Diversity: State, Capital and Social Movements in the Pacific Coast of Colombia," in Orin Starn and Richard Fox, eds, *Between Resistance and Revolution: Culture and Social Protest*, New Brunswick, NJ: Rutgers University Press, 1998, pp. 40–64.

17. Murray Bookchin, the principal theorist associated with social ecology, writes about the need to reconstitute citizenship in a post-state context by creating direct democracy. See *Urbanization Without Cities: The Rise and Decline of Citizenship*. Montreal: Black Rose Books, 1992.

18. Ibid., pp. 123–75.

19. I discuss the limits of such projects in my book *Ecology of Everyday Life: Rethinking the Desire for Nature*, Montreal: Black Rose Books, 1999.

20. See, for example, "Libertarian Municipalism," in Janet Biehl, ed., *The Murray Bookchin Reader*, London: Cassell, 1997, pp. 172–3. Biehl also provides a discussion of Bookchin's notion of confederalism in *The Politics of Social Ecology: Libertarian Municipalism*, Montreal: Black Rose, 1998.

21. Bookchin, "Libertarian Municipalism."

Resources for Information and Action

In each category, US groups are listed first, followed by Canadian, UK and international contacts. For more detailed descriptions and contacts in your region, see Luke Anderson's book *Genetic Engineering, Food and Our Environment*, published by Green Books in the UK, Chelsea Green Publishers in the US, Scribe Publications in Australia, Lilliput Press in Ireland, Telesma in France and Québec, and Gaia Proyecto in the Spanish-speaking world.

REGIONAL AND NATIONAL ACTIVIST NETWORKS

Northeast Resistance Against Genetic Engineering (Northeast RAGE)

c/o ISE, 1118 Maple Hill Road, Plainfield, VT 05667, USA
802–454–9957, nerage@sover.net, www.nativeforest.org/nerage, www.biodev.org

The first of the regional grassroots networks in the US, with affiliates throughout New England and New York state; organizers of Biodevastation 2000 in Boston and other events and regional campaigns.

Upper Midwest Resistance Against Genetic Engineering (GrainRAGE)

P.O. Box 580444, Minneapolis, MN 55458, USA, 651–213–6131, grainrage@visto.com, http://www.tao.ca/~ban/grainrage.htm

One of the most active RAGE groups; organizer of demonstrations at Minneapolis area corporate headquarters and major regional animal and plant genetics conferences.

Other RAGE groups in the US include:

Northwest RAGE, P.O. Box 15289, Portland, OR, 97293, 503–236–5772 x2, info@nwrage.org, www.nwrage.org

DownSouth RAGE, 787 Ellsworth, Memphis, TN 38111, denny@tao.ca

BayRAGE, c/o Long Haul Info Shop, 3124 Shattuck Avenue, Berkeley, CA 94705, 510–594–4000 x144

Desert RAGE, c/o P.O. Box 3412, Tucson, AZ 85722, DesertRAGE@visto.com

Missouri RAGE (MORAGE), c/o Confluence, P.O. Box 63232, St Louis, MO 63163, 573–721–3192, morage@unbounded.com, www.missouri.edu/~ldd0aa

Monterey Bay RAGE, 224A Walnut Ave, Santa Cruz, CA 95060, 831–515–4480 x2480

Ending Destructive Genetic Engineering (EDGE)

P.O. Box 21202, Santa Barbara, CA 93121, USA, 805–963–0583 x153, activist@edgecampaign.org, www.edgecampaign.org

Education and direct action on the central and southern California coast, aimed at halting the sale of genetically engineered foods and promoting organic agriculture.

Bioengineering Action Network (BAN)

ban@tao.ca, www.tao.ca/~ban

Organizes and publicizes direct actions against genetic engineering.

Genetic Engineering Action Network (GEAN)

2105 First Ave. S., Minneapolis, MN 55404, USA, 612–870–3423, rvanstaveren@iatp.org

Networks, activists and NGOs working to address the widespread consequences of genetic engineering.

The Campaign to Label Genetically Engineered Foods

P.O. Box 55699, Seattle, WA 98155, USA
425–771–4049, label@thecampaign.org,
www.thecampaign.org

National consumer campaign for legislation to label engineered foods.

Genetic Engineering Network

P.O. Box 9656, London, N4 4JY, UK,
020–8374 9516,
info@genetix.freeserve.co.uk,
www.visitweb.com/totnes

Links local activists with national and international campaigns and supports formation of local groups.

GenetiX Snowball

One World Centre, 6 Mount Street,
Manchester M2 5NS, UK,
0161–834–0295,
genetixsnowball@onet.co.uk,
www.gn.apc.org/pmhp/gs

Nonviolent direct action campaign against experimental genetically engineered crops.

Earth First! UK

c/o Cornerstone Resource Centre, 16 Sholebroke Avenue, Leeds LS7 3HB, UK
01132–629–365, actionupdate@gn.apc.org,
www.snet.co.uk/ef

Local groups throughout the UK focus on grassroots organizing, civil disobedience and direct action.

Totnes Genetics Group

P.O. Box 77, Totnes, TQ9 5ZJ, UK
01803–840098, info@togg.freeserve.co.uk,
www.togg.org.uk

Community-based campaign in Devon, providing speakers, info, resources, support, street theatre, and other ideas for local campaigning.

Resistance is Fertile

c/o TOGG, P.O. Box 77, Totnes, Devon,
TQ9 5J, UK, 01870–122–1403,
info@resistanceisfertile.com,
www.resistanceisfertile.com

International weeks of action against genetic engineering and patenting of life.

ASEED Europe

P.O. Box 92066, 1090 AB Amsterdam,
The Netherlands, 31–20–66–82–236,
aseedeur@antenna.nl,
www.antenna.nl/aseed

Activist network committed to ecological and social justice, offering corporate profiles and other information tools on genetic engineering, and coordinated actions and campaigns.

Australian GeneEthics Network

340 Gore St, Fitzroy 3065, Victoria,
Australia
61–3–9416–0767, acfgenet@peg.apc.org,
www.zero.com.au/agen

Promotes critical discussion and debate on the environmental, social and ethical impacts of genetic engineering; campaigns for a minimum five-year freeze on GMO releases, GE foods, and life patents.

GE Free New Zealand/RAGE

Contact via Greenpeace NZ (see below)

New Zealand GE-free zone campaign, offering GE-free foods database in collaboration with the Green Party.

ACTIVIST NGOs

Greenpeace International

USA   702 H Street NW, Washington DC 20001, USA, 1–800–326–0959,
www.greenpeaceusa.org
UK   Canonbury Villas, Islington, London, N1 2PN, UK, 020–7865–8100,
info@greenpeace.org.uk,
www.truefood.org
New Zealand   113 Valley Rd, Mt Eden, Private Bag 92507, Wellesley St, Auckland 1, New Zealand, 64–9–630–6317,
greenpeace.new.zealand@dialb.greenpeace.org,
www.greenpeace.org.nz
International HQ   Chaussesstr. 131–10115, Berlin, Germany, 49–30–30–889914,
www.greenpeace.org/~geneng

International campaigns of lobbying and direct action, including work with consumers, gardeners, restaurants, and the food industry in various countries, aimed to stop the genetic engineering of food and promote organics.

Center for Food Safety/International Center for Technology Assessment

310 D Street NE, Washington DC 20002,
USA, 202–547–9359,
office@icta.org, www.icta.org

Education and legal interventions for full testing and labelling of engineered foods.

BioDemocracy Campaign/Organic Consumers Association

860 Highway 61, Little Marais, MN 55614, USA, 218–226–4164, alliance@mr.net, www.purefood.org

West Coast office: 415–643–9592, simon@organicconsumers.org

Information and organizing resources on engineered foods and US organic food standards.

Friends of the Earth

USA 1025 Vermont Ave NW, Washington, DC 20005, USA, 877–843–8687, foe@foe.org, www.foe.org/safefood

UK 26–28 Underwood Street, London N1 7JQ, 020–7490–1555, info@foe.co.uk, www.foe.org.uk/camps/foodbio/index.htm

Australia P.O. Box 8212, Alice Springs NT 0871, 61–8–9522–497, www.metropolis.net.au/foe/foeOz.html

US Safe Food, Safer Farms Campaign seeks a ban on Bt crops and a moratorium on the release of genetically engineered organisms into the environment. UK Food and Biotechnology Campaign seeks to end importation, growing and sale of modified foods.

National Family Farm Coalition

110 Maryland Ave, N.E., Suite 307, Washington DC, 20002, USA 202–543–5675, nffc@nffc.net, www.nffc.net

Alliance of grassroots farm, resource conservation, and rural advocacy groups, focusing on the effects of engineered crops on farmers, among other issues.

Institute for Social Ecology

1118 Maple Hill Rd, Plainfield, VT 05667, USA, 802–454–8493, ise@sover.net, ise.rootmedia.org

Activism and educational programmes on biotechnology, along with education on ecological politics, philosophy, design and building, rooted in a politics of direct democracy and moral economy.

Genetically Engineered Food Alert

c/o PIRG, 3435 Wilshire Blvd # 380, Los Angeles, CA 90010, USA 213–251–3680, www.gefa.org

National NGO campaign for labelling, safety testing and liability for GE foods. An alliance of seven national groups.

Native Forest Network/ACERCA

P.O. Box 57, Burlington, VT 05402, USA 802–863–0571, nfnena@sover.net, acerca@sover.net, www.nativeforest.org

International forest protection network, with a focus on opposing genetic engineering in forestry. ACERCA highlights issues in the Central American region.

Pesticide Action Network North America

49 Powell St, Suite 500, San Francisco, CA 94102, USA, 415–981–1771, panna@panna.org, www.panna.org/panna

US office for international anti-pesticide network, with offices also in Africa, Asia/Pacific, Latin America and Europe; offers weekly PANUPS news updates.

Alliance for Bio-Integrity

P.O. Box 110, Iowa City, IA 52244, USA 515–472–5554, info@bio-integrity.org, www.bio-integrity.org

Organizer of lawsuit against FDA's lax regulation of engineered foods.

Northeast Organic Farming Association

411 Sheldon Rd, Barre, MA 01005, USA P.O. Box 697, Richmond, VT 05477, USA, 802–434–4122, novavt@together.net, www.nofavt.org

Assistance and advocacy for organic growers in the Northeastern US, including active opposition to genetic engineering in agriculture, local and regional conferences, etc.

California Certified Organic Farmers

1115 Mission St, Santa Cruz, CA 95060, USA, 831–423–2263, www.ccof.org

One of the leading organic farming groups in the US; active on GE issues since the late 1980s.

Campaign for Responsible Transplantation

P.O. Box 2751, New York, NY 10163, USA, 212–579–3477, alixfano@mindspring.com, www.crt-online.org

Research and activism exposing the myths of xenotransplantation, and its consequences for human and animal health.

Sierra Club of Canada

412–1 Nicholas St, Ottawa, Ontario, K1N 7B7, Canada, 613–241–4611, sierra@web.net, www.sierraclub.ca/national/genetic

Safe Food/Sustainable Agriculture campaign distributes information at grocery stores and exposes the environmental and human health risks of genetic engineering.

Council of Canadians

502–151 Slater Street, Ottawa, Ontario, K1P 5H3, Canada, 1–800–387–7177, inquiries@canadians.org, www.canadians.org

Independent, non-partisan citizens' interest group active on key national issues, including economic justice, genetic engineering and international trade.

Genetics Forum

94 White Lion Street, London N1 9PF, UK, 020–7837–9229, geneticsforum@gn.apc.org, www.geneticsforum.org.uk

Policy development, campaigns and publications on genetic engineering from a social, environmental and ethical perspective; publishes bimonthly magazine, SPLICE.

Greenpeace UK, Friends of the Earth UK

(See above)

Campaign Against Human Genetic Engineering

P.O. Box 6313, London N16 0DY, UK cahge@globalnet.co.uk, www.users.globalnet.co.uk/~cahge

Opposes human genetic engineering and works on related issues such as genetic discrimination and eugenics.

Gaia Foundation

18 Well Walk, Hampstead, London NW3 1LD, UK, 020–7435–5000, gaia@gaianet.org,

Links with farmers, scientists and grassroots organizations in the Third World to address the impacts of genetic engineering and patenting.

Women's Environmental Network

87 Worship St, London EC2A 2BE, UK, 020–7247–3327, testtube@gn.apc.org, www.gn.apc.org/wen

Information and science-based campaign resources on genetics issues.

EarthRights Environmental Law & Resource Centre

26 Field Lane, Teddington, TW11 9AW, UK, 020–8977–2248, www.gn.apc.org/earthrights

Offers free and low-cost legal resources to individuals and groups seeking to protect the environment; an affiliate of the Environmental Law Alliance Worldwide (E-LAW).

The Soil Association

Bristol House, 40–56 Victoria Street, Bristol, BS1 6BY, UK, 0117–929–0661, info@soilassociation.org, www.soilassociation.org

Campaigning for and certification of organic farms, seeking to combine food production with environmental protection and human health.

Primal Seeds

mail@primalseeds.org, www.primalseeds.org

Web-based campaign to protect biodiversity and support local food security, in response to biopiracy, and corporate control of seeds and food.

ActionAid

Hamlyn House, Macdonald Road, Archway, London, N19 5PG, UK, 020–7561–7611, campaigns@actionaid.org.uk, www.actionaid.org

Campaigns against hunger and poverty in thirty countries, increasingly focused on genetic engineering and patenting.

Friends of the Earth Australia, Greenpeace New Zealand, Greenpeace International

(see above)

RESEARCH, EDUCATION AND POLICY ORGANIZATIONS

Rural Advancement Foundation International

110 Osborne St, Suite 202, Winnipeg, MB R3L 1Y5, Canada, 204–453–5259, rafi@rafi.org, www.rafi.org

US office P.O. Box 640, Pittsboro NC 27312, USA, 919–542–1396

Focuses on genetic diversity, intellectual property rights, world food security and human genomics. Ongoing investigation of patents and corporate policies, with frequent detailed communiqués posted to the Internet.

Edmonds Institute

20319 92nd Avenue W., Edmonds, WA 98020, USA
425–775–5383, beb@igc.org,
www.edmonds-institute.org

Research and activism focusing on biosafety, patenting, and technology assessment. Publisher of peer-reviewed *Manual for Assessing the Ecological and Human Health Effects of Genetically Engineered Organisms*.

Union of Concerned Scientists

2 Brattle Square, Cambridge, MA 02238, USA, 617–547–5552, ucs@ucsusa.org,
www.ucsusa.org

Studies risks and benefits of genetic engineering and supports sustainable alternatives. Publishes quarterly *FoodWeb* electronic newsletter and ongoing survey of approved engineered crops in the US.

Council for Responsible Genetics

5 Upland Road, Suite 3, Cambridge, MA 02140, USA, 617–868–0870,
marty@gene-watch.org,
www.gene-watch.org

Focuses on patenting, genetic discrimination and food safety; publishes monthly *Gene-Watch*.

Institute for Food and Development Policy

398 60th St, Oakland, CA 94618, USA
510–654–4400, info@foodfirst.org,
www.foodfirst.org/progs/global/biotech

Focus on international development, exposing causes of poverty and hunger and the consequences of agribusiness practices for the global South.

Institute for Agriculture and Trade Policy

2105 1st Avenue South, Minneapolis, MN 55404, USA, 612–870–0453,
support@iatp.org, www.iatp.org

Research, activism, networking and technical assistance on diverse aspects of agriculture and trade policy.

Ag Biotech InfoNet

www.biotech-info.net

Comprehensive Internet resource on biotechnology and genetic engineering in agricultural production and food processing and marketing, developed by Idaho-based crop consultant Charles Benbrook.

Environmental Research Foundation

P.O. Box 5036, Annapolis, MD 21403, USA, 1–888–2RACHEL, erf@rachel.org,
www.rachel.org

Weekly updates on environmental toxins, corporate power and genetic engineering.

GeneWatch UK

The Courtyard, Whitecross Road, Tideswell, Buxton, 5K17 8NY, UK, 01298–871–898,
gene-watch@dial.pipex.com,
www.genewatch.org

Research and analysis on the science, ethics, risks and regulation of genetic engineering, including corporate database and details of crop trials.

Econexus

P.O. Box 3279, Brighton, BN1 1TL, UK, 01273–625–173, info@web-econexus.org,
www.web-econexus.org

Science-based screening and analysis of scientific documents and patents to assess the impacts of genetic engineering on the environment, farming, food security, health and medicine.

CornerHouse

P.O. Box 3137, Station Road, Sturminster Newton, Dorset, DT10 1YJ, UK,
cornerhouse@gn.apc.org

Monthly in-depth briefings on environment and development issues, including genetic engineering, from an international perspective.

Compassion in World Farming

Charles House, 5A Charles Street, Petersfield, GU32 3EH, UK
01730–264208/268863,
compassion@ciwf.co.uk, www.ciwf.co.uk

Provides videos and other resources on the genetic engineering of animals and xenotransplantation.

Genetic Resources Action International (GRAIN)

Girona 25, pral., E-08010 Barcelona, Spain, 093–301–1627,
grain@bcn.servicom.es, www.grain.org

International research and action against erosion of genetic diversity. Publishes quarterly newsletter, *Seedling*.

Consumer Food Network

223 Logan Road, Buranda QLD 4102, Australia, 61–7–3217–3187,
eco-cons@bit.net.au,
www.ozemail.com.au/~confoodnet

Addresses food issues and policies from a consumer perspective throughout Australia and New Zealand, coordinating consumer input to government and industry bodies; monthly print and electronic newsletter.

Research Foundation for Science, Technology and Ecology

A-60 Hauz Khas, New Delhi, 110016, India, 91–11–696–8077, vshiva@giasdl01.vsnl.net.in, www.ipsil.com/vshiva/

Research on genetic engineering, biopiracy, globalization and intellectual property, and support for traditional farming communities in rural India.

Third World Network

228 Macalister Rd, 10400 Penang, Malaysia, 60–4–226–6728, twn@igc.apc.org, www.twnside.org.sg

Focus on development and North–South issues; intervenes at the UN and other international fora. Publishes monthly *Third World Resurgence* and bimonthly *Third World Economics*.

OTHER PERIODICALS

The Ecologist

Unit 18, Chelsea Wharf, 15 Lots Road, London, SW10 0QJ, UK, 020–7351–3578, ecologist@gn.apc.org, www.theecologist.com
US Subscriptions   1920 Martin Luther King Jr. Way, Berkeley, CA 94704, USA, 510–548–2032

Earth First! Journal

P.O. Box 1415, Eugene, OR 97440, USA 541–344–8004, earthfirst@igc.org

GenEthics News

P.O. Box 6313, London N16 0DY, UK genethicsnews@compuserve.com

The Ram's Horn

S-6, C-27, RR 1, Sorrento, BC V0E 2W0 Canada
250–675–4866, ramshorn@ramshorn.bc.ca, www.ramshorn.bc.ca

GM Free

KHI Publications, Beacon House, Woodley Park, Skelmersdale, Lancashire WN8 6UR, UK, 01695–50504, gmfree@cableinet.co.uk, wkweb4.cableinet.co.uk/pbrown/avoidgm/FS_Intro.htm

ELECTRONIC LISTS
(a representative sample)

ban@tao.ca
(Bioengineering Action Network, US)

Ban-GEF@lists.txinfinet.com
(Discussion list on GE foods)

bioethics@egroups.com
(Discussion list on bioethics and human genetic engineering)

bio-ipr@cuenet.com (GRAIN list on patenting and biodiversity issues)

biotech_activists@iatp.org
(Institute for Agriculture and Trade Policy)

frankentrees-NFN@egroups.com (Native Forest Network GE Trees Campaign)

genet@agoranet.be (GENET Europe)

genetics@gn.apc.org
(Genetic Engineering Network, UK)

organicview@organicconsumers.org
(Organic Consumers Association, US)

teel@adax.com
(Techno-Eugenics newsletter)

Contributors

Brian Tokar has been an activist since the 1970s in the peace, anti-nuclear, environmental and Green politics movements, and is currently a faculty member at the Institute for Social Ecology and Goddard College in Vermont. He is the author of *The Green Alternative: Creating an Ecological Future* (1987; revised edition, 1992) and *Earth for Sale: Reclaiming Ecology in the Age of Corporate Greenwash* (1997), and was the recipient of a 1999 Project Censored award for his investigative history of the Monsanto company (*The Ecologist*, September/October 1998). Tokar's articles on environmental politics and emerging ecological movements appear frequently in *Z Magazine*, *The Ecologist*, *Food & Water Journal*, *Synthesis/Regeneration*, *Toward Freedom*, and numerous other publications. He has lectured throughout the US, as well as internationally, is a founding member of Northeast Resistance Against Genetic Engineering, and serves on the national boards of the Native Forest Network and the Edmonds Institute. He graduated from MIT with degrees in biology and physics and received a Masters degree in biophysics from Harvard University.

Beth Burrows is the founder, president and director of the Edmonds Institute, an award-winning, public interest, non-profit organization concerned with the environmental, social, and ethical implications of new technologies. She has lectured and published internationally on trade and technology issues, and for many years has been a involved in international deliberations on biosafety and patenting.

Mitchel Cohen is a freelance writer, poet and activist with the Green Party of New York State, the Red Balloon Collective, and the Direct Action Network to Free Mumia Abu Jamal and Leonard Peltier. The Coalition Against the Violence Initiative can be contacted c/o Social Justice Ministries, The Riverside Church, 490 Riverside Drive, New York, NY 10027, USA.

Dr Martha L. Crouch did pioneering research in plant genetics at Indiana University until she discovered that her results were being used to advance industrial agriculture at the expense of the environment. She has over the past twenty years been a naturalist, genetic engineer, university professor, and consultant on issues related to food and agriculture.

Dr Marcy Darnovsky is the co-editor of the *Techno-Eugenics Newsletter*, available electronically from teel@adax.com. She teaches courses on the politics of science and the environment in the Hutchins School of Liberal Studies at Sonoma State University in California.

Michael K. Dorsey is a long-time environmental justice activist, currently studying the political economy of bioprospecting in the rainforests of Ecuador. He is finishing his PhD in natural resource and environmental policy at the

University of Michigan and is a member of the US Board of Directors of the Sierra Club.

Steve Emmott is the Policy Advisor on Genetic Engineering for the Green Group of Members of the European Parliament (MEPs), and has for the last ten years been at the centre of the biotechnology patent debate. Portions of his chapter are adapted with permission from GRAIN's journal, *Seedling*.

Alix Fano is director of the New York City-based Campaign for Responsible Transplantation, an international coalition of physicians, scientists and eighty public interest groups opposed to xenotransplantation (http://www.crt-online.org).

Jennifer Ferrara is a freelance writer and English teacher in California, and the former associate director of the US food safety organization, Food & Water. An earlier version of her chapter on regulation appeared in the *Ecologist* (vol. 28, no. 5, September/October 1998).

Chaia Heller teaches ecological philosophy at the Institute for Social Ecology in Vermont and is currently completing her doctoral dissertation, on risk discourse and the GMO debate in France, at the University of Massachusetts, Department of Anthropology. She is the author of *Ecology of Everyday Life: Rethinking the Desire for Nature* (Montreal: Black Rose, 1999).

Dr David King, editor of the London-based *GenEthics News*, has a PhD in molecular biology from Edinburgh University. Since 1990 he has spent his time campaigning and writing on genetics issues and in 1998 founded the Campaign Against Human Genetic Engineering.

Dr Jack Kloppenburg is a member of the Department of Rural Sociology at the University of Wisconsin at Madison, and is the author of two books on biotechnology and agriculture. His chapter is excerpted with permission from *Does Technology Know Where it is Going? 12 Reasons to Stop Expecting Modern Biotechnology to Create a Sustainable Agriculture and What to Do After the Expectation has Ceased* (Edmonds, WA: Edmonds Institute, 1996). An earlier version appeared in *The Ecologist* (vol. 26, no. 2, March/April 1996).

Orin Langelle is a freelance writer and photojournalist, the coordinator of ACERCA (Action for Community and Ecology in the Regions of Central America), and a founding board member of the international Native Forest Network.

Zoë C. Meleo-Erwin is a long-time anti-racist, feminist and radical activist, and a graduate of the Institute for Social Ecology's Masters program at Goddard College in Vermont. She is currently living in Philadelphia, where she is working to free US political prisoners. Her research for this book was supported by a grant from the Institute for Anarchist Studies in New York City.

Dr Barbara Katz Rothman is on the faculty of the Department of Sociology at Baruch College of the City University of New York. She is the author of *Genetic Maps and Human Imaginations* (New York: W.W. Norton, 1998), from which her chapter here is excerpted with permission.

Sonja A. Schmitz has a Masters degree in molecular biology. She worked for DuPont in their Agricultural Biotechnology department for many years. She is presently a doctoral candidate in plant systematics and evolution at the University of Vermont.

Thomas G. Schweiger, a native of Vienna, has been an environmental campaigner since the mid-1980s. Since 1995 he has been based in Brussels as a lobbyist against genetic engineering, most recently for Greenpeace International.

Sarah Sexton works with The CornerHouse, a UK-based research and solidarity group which publishes regular briefing papers on a range of social and environmental justice issues. Copies of the longer briefing paper from which her chapter was extracted with permission can be obtained from The CornerHouse, PO Box 3137, Station Road, Sturminster Newton, Dorset DT10 1YJ, UK. Email: cornerhouse@gn.apc.org.

Robin Seydel is a community organizer who has worked in cooperatives in Ohio, Vermont, Arizona and, for the past thirteen years, at La Montañita Coop in New Mexico, where she serves as the member services coordinator and newsletter editor. She is a Clinical Medical Herbalist, teaches herbal medicine, has a small private practice, and would like to spend more time in her organic garden.

Hope Shand is research director of the Rural Advancement Foundation International, an international civil society organization based in Canada. RAFI is dedicated to the conservation and sustainable use of biodiversity and the socially responsible development of technologies useful to rural societies. http://www.rafi.org.

Lucy Sharratt is coordinator of the Safe Food, Sustainable Agriculture Campaign at the Sierra Club of Canada. She was active in the campaign against recombinant Bovine Growth Hormone and works locally against genetic engineering. She is a long-time social-justice activist and human-rights educator.

Dr Vandana Shiva is the director of the Research Foundation for Science, Technology and Ecology in New Delhi, India. She is internationally renowned as an environmentalist, feminist and widely published author, and a 1993 recipient of the Right Livelihood Award.

Dr Ricarda A. Steinbrecher is a biologist and genetic scientist and an outspoken critic of genetic engineering in food and farming, and patents on life. Based in England, she is an advisor to many national and international NGOs and university research programmes and is the co-director of Econexus (http://www.web-econexus.org).

Victoria Tauli-Corpuz is the director of the Philippines-based Tebtebba Foundation (Indigenous Peoples' International Centre for Policy Research and Education), and has written widely and spoken at numerous international conferences on human genetics and indigenous concerns. She is an indigenous person belonging to the Kankana-ey-Igorot peoples in the Cordillera Region.

Jim Thomas is a Food Campaigner with Greenpeace UK, and has been closely associated with the Genetic Engineering Network since it was established in 1996. He has written for the *Guardian*, *BBC Online*, *Resurgence* magazine, *GenEthics News*, *Splice*, *The Ecologist* and other publications.

Kimberly A. Wilson is the co-author (with Martin Teitel) of *Genetically Engineered Food: Changing the Nature of Nature* (Rochester, VT: Park Street Press, 1999).

Index

Abir-Am, Pnina, 322
aborted fetuses *see* fetuses
abortion, 154, 165, 173, 175, 176, 199; as a
  result of testing, 164; for sex selection, 135;
  justification for, 179
Action Aid (UK), 346
Action for Solidarity, Equality, Environment and
  Development (ASEED), 112
Action for Community and Ecology in the
  Regions of Central America (ACERCA),
  115
adaptive crops, hazards of, 89–91
adenosine deaminase (ADA) deficiency, 139
Advanced Genetic Sciences company (AGS),
  322–4
Aeta people, gene collection from, 262
ageing, 161; slowing of, 158
Agfa company, 228
Agracetus company, 87
AgroEvo company, 47, 82, 346; field test
  sabotaged, 340–41
agriceuticals, importance of, 230
Agricultural Renewal in India for a Sustainable
  Environment (ARISE), 352
agriculture: as war by other means, 19;
  biotechnology in, 22–39; commercialization
  of, 25; globalization of, 43; idea of
  revolution in, 49, 77; industrial, 60;
  sustainable, 49
agrochemicals, 230–31
AIDS, 150, 185, 190
Akre, Jane, 303
alar, ban on, 17
Albert, Bill, 165
alcohol: health problems of, 161, 191; produced
  from plant debris, 30–31, 70, 94
alcoholism, genetic aspect to, 162
Allan, Jonathan, 183
Allen, Darina, 346
allergies to food, 54, 57–8
Alliance for Bio-Integrity (USA), 332
Aloha Medical Mission, 262
Alzheimer's disease, 159, 245
Ambridge Against Genetix group (UK), 343
American Corn Growers Association, 332
American Cyanamid company, 83
American Home Products company, 231
American Society of Transplant Surgeons, 188
Amish farmers, 105
amniocentesis, 173, 199; for sex determination,
  263
Anderson, Per Pinstrup, 41, 43
Andy, O.J., 209–10

animal pharmaceuticals sector, 231–2
animal welfare concerns, in animal modification,
  380, 382
Annas, George, 245–6
anti-Americanism in resistance to GMOs, 368–9
anti-freeze genes, incorporation of, 90
anti-nuclear movement, 1, 321
antibiotic resistance, 32, 51, 285; marker genes, 4,
  59–60, 92, 95, 366
antibiotics: allergy to, 54; banned in animal feeds,
  366; in milk, 54
Archer Daniels Midland company, 11, 332;
  genetically engineered grain, 219
Archer, Tommy, 343
Arewete tribe, DNA collected from, 108
Aristotle, 416
Armstrong, Jeannette, 244
Asgrow company, 9, 62, 89
Aspergillus niger fungus, picks up engineered
  genes, 94
assay technology, improved efficiency of, 273
asthma, research into, 219
Astra company, 229, 231
AstraZeneca company, 8, 87, 104, 119, 229, 230,
  231; demergers in, 229; genetically modified
  trees destroyed, 121
Atomic Energy Act (1954) (USA), 295
Audubon Ballroom, planned biotechnology
  centre, 329
Aventis company, 8, 47, 62, 82, 104, 106, 228,
  229, 231
Aviles, Jaime, 116
Axis Genetics company, 228
Axys Pharmaceuticals company, 219
ayahuasca plant, 259
Ayurveda medicine system, 284, 285, 286
Azadirachta indica see neem
Aztecs, medicine of, 272

Babagouda, a farmers' leader, 353
baboon cytomegalovirus, 183
baboon heart transplants, 183
Baby Fae, heart transplant, 183
Bacillus thuringiensis (Bt), 34, 56, 96, 118,
  299–300, 403; Bt corn, 332 (sabotage of,
  316); Bt cotton, 84, 96, 356 (field trials in
  India, 355); Bt-producing crops, 84–5;
  regulation of, 325; resistance to, 86; side
  effects of, 29–30
Bacon, Francis, 129, 172
Bailey, Britt, 61–2
Bailey, Leonard, 275
Baltimore, David, 233